FUNDAMENTALS OF
QUANTUM ELECTRONICS

FUNDAMENTALS OF QUANTUM ELECTRONICS

Richard H. Pantell,
Professor, Electrical Engineering
Stanford University,
Stanford, California

Harold E. Puthoff,
Research Associate
Stanford University,
Stanford, California

JOHN WILEY & SONS, INC., NEW YORK, LONDON, SYDNEY, TORONTO

Copyright © 1969
by John Wiley & Sons, Inc.
All rights reserved. No part of this book may be reproduced by any means, nor transmitted, nor translated into a machine language without the written permission of the publisher. Library of Congress Catalog Card Number: 70-76060 SBN 471 65790 5 Printed in the United States of America.

1 2 3 4 5 6 7 8 9 10

PREFACE

With the many new applications for quantum electronics over the past few years, a need has arisen in the academic and research community for a textbook that approaches the analysis and description of quantum electronic devices from a unified point of view. Unified points of view are of course desirable in any newly emerging field, but this need is especially accented in the area of quantum electronics because of the rapid development of this field as it has passed from the hands of the theoretical physicist to the applied physicist to the engineer. As a result it has become necessary for device-oriented physicists and engineers to assimilate a large background of material that ranges from the basic elements of the theory of quantum mechanics to the behavior of such complex macroscopic devices as lasers and parametric amplifiers.

We hope that the present text will help the applied physicist or engineer to bridge the gap, starting with first principles (that is, Schrödinger's equation or its equivalent) and carrying through a somewhat orderly step-by-step analysis to the level of describing the macroscopic behavior of a useful quantum electronic device. The text has been used as the basis of a graduate course at Stanford University and is intended primarily for the first-year graduate student in applied physics or electrical engineering who has had the usual introductory quantum mechanics course. The student is thus ready to take the basic tools provided by such a course and from them forge more powerful techniques for the analysis of macroscopic systems based on such processes as stimulated emission, multiple photon effects, and photon-phonon interactions. The material we present is designed to meet this need and therefore serves as an introduction to a more detailed study of quantum electronics, semiconductor theory, and the optical properties of solids. The techniques developed in the text are illustrated by applications to problems of current interest. The emphasis, however, is on the development of problem-solving techniques appropriate to the analysis of such problems, and not on the actual devices analyzed.

The purpose of the text is to make the transition from the mathematical postulates and formalisms of quantum mechanics to equations involving the macroscopic variables of interest. Macroscopic observables are determined from the expectation value of the operator corresponding to the observable, and the expectation value is obtained by means of the density operator. The equations are then reformulated solely in terms of expectation values by substituting equivalent expressions for the density operator. This procedure results in a set of simultaneous equations in terms of macroscopic variables such as polarization and electric field. The semiclassical approach, in which the electromagnetic field is not quantized, requires the addition of Maxwell's electromagnetic field equations to the equations for the expectation values. This whole then constitutes a self-consistent set of descriptive equations.

Considerable thought was given to the question of units, that is, whether CGS or MKS units should be used. Although much of the literature in the field of quantum electronics employs CGS units, in preparing a text that will find its primary use in the graduate curricula of applied physics and engineering programs one must take into account the growing trend toward the use of MKS units. Based upon requests by the students themselves, the conflict between reading the literature, on the one hand, and ease of learning in notation that the student is comfortable with, on the other, was finally resolved in favor of the latter and MKS units adopted.

The text begins with a review, in Chapter 1, of the quantum mechanics pertinent to the material presented in subsequent chapters. The density operator formulation is developed, and its importance is stressed because it forms the basis for relating quantum mechanical operators to observables. A treatment of the interaction between a radiation field and a two-level quantized medium is presented in Chapter 2. Semiclassical equations are developed in terms of polarization, electromagnetic field, and energy level populations for both the electric dipole and spin $\frac{1}{2}$ transitions. Chapter 3 is concerned with resonance absorption, saturation, and dispersion. The parameters of the dipole transition are expressed in terms of quantities that are experimentally obtained, and in this manner it is possible to evaluate matrix elements and relaxation times. Frequency dependence of the complex refractive index is discussed from the standpoint of the effect produced by the presence of a dipole interaction. A description of the laser including threshold requirements, steady-state conditions, transient behavior, and Q-switched operation is considered in Chapter 4. If the laser is homogeneously broadened, the results follow from the rate equation approach of the preceding chapter. For an inhomogeneously broadened resonance the semiclassical equations are used. In Chapter 5 we consider the interaction of nonresonant fields with a quantized medium; that is, the frequency of the radiation field differs from that associated with the transition. Multiple quanta absorption and harmonic generation are analyzed, and transition probabilities are evaluated for these higher-order processes. Chapter 6 is devoted to a comparison of the semiclassical and quantum electrodynamic approaches to the interaction of radiation and matter. In particular, this chapter discusses the manner in which spontaneous emission terms enter and bring about modification of previous results. The interaction of radiation with molecular vibrations is considered in Chapter 7. The stimulated and nonstimulated Raman and Brillouin effects are discussed, with derivations presented for the threshold and phase-matching conditions. Optical phenomena in semiconductors are explored in Chapter 8, where there is an interaction of radiation with a continuum of energy states. The topics that are discussed include absorption, photoconductivity, and semiconductor lasers.

Throughout the text, lengthy derivations that would interrupt the natural flow of ideas are relegated either to an appendix or to problems at the end of the chapter. The problems should be considered a necessary adjunct to the text for the development of certain concepts that are illustrated best by the working of an example. Some of the problems are intended to be minor research projects rather than merely a reproduction of material in the text. In practice, the researcher must establish appropriate assumptions and utilize a variety of references. In solving some research problems, perhaps the student will obtain a foretaste of what might be expected from him after graduation or in the performance of his thesis work.

The authors wish to express their gratitude to Professor P. D. Coleman at the University of Illinois and Professors S. E. Schwarz and J. R. Whinnery at the University of California in Berkeley for reviewing the manuscript prior to publication, and to Mrs. Suzanne Wise for her secretarial assistance. We are grateful to Mr. D. Hecht for contributing the Nomograph of Electromagnetic Conversions.

Richard H. Pantell
Harold E. Puthoff

Stanford, California
June 1969

LIST OF SYMBOLS[‡]

A	spontaneous emission rate
$A^\dagger$	adjoint operator of A
$\langle A \rangle$	expectation value of operator A
A	vector potential
a	lattice constant
$a^\dagger, a$	creation and annihilation operators, boson fields easily distinguished from above by context
$[A, B]$	commutator of operators A and B
$\mathscr{A}$	power attenuation constant
B	magnetic flux density
B_{ij}	relative dielectric impermeability tensor
b^*	complex conjugate of b
C_T, C_L	crystal force constant for transverse, longitudinal waves
c	velocity of light in vacuum
D	electric flux density
d^3k	volume element in **k** space
E_k	energy eigenvalue of kth eigenstate
E, $\mathscr{E}$	electric field
e	electronic charge
f	force
f_{ji}	oscillator strength associated with transition $i \rightarrow j$
F	filling factor
G	reciprocal lattice vector
g	spectroscopic splitting factor
$g_G(\Omega_i, \omega_0)$	Gaussian lineshape function
g_i	degeneracy of ith level
$g_L(\omega, \Omega)$	Lorentzian lineshape function (real part)

‡ Other symbols used locally in the text are defined as needed.

LIST OF SYMBOLS

$\tilde{g}_L(\omega, \Omega)$	complex Lorentzian lineshape function
g_s	Stokes power gain per unit length
H	magnetic field
$\hbar$	$(1/2\pi) \times$ Planck constant
$\hbar \mathbf{J}$	angular momentum
$\hbar \Omega$	energy separation between a pair of eigenstates
$\mathcal{H}$	Hamiltonian
$\mathcal{H}'$	interaction or perturbation Hamiltonian
$\mathcal{H}_0$	unperturbed Hamiltonian
$\mathcal{H}^r$	randomizing Hamiltonian
I	intensity (power per unit area)
i	imaginary unit, $\sqrt{-1}$
J	current density
k or $|\mathbf{k}|$	magnitude of wavevector
k	wavevector
L	Lorentz correction factor
L	angular momentum
$\mathcal{L}$	Lagrangian
M	atomic mass
M	magnetization
m	electron mass
m^*	effective electron mass
m	magnetic dipole moment
m_h^*	effective hole mass
N	population difference per unit volume, $N_2 - N_1$
$\bar{N}$	population difference per unit volume normalized to steady-state value
N^e	population difference per unit volume under equilibrium conditions
N_V	number per unit volume
n	photon number
$\mathcal{N}$	number
P	power
P	polarization
p	canonical momentum conjugate to coordinate q
p_n	probability that a quantum mechanical system is characterized by $|\psi_n\rangle$
$p(\omega)$	mode density (modes/unit volume-unit frequency range)
p_{ijrs}	elasto-optical coefficient
$\mathcal{P}$	power per unit volume
Q	relative atomic or molecular displacement, $q_2 - q_1$

LIST OF SYMBOLS

Q_c, Q_l	cavity, linewidth quality factors
q	generalized coordinate
$\mathscr{R}$	power reflectivity
T	temperature
T_1	longitudinal relaxation time
T_2	transverse relaxation time
$\mathbf{T}$	torque
Tr	trace
U	energy/unit volume
V	volume
$\mathscr{V}$	potential
v	velocity
W_{jk}	transition rate for transitions from state j to state k
$\mathscr{W}$	energy
Z	effective charge
α	polarizability
β	Bohr magneton
Γ	power absorption constant
γ	gyromagnetic ratio
$\Delta\omega_G$	linewidth of Gaussian line
$\Delta\omega_L$	linewidth of Lorentzian line
δ_{ij}	Kronecker delta
$\delta(x - x')$	Dirac delta function
ϵ	permittivity
ϵ_0	permittivity of free space
ζ	solid angle
η	refractive index
κ	Boltzmann constant or dielectric constant (the distinction is obvious in the text)
λ	wavelength
μ	permeability
μ_e	electron mobility
μ_h	hole mobility
μ_0	permeability of free space
$\boldsymbol{\mu}$	electric dipole moment
$|\mu_k\rangle$	time-independent eigenstate state vector
$|\boldsymbol{\mu}_{12}|$	matrix element of dipole moment operator coupling states $|1\rangle$ and $|2\rangle$
ν	frequency
ρ	density operator
ρ_{ij}	i, j matrix element of density operator ρ

ρ_m	mass density
$\rho(\mathbf{k})$	probability that an electron will be found in unit volume in $\mathbf{k}$ space
$\rho(\nu)$	electromagnetic energy per unit volume per unit frequency interval
σ	conductivity
σ_c	cross section
$\boldsymbol{\sigma}, \sigma_x, \sigma_y, \sigma_z$	Pauli spin operators
τ	relaxation time constant
τ_c	cavity lifetime
τ_d	lifetime associated with cavity diffraction losses
τ_m	lifetime associated with cavity mirror losses
τ_s	lifetime associated with cavity scattering and absorption losses
τ_{sp}	spontaneous emission time
Φ	scalar potential
$\lvert \Phi_k \rangle$	eigenstate vector
φ	photon density (electromagnetic energy density$/\hbar\omega$)
$\bar{\varphi}$	photon density normalized to steady-state value
χ_B	Brillouin susceptibility
χ_R	Raman susceptibility
$\chi = \chi' + i\chi''$	susceptibility
ψ	wavefunction
$\lvert \psi \rangle$	arbitrary state vector
ω	angular frequency
ω_c	cavity resonant angular frequency
$\mathbf{1}_x, \mathbf{1}_y, \mathbf{1}_z$	unit vectors in the x, y, z directions, respectively

CONTENTS

1 QUANTUM THEORY

1.1	INTRODUCTION	1
1.2	SOME BASIC CONCEPTS	2
1.3	SOME PROPERTIES OF OPERATORS	5
1.4	THE DENSITY OPERATOR	6
	1.4.1 Some Properties of ρ	9
	1.4.2 Time Dependence of ρ	10
	1.4.3 Relaxation Terms	11
	1.4.4 The Density Operator for a Continuum of Eigenstates	14
	1.4.5 Time Derivatives of the Expectation Value of an Operator	15

2 DIPOLE TRANSITIONS

2.1	INTRODUCTION	20
2.2	HAMILTONIAN OF AN ATOM IN AN ELECTROMAGNETIC FIELD	20
	2.2.1 Multipole Expansion	21
	2.2.2 Electric Dipole Interaction	22
	2.2.3 Electric Quadrupole, Magnetic Dipole, and Diamagnetic Interactions	23
	2.2.4 Multiparticle Systems	25
2.3	MATRIX ELEMENTS AND THE CONCEPT OF PARITY	26
	2.3.1 Electric Dipole Transition	26
	2.3.2 Electric Quadrupole and Magnetic Dipole Transitions	29
2.4	EQUATIONS OF MOTION FOR THE ELECTRIC DIPOLE TRANSITION	30
	2.4.1 Hamiltonian	30
	2.4.2 Electric Dipole Moment	31

	2.4.3 Molecular Averaging	33
	2.4.4 Population Difference	35
	2.4.5 Field Equations	37
	2.4.6 Electric Dipole Transition—Summary	40
	2.4.7 Local Field Correction Factors	40
2.5	EQUATIONS OF MOTION FOR THE MAGNETIC DIPOLE SPIN $\tfrac{1}{2}$ SYSTEM	42
	2.5.1 Hamiltonian	45
	2.5.2 Equations of Motion	47
	2.5.3 Field Equations for Magnetic Dipole Spin $\tfrac{1}{2}$ Case	51

3 RESONANT PROCESSES

3.1	INTRODUCTION	55
3.2	STEADY-STATE BEHAVIOR OF THE ELECTRIC DIPOLE TRANSITION; ABSORPTION, DISPERSION, AND SATURATION	56
	3.2.1 Susceptibility	57
	3.2.2 Absorption and Dispersion	58
	3.2.3 Oscillator Strength and Sum Rules	63
	3.2.4 Lorentzian Lineshape—Homogeneous Broadening	65
	3.2.5 Gaussian Lineshape—Inhomogeneous Broadening	66
	3.2.6 Saturation	71
	3.2.7 Degeneracy	75
3.3	TENSOR PROPERTIES OF THE SUSCEPTIBILITY	77
3.4	TRANSIENT BEHAVIOR OF THE ELECTRIC DIPOLE TRANSITION—COUPLED AMPLITUDE EQUATIONS AND RATE EQUATIONS	85
	3.4.1 Traveling-Wave Amplification	86
	3.4.2 Cavity Amplification	89
	3.4.3 Laser Cavity Oscillator	91
	3.4.4 Maser Cavity Oscillator	93
	3.4.5 Transition Probability	94
3.5	PARAMAGNETIC RESONANCE ABSORPTION	95

4 LASERS

4.1	INTRODUCTION	101
4.2	POPULATION INVERSION	102
4.3	THRESHOLD REQUIREMENTS	107
4.4	STEADY-STATE POWER OUTPUT	111
4.5	TRANSIENT BEHAVIOR	112

4.6	Q-SWITCHING	117
4.7	THE FOUR-LEVEL LASER	120
4.8	THE HELIUM-NEON GAS LASER	121

5 NONLINEAR EFFECTS IN QUANTIZED MEDIA

5.1	INTRODUCTION	132
5.2	PARITY CONSIDERATIONS AND TRANSITION PROBABILITIES	135
5.3	SCATTERING CROSS-SECTIONS FOR MULTIPLE-PHOTON ABSORPTIONS	143
5.4	HARMONIC GENERATION	145
5.5	STARK SHIFT	150
5.6	STIMULATED RAMAN OSCILLATIONS	153

6 FIELD QUANTIZATION

6.1	INTRODUCTION		158
6.2	QUANTIZATION OF CAVITY FIELDS		159
	6.2.1	Energy Decay in a Cavity Mode	166
	6.2.2	Field Equation for a Cavity Mode	167
6.3	QUANTIZATION OF PLANE WAVES		169
6.4	INTERACTION OF RADIATION WITH MATTER WHERE BOTH FIELDS AND MATTER ARE QUANTIZED		171
	6.4.1	Equations of Motion for the Density Operator	173
	6.4.2	Rate Equations	174
	6.4.3	Mode Density	179
	6.4.4	Spontaneous Emission	182
	6.4.5	Modified Rate Equations for the Single-Mode Case	183
	6.4.6	Interaction Between Matter and a Broadband Radiation Field	184
	6.4.7	Blackbody Radiation	186
	6.4.8	Einstein Treatment of Induced and Spontaneous Transitions	187

7 INTERACTIONS BETWEEN RADIATION AND PHONONS

7.1	INTRODUCTION	191
7.2	CRYSTAL LATTICE VIBRATIONS	192

      7.2.1 Modes of a Monatomic Crystal Lattice   192
      7.2.2 Modes of a Diatomic Crystal Lattice   196
7.3 QUANTIZATION OF LATTICE VIBRATIONS   200
      7.3.1 Periodic Boundary Conditions and Normal Modes   200
      7.3.2 Quantization of the Acoustic Modes of a Monatomic Crystal   204
      7.3.3 Quantization of the Optical Modes of a Diatomic Crystal   207
7.4 CONSERVATION OF ENERGY AND MOMENTUM IN PROCESSES INVOLVING PHONONS   209
      7.4.1 Conservation Conditions   209
      7.4.2 Conservation Conditions for Brillouin Scattering   212
      7.4.3 Conservation Conditions for Raman Scattering   214
7.5 INFRARED PROPERTIES OF OPTICAL PHONONS   217
7.6 RAMAN EFFECT   225
      7.6.1 General Considerations—Classical Approach   229
      7.6.2 Quantum Mechanical Treatment   232
      7.6.3 Raman Susceptibility   234
      7.6.4 Raman Gain   235
      7.6.5 Spontaneous Raman Scattering   239
7.7 BRILLOUIN EFFECT   243
      7.7.1 Hamiltonian   246
      7.7.2 Quantum Mechanical Treatment   247
7.8 SELF-FOCUSING   255

8 ELECTRONS IN CRYSTALS

8.1 INTRODUCTION   266
8.2 ELECTRONS IN CRYSTALS WITHOUT AN APPLIED FIELD   266
      8.2.1 Energy Bands   267
      8.2.2 Bloch Functions   271
      8.2.3 Momentum   273
      8.2.4 Reflection by Impurities   275
      8.2.5 Trapping   278
8.3 INTRABAND EFFECTS   279
      8.3.1 Intraband Motion in an Ideal Crystal   280
      8.3.2 Effective Mass   282
      8.3.3 Electron Motion Including Collisions   284
      8.3.4 Current Density   288
      8.3.5 Holes   290
      8.3.6 Current Density from the Boltzmann Equation   291

8.4	INTERBAND EFFECTS	294
	8.4.1 Direct Transitions	294
	8.4.2 Indirect Transitions	302
8.5	PHOTOCONDUCTIVITY	307
8.6	SEMICONDUCTOR INJECTION LASERS	310
	8.6.1 A General Description	310
	8.6.2 Threshold Current	313

APPENDICES

1	THE EFFECT OF $\mathscr{H}^r$ ON ρ_{11}	319
2	TRANSITION RATES IN EQUILIBRIUM	321
3	TO PROVE THAT $\mu_\alpha \mu_\beta^*$ IS REAL	323
4	LORENTZ LOCAL FIELD CORRECTION FACTOR	325
5	KRAMERS-KRONIG RELATIONSHIPS	328
6	SPATIAL AVERAGING	330
7	DERIVATION OF PHOTON RATE EQUATION	332
8	SUBSTANCES FOUND TO EXHIBIT STIMULATED RAMAN EFFECT	335
9	RELATIONSHIP BETWEEN POLARIZABILITY AND DISPLACEMENT	340
10	AN EXPRESSION FOR d^3k IN TERMS OF ω_{21}	342
11	EVALUATION OF THE INTEGRALS IN EQUATION 8.54	344
12	THE DERIVATION OF $\left(\dfrac{\partial E}{\partial k_\alpha}\right)^2 = \dfrac{2\hbar^2}{3m_\alpha^*} E$	346
13	THE DERIVATION OF (8.99) FROM (8.98).	347
	INDEX	351
	PHYSICAL CONSTANTS	359
	USE OF THE NOMOGRAPH	360
	NOMOGRAPH OF ELECTROMAGNETIC CONVERSIONS	361

QUANTUM THEORY

1

1.1 INTRODUCTION

Introductory courses in quantum mechanics generally begin with a review of the progressive failure of classical physics to account for the outcome of certain physical experiments. This failure is then traced to the wave-particle duality and the inherent property of discreteness found in nature, concepts not fully developed in the classical view. To account more fully for these concepts, the fundamental postulates on which physics is to be based must be reformulated. The concept of a wave function ψ that is postulated to contain all the information that can be known about a system, and that satisfies, in the nonrelativistic case, the Schrödinger wave equation

$$\mathcal{H}\psi = i\hbar \frac{\partial \psi}{\partial t} \tag{1.1}$$

is introduced in this reformulation. The theory of wave mechanics is then presented, and in the course of its development a large number of relatively simple problems such as the harmonic oscillator and hydrogenic atom are solved in detail.

Unfortunately, even these relatively simple problems require a considerable degree of mathematical detail, and the student can be left with the feeling that the solution to a problem of more than just academic interest must be hopelessly complex. In reality, the application of quantum mechanics to some important large-scale problems is often simpler in many respects.

In initial studies of quantum mechanics much of the effort is directed toward determining the explicit form of the wave function ψ, given a Hamiltonian describing the system of interest. However, in the application of quantum mechanics to macroscopic processes it is not generally necessary to determine the exact form of the wave function. Instead, we can more usefully approach the problem from a viewpoint similar in spirit to that used in deriving Ehrenfest's theorem. Ehrenfest's theorem, it will be recalled, is a basic derivation in quantum mechanics used to show that the equations of motion for expectation values are closely analogous to classical equations. The derivation is important because the resulting equations are independent of the precise form of the wave functions. Such a technique therefore provides a useful bridge for making the transition from quantum mechanics to equations similar in form to a classical description.

The density matrix formulation, which allows us to obtain equations of motion for the expectation values of quantum mechanical variables without the necessity of determining the actual wave functions involved, provides a comprehensive yet concise way of applying the approach just described. In the density matrix approach we need only assume that a complete set of known but unspecified wave functions exists, and then straightforward application of the formalism proceeds to yield the desired equations.

Chapter 1 is devoted to a review of the quantum mechanics relevant to the development of the density matrix approach. Sections 1.2 and 1.3 summarize the postulates and mathematics of this subject that are pertinent to the text. In Section 1.4 the density operator is defined and its properties developed.

1.2 SOME BASIC CONCEPTS

It is assumed that whatever can be known about a system can be obtained from a *state* function ψ that satisfies Schrödinger's wave equation (1.1), where $\mathcal{H}$ is the Hamiltonian operator and $2\pi\hbar$ is Planck's constant. It is convenient to normalize the function so that

$$\int dV \psi^* \psi = 1$$

where the asterisk denotes a complex conjugate and the integral is over all the coordinates upon which ψ depends. In general, ψ may be expressed in terms of a variety of independent variables such as spatial coordinates, spin coordinates, or momentum coordinates.

The notation may be simplified by adopting the bra and ket designation of Dirac.[1] The state vector ψ is represented by $|\psi\rangle$ and is termed a *ket* vector. There is a one-to-one correspondence between each ket vector and another vector $\langle\psi|$, termed a *bra* vector, where the product of a bra and ket $\langle\psi|\psi\rangle$ represents the integral $\int \psi^*\psi \, dV$. In general, the product between any bra $\langle u|$ and ket $|v\rangle$ represents $\int u^*v \, dV$. This product, termed a *scalar* product, is postulated to have the following properties:

$$\langle u | v \rangle = \langle v | u \rangle^*$$

$$\langle u | u \rangle \geqq 0$$

These properties are consistent with the integral representation for the scalar product.

With Dirac's notation, the wave equation for the ket vector becomes

$$\mathscr{H}|\psi\rangle = i\hbar \frac{\partial |\psi\rangle}{\partial t}$$

and the corresponding normalization condition is

$$\langle \psi | \psi \rangle = 1$$

We shall first consider an isolated system free of any interaction or perturbation. For such a system there exists a set of states, designated by $|\Phi_k\rangle$, that are in principle found by solving Schrödinger's equation in the form

$$\mathscr{H}_0 |\Phi_k\rangle = i\hbar \frac{\partial |\Phi_k\rangle}{\partial t} \qquad (1.2)$$

where $\mathscr{H}_0$ is the unperturbed Hamiltonian operator for the system, assumed time-independent, and k labels the kth quantum state. If we now consider a complete set of time-independent state vectors $|u_k\rangle$ that individually satisfy the eigenvalue equation

$$\mathscr{H}_0 |u_k\rangle = E_k |u_k\rangle \qquad (1.3)$$

and the orthonormality condition involving the Kronecker delta function, δ_{kl},

$$\langle u_k | u_l \rangle = \delta_{kl} = 0 \qquad k \neq l$$
$$= 1 \qquad k = l$$

we find that (1.2) can be solved by assuming a product solution of the form

$$|\Phi_k\rangle = a_k(t) |u_k\rangle \qquad (1.4)$$

Substitution of (1.4) and (1.3) into (1.2) yields

$$\frac{da_k}{dt} = -\frac{i}{\hbar} E_k a_k$$

with the solutions

$$a_k = a_k(0) e^{-i\omega_k t}$$

where the frequency ω_k is given by $\omega_k = E_k/\hbar$. If the normalization condition $\langle \psi | \psi \rangle = 1$ is applied to each of the eigenvectors, then, apart from an arbitrary phase factor, the $|\Phi_k\rangle$ are given by†

$$|\Phi_k\rangle = e^{-i\omega_k t} |u_k\rangle \qquad (1.5)$$

† As a result of the scalar product requirement $\langle u | v \rangle = \langle v | u \rangle^*$, the bra corresponding to the above ket is

$$\langle \phi_k | = e^{i\omega_k t} \langle u_k |$$

Thus, in passing from bras to kets and vice versa, constants are replaced by their complex conjugates, and the bras and kets are interchanged.

If the isolated system is not in one of its eigenstates, but is instead in a state $|\psi\rangle$ that is a linear combination of eigenstates, the state vector is given by a superposition of eigenvectors,

$$|\psi\rangle = \sum_k c_k |\Phi_k\rangle = \sum_k c_k e^{-i\omega_k t} |u_k\rangle \tag{1.6}$$

where $|c_k|^2$ is the probability that a measurement will find the system in its kth eigenstate. The expansion coefficients c_k are independent of time because we assume that no interaction or perturbation is present to alter the system. Any given measurement of the energy of the system yields one of the eigenvalues of the Hamiltonian, and the distribution of a series of such measurements is required to establish the complete form of the total state vector $|\psi\rangle$. The normalization condition $\langle \psi | \psi \rangle = 1$ implies $\sum_k |c_k|^2 = 1$, which is a statement that the sum of the probabilities of finding the system in one or another of its eigenstates equals unity.

In Equation 1.6 the state vector $|\psi\rangle$ is written as a sum of terms involving the eigenvectors of the Hamiltonian operator. At times it is useful to select the eigenvectors of another operator, such as momentum, for the summation representation of $|\psi\rangle$. The particular set of functions that are used in the expansion are termed the *basis* vectors. In general, orthonormal functions are chosen as the basis vectors because this simplifies most mathematical operations. For example, a direct consequence of the orthonormal property of $|u_k\rangle$ is that $c_k = \langle \Phi_k | \psi \rangle$. The state vector $|\psi\rangle$ is often written as a column array of the coefficients of the orthonormal basis vectors:

$$|\psi\rangle \rightarrow \begin{pmatrix} c_1 \\ c_2 \\ c_3 \\ \vdots \end{pmatrix}$$

which is the matrix notation for the state vector.

If a perturbation or interaction $\mathscr{H}_1$ acts on the system to alter its dynamical state, then the complete Hamiltonian $\mathscr{H} = \mathscr{H}_0 + \mathscr{H}_1$ must be used in the solution of Schrödinger's equation (1.1). Even in the presence of an interaction, however, it is still convenient to consider an expansion in terms of the basis vectors $|u_k\rangle$, which are eigenvectors of the unperturbed Hamiltonian provided by the solution to (1.3). In this case the solution can be written as in (1.5) with the exception that the expansion coefficients are now taken to be functions of time,

$$|\psi\rangle = \sum_k c_k(t) |\Phi_k\rangle = \sum_k c_k(t) e^{-i\omega_k t} |u_k\rangle$$

At this point the assumed expansion above can be substituted into Schrödinger's equation (1.1) to obtain a set of differential equations for the expansion coefficients $c_k(t)$ that can then be solved by approximation techniques such as perturbation theory. However, we shall instead develop the mathematics appropriate to the density matrix approach, which provides an alternative and, in many respects, more useful path to the solution.

1.3 SOME PROPERTIES OF OPERATORS

An operator is an instruction or set of instructions that transforms one vector into another. The instruction might be: "Take the time derivative," or "Differentiate with respect to x." The equation is

$$|w\rangle = A\,|v\rangle$$

where $|v\rangle$ and $|w\rangle$ are vectors and A is the operator.

A product of a bra and ket $\langle u \mid y \rangle$ is a scalar; the product of a ket and bra $|y\rangle\langle u|$ may be shown to be an operator. If A is defined as $|y\rangle\langle u|$, we see that

$$A\,|v\rangle = |y\rangle\langle u \mid v \rangle$$

Therefore, $A\,|v\rangle$ is a constant times the vector $|y\rangle$, and so $|y\rangle\langle u|$ has been shown to be an operator.

Just as we may expand a vector in terms of an orthonormal set of basis vectors $|u_i\rangle$, we may expand an operator in terms of the set of operators $|u_i\rangle\langle u_j|$:

$$A = \sum_{i,j} A_{ij}\,|u_i\rangle\langle u_j|$$

From the orthonormal property of $|u_i\rangle$ it is found that

$$A_{ij} = \langle u_i|\,A\,|u_j\rangle \tag{1.7}$$

and A_{ij} is termed the *ij* matrix element of the operator A. The diagonal matrix elements are those for which $i = j$. In matrix notation an operator is written as a two-dimensional array of the elements A_{ij},

$$A \rightarrow \begin{pmatrix} A_{11} & A_{12} & A_{13} & \cdots \\ A_{21} & A_{22} & A_{23} & \cdots \\ A_{31} & A_{32} & A_{33} & \cdots \\ \cdot & \cdot & \cdot & \\ \cdot & \cdot & \cdot & \\ \cdot & \cdot & \cdot & \end{pmatrix}$$

A useful operator is the identity operator I, which satisfies the condition that $IA = AI = A$, where A is any operator. In terms of an orthonormal set of basis vectors,

$$I = \sum_k |u_k\rangle\langle u_k|$$

That this expansion is an identity operator may be seen by writing

$$IA = \left[\sum_k |u_k\rangle\langle u_k|\right]\left[\sum_{ij} A_{ij} |u_i\rangle\langle u_j|\right]$$

$$= \sum_{i,j,k} A_{ij} |u_k\rangle\langle u_k | u_i\rangle\langle u_j|$$

$$= \sum_{i,j,k} A_{ij} |u_k\rangle \delta_{ki} \langle u_j|$$

$$= \sum_{ij} A_{ij} |u_i\rangle\langle u_j| = A$$

The matrix elements for I are

$$\langle u_i| I |u_j\rangle = \sum_k \langle u_i | u_k\rangle\langle u_k | u_j\rangle$$

$$= \delta_{ij}$$

That is, the diagonal elements are unity and the off-diagonal elements are zero.

If the operator A corresponds to an observable, then the expectation value of this observable is given by $\langle A \rangle$, where

$$\langle A \rangle = \langle \psi| A |\psi\rangle \tag{1.8}$$

It is this postulate that enables us to go from the mathematical formalism of quantum mechanics to the equations of motion for the variables of interest associated with a specific device or experiment. For example, the measured dipole moment for a molecule is taken to be $\langle \mu \rangle$, where μ is the dipole moment operator; the measured electric field is taken to be $\langle E \rangle$, where E is the electric field operator; and so on.

Operator B is termed the *adjoint* operator of A if the following equality holds:

$$\langle \varphi| A |\psi\rangle = \langle \psi| B |\varphi\rangle^* \tag{1.9}$$

and B is written as $A^\dagger$. If $A = A^\dagger$, then A is a self-adjoint or *Hermitian* operator. An operator A corresponding to an observable is Hermitian. This follows from the fact that $\langle A \rangle$ represents a measurable quantity and so must be real, which in turn requires A to be Hermitian. In terms of the matrix elements of a Hermitian operator, Equation 1.9 yields the condition that $A_{ij} = A_{ji}^*$.

1.4 THE DENSITY OPERATOR

In general, there is not sufficient information to say that a system is characterized by a specific state function $|\psi\rangle$. For example, suppose that we have two groups of harmonic oscillators each prepared in a different way so that, at $t = t_0$,

$$|\psi_a(t_0)\rangle = \sum_k a_k |u_k\rangle$$

$$|\psi_b(t_0)\rangle = \sum_k b_k |u_k\rangle$$

where the $|u_k\rangle$ are an orthonormal set, $|\psi_a(t_0)\rangle$ is the state function for one group of oscillators, and $|\psi_b(t_0)\rangle$ is the state function for the other group at $t = t_0$. The expectation values for an operator A are given by $\langle \psi_a(t)| A |\psi_a(t)\rangle$ and $\langle \psi_b(t)| A |\psi_b(t)\rangle$ for the two separate groups, respectively. If N_a oscillators characterized by the state function $|\psi_a(t)\rangle$ are mixed with N_b oscillators characterized by $|\psi_b(t)\rangle$, assuming that the particles in the two groups do not interact with each other, then $\langle A \rangle$ for the mixture is

$$\langle A \rangle = p_a \langle \psi_a(t)| A |\psi_a(t)\rangle + p_b \langle \psi_b(t)| A |\psi_b(t)\rangle$$

where $p_a = N_a/(N_a + N_b)$ is the probability that an oscillator is characterized by $|\psi_a(t)\rangle$. Therefore, when it is necessary to describe a system by more than one state function, the expectation value of an operator is given by

$$\langle A \rangle = \sum_n p_n \langle \psi_n| A |\psi_n\rangle \tag{1.10}$$

where p_n is the probability that the system is characterized by the function $|\psi_n\rangle$. The various state functions may or may not be orthogonal. If a sufficient number of measurements are made on a system, the p_n are determined, but if the set of measurements is incomplete, it is necessary to include statistical distribution theory to evaluate the probabilities.

The probability p_n introduced here differs from the probability $|c_n|^2$ discussed earlier in connection with the distribution of eigenstates. In the case discussed earlier, the system was assumed to be in a known state $|\psi\rangle$ made up of a particular linear combination of eigenstates $|u_k\rangle$. Such a state is known as a pure state. There the probability $|c_n|^2$ of finding the system in its nth eigenstate arose not because of any lack of knowledge of the state of the system, but rather from the inherent nature of the quantum mechanical measurement process, which disturbs the system in such a way as to place it in one of its eigenstates.

The probability p_n, on the other hand, results from a lack of information as to which of several possible states $|\psi_i\rangle$ the system may be in. This may be due simply to a lack of complete measurement or, equivalently, a lack of knowledge about the preparation of the system. Therefore a need for statistical averaging arises in the same sense as in classical physics. The quantum state must therefore be described by a suitably weighted mixture of states and is termed a mixed state. Table 1.1 summarizes the hierarchy of the states and their relation to expectation values. For further discussion of these points, see Chapter 6 of Ref. 3.

TABLE 1.1

System in	Expectation value of operator A
Eigenstate $\|u_k\rangle$	$\langle A \rangle = \langle u_k\| A \|u_k\rangle$
Pure state $\|\psi\rangle$	$\langle A \rangle = \langle \psi\| A \|\psi\rangle$
Mixed state $\|\psi_a\rangle, \|\psi_b\rangle, \ldots$	$\langle A \rangle = \sum_n p_n \langle \psi_n\| A \|\psi_n\rangle$

The density operator may be introduced by rewriting (1.10) in the form

$$\langle A \rangle = \sum_{n,k} p_n \langle \psi_n| A |u_k\rangle\langle u_k | \psi_n \rangle \tag{1.11}$$

which is the same as (1.10) since $\sum_k |u_k\rangle\langle u_k|$ is the identity operator. Transposing the two scalar products on the right-hand side of (1.11), we have

$$\langle A \rangle = \sum_{n,k} p_n \langle u_k | \psi_n\rangle\langle \psi_n| A |u_k\rangle$$

$$\equiv \sum_k \langle u_k| \rho A |u_k\rangle$$

where

$$\rho \equiv \sum_n p_n |\psi_n\rangle\langle \psi_n| \tag{1.12}$$

With this definition of ρ, which is termed the density operator, we see that the expectation value for any operator A is the sum of the diagonal matrix elements of the product ρA. The trace of a matrix, abbreviated as Tr, is defined as the sum of the diagonal elements, so that we have

$$\langle A \rangle = \text{Tr}\,(\rho A) \tag{1.13}$$

Since the expectation value of any observable can be obtained by use of the above expression, the density operator ρ contains all the physically significant information that can be known about the system. Therefore the reformulation of quantum mechanics in terms of the density operator ρ is extremely useful in the application of quantum mechanics to physical problems.

If the state vectors are expressed as a linear combination of the eigenfunctions of the Hamiltonian, the matrix elements of ρ are simply related to the coefficients that appear in the expansion. As given by (1.6), we may write $|\psi_n\rangle$ as

$$|\psi_n\rangle = \sum_k c_{nk} e^{-i\omega_k t} |u_k\rangle \tag{1.14}$$

If (1.14) is substituted into (1.12), we obtain

$$\rho = \sum_{n,k,l} p_n c_{nk} c_{nl}^* e^{-i(\omega_k - \omega_l)t} |u_k\rangle\langle u_l| \tag{1.15}$$

The matrix elements ρ_{mj} are given by

$$\rho_{mj} = \langle u_m | \rho | u_j \rangle$$

so that from (1.15) we have

$$\rho_{mj} = \sum_n p_n c_{nm} c_{nj}^* e^{-i(\omega_m - \omega_j)t} \qquad (1.16)$$

If $m = j$, then (1.16) gives an expression for the diagonal elements of ρ:

$$\rho_{jj} = \sum_n p_n |c_{nj}|^2 \qquad (1.17)$$

The diagonal element ρ_{jj} is the probability that the system is characterized by the eigenstate $|u_j\rangle$. Since the probability of finding the system in the state $|\psi_n\rangle$ is p_n, and since the probability that $|\psi_n\rangle$ is in state $|u_j\rangle$ is $|c_{nj}|^2$, the product $p_n |c_{nj}|^2$ is the probability that the system is in state $|\psi_n\rangle$ and eigenstate $|u_j\rangle$. Therefore the total probability that the system is characterized by $|u_j\rangle$ is given by $\sum_n p_n |c_{nj}|^2$, which is ρ_{jj}. In a pure state only one $|\psi_n\rangle$ is present, and thus the subscript n may be dropped and

$$\rho_{jj} = |c_j|^2 \qquad (1.18)$$

1.4.1 Some Properties of ρ

Two properties of ρ will be useful for later derivations:

1. ρ is Hermitian
2. Tr $(\rho) = 1$

The Hermitian property is proved as follows:

$$\begin{aligned}
\langle \varphi | \rho | \phi \rangle &= \sum_n p_n \langle \varphi | \psi_n \rangle \langle \psi_n | \phi \rangle \\
&= \sum_n p_n (\langle \phi | \psi_n \rangle \langle \psi_n | \varphi \rangle)^* \\
&= \langle \phi | \rho | \varphi \rangle^*
\end{aligned}$$

To prove that Tr $(\rho) = 1$, we assume that the $|\psi_n\rangle$ have been normalized so that $\langle \psi_n | \psi_n \rangle = 1$, and that $\sum_n p_n = 1$. Since p_n is a probability, the condition $\sum_n p_n = 1$ means that there is unity probability of locating the system within the span of state functions. From the definition of a trace,

$$\begin{aligned}
\text{Tr } (\rho) &= \sum_k \langle u_k | \rho | u_k \rangle \\
&= \sum_{k,n} p_n \langle u_k | \psi_n \rangle \langle \psi_n | u_k \rangle \\
&= \sum_n p_n \langle \psi_n | \psi_n \rangle = 1 \qquad (1.19)
\end{aligned}$$

The summation over k was eliminated by noting that $\sum_n |u_k\rangle\langle u_k|$ is the identity operator.

1.4.2 Time Dependence of ρ

With the aid of Equation 1.13 we can determine the expectation value of any observable, provided that the density operator is known. The next step, therefore, is to obtain an expression for ρ. Differentiating ρ with respect to time, we have

$$i\hbar \frac{\partial \rho}{\partial t} = i\hbar \sum_n p_n \left[\frac{\partial |\psi_n\rangle}{\partial t} \langle \psi_n| + |\psi_n\rangle \frac{\partial \langle \psi_n|}{\partial t} \right] \qquad (1.20)$$

From the wave equation it is possible to substitute equivalent expressions for the time derivatives of the bra and ket vectors in Equation 1.20. The wave equation for the ket vector is

$$\mathcal{H} |\psi_n\rangle = i\hbar \frac{\partial |\psi_n\rangle}{\partial t}$$

A wave equation for the bra vector may be obtained from the previous equation by premultiplying both sides by $\langle \psi|$, taking the complex conjugate, and using the fact that $\mathcal{H}$ is an Hermitian operator. This leads to

$$\langle \psi_n| \mathcal{H} = -i\hbar \frac{\partial \langle \psi_n|}{\partial t}$$

Substitution into (1.20) for the time derivatives of the vectors as given by the wave equation yields

$$i\hbar \frac{\partial \rho}{\partial t} = \sum_n p_n [\mathcal{H} |\psi_n\rangle\langle \psi_n| - |\psi_n\rangle\langle \psi_n| \mathcal{H}]$$

$$= [\mathcal{H} \rho - \rho \mathcal{H}]$$

The operator $\mathcal{H}\rho - \rho\mathcal{H}$ is termed the *commutator* and is written as $[\mathcal{H}, \rho]$. Thus

$$i\hbar \frac{\partial \rho}{\partial t} = [\mathcal{H}, \rho] \qquad (1.21)$$

Equations 1.13 and 1.21 are the basic equations of the text, for it is the simultaneous solution of these equations that leads to equations of motion for the observables.

It is useful to obtain differential equations for the matrix elements of ρ using the eigenvectors of the unperturbed Hamiltonian for the medium $\mathcal{H}_0$ as the basis vectors. Let the Hamiltonian be written as $\mathcal{H} = \mathcal{H}_0 + \mathcal{H}_1$, where $\mathcal{H}_0 |u_k\rangle = E_k |u_k\rangle$ and $\mathcal{H}_1$ is the energy operator for the interaction between the medium and a perturbation. From (1.21) it may be shown that

$$i\hbar \frac{\partial \rho_{ij}}{\partial t} = (E_i - E_j)\rho_{ij} + [\mathcal{H}_1, \rho]_{ij} \qquad (1.22)$$

where the *ij* subscript indicates the *ij* matrix element.

1.4.3 Relaxation Terms

It is convenient to rewrite $\mathcal{H}_1$ as a sum of two terms, the first corresponding to the interaction between the medium and an applied perturbation such as an electromagnetic field, and the second corresponding to internal interaction energies. The second term, for example, might include interactions between an atom and its lattice or exchanges of energy resulting from collisions between molecules. These terms in the Hamiltonian are generally referred to as randomizing terms because they produce changes in the elements of the density operator without an applied perturbation. Thus we write

$$\mathcal{H}_1 = \mathcal{H}' + \mathcal{H}^r \qquad (1.23)$$

where $\mathcal{H}'$ is the applied interaction term and $\mathcal{H}^r$ corresponds to the randomizing terms. If (1.23) is substituted into (1.22), we obtain

$$i\hbar \frac{\partial \rho_{ij}}{\partial t} = (E_i - E_j)\rho_{ij} + [\mathcal{H}', \rho]_{ij} + [\mathcal{H}^r, \rho]_{ij} \qquad (1.24)$$

We may show that if the system is characterized by $|u_1\rangle$ at $t = 0$, so that at $t = 0$ we have $\rho_{ij} = \delta_{i1}\delta_{j1}$, then the presence of a time-independent $\mathcal{H}^r$ causes ρ_{11} to begin to decay exponentially with time. This result was first given by Weisskopf and Wigner.[2] For small values of t, ρ_{11} is the largest element of the density matrix, and thus from (1.24) with $\mathcal{H}' = 0$ we have

$$i\hbar \frac{\partial \rho_{1j}}{\partial t} \cong (E_1 - E_j)\rho_{1j} - \rho_{11}\mathcal{H}^r_{1j} \qquad \text{for} \quad j \neq 1 \qquad (1.25)$$

and

$$i\hbar \frac{\partial \rho_{11}}{\partial t} = \sum_k [\mathcal{H}^r_{1k} \rho_{k1} - \rho_{1k}\mathcal{H}^r_{k1}] \qquad (1.26)$$

As shown in Appendix 1, we may solve for $\rho_{11}(t)$ by means of the Laplace transform, with the result

$$\rho_{11}(t) = e^{-t/\tau} \qquad (1.27)$$

where τ is defined by Equation A1.7. On the basis of this result the effect of the randomizing Hamiltonian may be approximated by replacing the last term in (1.24) with a relaxation term. The utility of any approximation is dependent upon how well the replacement predicts the actual event, and on a pragmatic basis the approximation is a good one.

In thermal equilibrium the diagonal elements of ρ, which are the occupation probabilities, are given by the Boltzmann distribution

$$\rho_{ii} = \frac{e^{-E_i/\kappa T}}{\sum_m e^{-E_m/\kappa T}} \qquad (1.28)$$

where κ is Boltzmann's constant and T is temperature. Also, with the aid of statistical mechanics, it is shown that in thermal equilibrium the off-diagonal elements of ρ are zero:[3]

$$\rho_{ij} = 0 \quad \text{for} \quad i \neq j \qquad (1.29)$$

The effect of $\mathscr{H}^r$ is to produce changes in ρ_{ij} in the absence of any applied perturbation so that a disturbed system can reach equilibrium with its environment.

For $\mathscr{H}' = 0$ and $i \neq j$, ρ_{ij} becomes zero in equilibrium, so that we modify (1.24) to obtain

$$i\hbar \frac{\partial \rho_{ij}}{\partial t} = \hbar \omega_{ij} \rho_{ij} + [\mathscr{H}', \rho]_{ij} - \frac{i\hbar}{\tau_{ij}} \rho_{ij} \qquad (1.30)$$

where $\omega_{ij} = (E_i - E_j)/\hbar$, and $[\mathscr{H}^r, \rho]_{ij}$ has been replaced by a relaxation term. The constant τ_{ij} is taken to be real and positive so that $\rho_{ij} \to 0$ for $\mathscr{H}' = 0$. Since ρ is Hermitian, it is necessary that $\tau_{ij} = \tau_{ji}$.

The diagonal elements ρ_{jj} are to approach time-independent equilibrium values ρ_{jj}^e for $\mathscr{H}' \to 0$. A transition probability W_{ij} is defined as the per-unit-time probability of a transition from the state characterized by energy E_i to the state characterized by energy E_j when $\mathscr{H}' = 0$. The inclusion of W_{ij} allows for a change of population of the different energy states in the absence of the applied perturbation $\mathscr{H}'$ in order to obtain the appropriate statistical distribution. With this modification,

$$i\hbar \frac{\partial \rho_{jj}}{\partial t} = [\mathscr{H}', \rho]_{jj} + i\hbar \sum_k (\rho_{kk} W_{kj} - \rho_{jj} W_{jk}) \qquad (1.31)$$

where the terms in (1.31) of the form $\rho_{kk} W_{kj}$ represent the per-unit-time increase in the probability of occupation of the j state resulting from transitions from the k to the j state, and the terms $-\rho_{jj} W_{jk}$ represent the per-unit-time decrease in the probability of occupation of the j state resulting from transitions from the j to the k state. With $\mathscr{H}' = 0$ and in equilibrium we require that the ρ_{jj} be time-independent and that there be no net emission or absorption of radiation at any transition frequency. As shown in Appendix 2, to satisfy these conditions the number of transitions per unit time from the j to the k state should equal the number of transitions per unit time from the k to the j state:

$$\rho_{kk}^e W_{kj} = \rho_{jj}^e W_{jk} \qquad (1.32)$$

It is useful to introduce a relaxation time T_{jk}, defined as

$$T_{jk} = \frac{\rho^e_{kk}}{W_{jk}} \qquad (1.33)$$

From Equation 1.32 we see that $T_{kj} = T_{jk}$. Therefore (1.31) can be written

$$i\hbar \frac{\partial \rho_{jj}}{\partial t} = [\mathcal{H}', \rho]_{jj} + i\hbar \sum_k \left(\rho_{kk} \frac{\rho^e_{jj}}{T_{jk}} - \rho_{jj} \frac{\rho^e_{kk}}{T_{jk}} \right) \qquad (1.34)$$

An interesting case to consider is where all T_{jk} are equal, for this results in a simplification of (1.34). Letting all $T_{jk} = T_1$, from Equation 1.34

$$i\hbar \frac{\partial \rho_{jj}}{\partial t} = [\mathcal{H}', \rho]_{jj} + \frac{i\hbar}{T_1} \sum_k (\rho_{kk} \rho^e_{jj} - \rho_{jj} \rho^e_{kk}) \qquad (1.35)$$

As proved previously, Tr $(\rho) = 1 = $ Tr (ρ^e), so that (1.35) becomes

$$i\hbar \frac{\partial \rho_{jj}}{\partial t} = [\mathcal{H}', \rho]_{jj} + \frac{i\hbar}{T_1} (\rho^e_{jj} - \rho_{jj}) \qquad (1.36)$$

With $\mathcal{H}' = 0$, ρ_{jj} decays with a time constant T_1 to the equilibrium value ρ^e_{jj}.

If we restrict our system to one in which only two eigenstates are involved with corresponding energy eigenvalues E_1 and E_2, then there are only two time constants to consider. The time constant for the diagonal matrix elements is T_{12}, and for the off-diagonal elements it is τ_{12}. Constant T_{12} is generally written as T_1 and is termed the longitudinal, spin-lattice, or dipole-lattice relaxation time. This constant is a measure of the time required for a perturbed system to reach an energy equilibrium with its environment. The terms in the randomizing Hamiltonian that produce T_1 could result from interactions between a molecule and its lattice, from collisions between molecules such as in a gas or liquid, or from spontaneous emission as discussed in Chapter 6.

Constant τ_{12} is generally written as T_2 and is termed the transverse, dipole-dipole, or spin-spin relaxation time. As we shall see in Chapter 3, this constant is associated with the linewidth of a transition. In general, we are interested in the electric or magnet moment resulting from an aggregate of molecules, and so the total moment depends upon the phase relationship among the moments of the individual molecules. If we initially start all the molecules vibrating in phase, T_2 is a measure of the amount of time required for the vibrating molecules to have arbitrary phases relative to each other. Any mechanism that contributes to T_1, such as a loss of energy to the lattice, also contributes to T_2, and thus $T_2 \leq T_1$. There can be additional randomizing interactions between molecules that contribute to T_2 and not T_1. For example, if there is an elastic collision in which one molecule makes the transition $|1\rangle \to |2\rangle$ and another makes the transition $|2\rangle \to |1\rangle$, the system does not alter its energy and so T_1 is unaffected by this interaction.

TABLE 1.2 RELAXATION MECHANISMS

Symbol	T_1	T_2
Nomenclature	Longitudinal relaxation; spin-lattice relaxation; dipole-lattice relaxation	Transverse relaxation; spin-spin relaxation; dipole-dipole relaxation
Contributing factors	Spontaneous emission; interactions with the lattice; inelastic collisions	All the factors contributing to T_1; elastic collisions

These transitions generally do result in a loss of phase information and thus T_2 is affected. Table 1.2 summarizes the two types of relaxation. The reason for the terms "transverse" and "longitudinal" relaxation will be discussed in the section of Chapter 2 dealing with spin interactions.

1.4.4 The Density Operator for a Continuum of Eigenstates

Thus far we have considered a system with a discrete set of energy eigenstates, but (as discussed in Chapter 8) at times we deal with a continuum of eigenstates. For example, the energy eigenvalue for an electron in a crystal is a continuous function of a vector **k**. A state vector $|\psi\rangle$ may be written as

$$|\psi\rangle = \int d^3k \, c(\mathbf{k}) \, |u(\mathbf{k})\rangle$$

where

$$d^3k = dk_x \, dk_y \, dk_z = \text{volume element in } \mathbf{k}\text{-space}$$

The $|u(\mathbf{k})\rangle$ are the basis functions satisfying the orthogonality relationship

$$\langle u(\mathbf{k}') \, | \, u(\mathbf{k}) \rangle = \delta(\mathbf{k}' - \mathbf{k})$$

where $\delta(\mathbf{k}' - \mathbf{k})$ is the Dirac delta function.[4]

For a pure state, ρ is defined as

$$\rho = |\psi\rangle\langle\psi|$$

and the matrix elements of ρ are

$$\rho(\mathbf{k}, \mathbf{k}') = \langle u(\mathbf{k})| \, \rho \, |u(\mathbf{k}')\rangle = c(\mathbf{k}) \, c^*(\mathbf{k}') \quad (1.37)$$

The diagonal elements $\rho(\mathbf{k}, \mathbf{k}) \equiv \rho(\mathbf{k})$ are probability densities, and thus $\rho(\mathbf{k})$ is the probability that an electron will be found in a unit volume in **k** space. The equation corresponding to $\text{Tr}(\rho) = 1$ now becomes

$$\int d^3k \, \rho(\mathbf{k}) = 1 \quad (1.38)$$

where the integral is over all **k** space.

For a set of discrete levels, the commutator equals

$$[\mathcal{H}', \rho]_{ij} = \sum_{l} [\mathcal{H}'_{il}\rho_{lj} - \rho_{il}\mathcal{H}'_{lj}]$$

whereas for the continuum the commutator is

$$[\mathcal{H}', \rho]_{\mathbf{kk}'} = \int d^3k'' [\mathcal{H}'_{\mathbf{kk}''}\rho(\mathbf{k}'', \mathbf{k}') - \rho(\mathbf{k}, \mathbf{k}'')\mathcal{H}'_{\mathbf{k}''\mathbf{k}'}] \quad (1.39)$$

The equations of motion for the matrix elements of the density operator, (1.30) and (1.36), are still applicable, but it is necessary to use the integral form of the commutator, Equation 1.39, for the continuum of eigenstates.

It is often convenient to express the probability density $\rho(\mathbf{k})$ as a product of three factors:

$$\rho(\mathbf{k}) = n(\mathbf{k})PN_V^{-1} \quad (1.40)$$

where

$n(\mathbf{k}) = $ number of eigenstates per unit volume in $\mathbf{k}$ space, per unit volume in coordinate space

$P = $ probability that an eigenstate is occupied by an electron

$N_V = $ number of electrons per unit volume in coordinate space.

Any operator A has an expectation value given by

$$\langle A \rangle = \text{Tr}(\rho A) = \int d^3k \int d^3k' \rho(\mathbf{k}, \mathbf{k}') A_{\mathbf{k}'\mathbf{k}} \quad (1.41)$$

where $A_{\mathbf{k}'\mathbf{k}}$ has the usual meaning

$$A_{\mathbf{k}'\mathbf{k}} = \langle u(\mathbf{k}')| A |u(\mathbf{k}) \rangle$$

The equations presented in this section will be used in Chapter 8, where we study the interaction between electromagnetic radiation and electrons in crystals.

1.4.5 Time Derivatives of the Expectation Value of an Operator

In succeeding chapters we shall be interested in the time derivatives of the expectation value for an operator. In particular, let us consider the case where all τ_{ij} are equal and all T_{ij} are equal, so that we may write

$$\tau_{ij} = T_2$$
$$T_{ij} = T_1$$

Since $\langle A \rangle = \text{Tr}(\rho A)$, for an operator that is not an explicit function of time we have

$$\langle \dot{A} \rangle = \text{Tr}(\dot{\rho} A) \quad (1.42)$$

where $\langle \dot{A} \rangle \equiv \partial \langle A \rangle / \partial t$. If (1.30) and (1.36) are substituted into the right-hand side of (1.42), we find that $\langle \dot{A} \rangle$ may be written as

$$\langle \dot{A} \rangle = \frac{1}{i\hbar} \sum_{i,j} [\mathscr{H}', \rho]_{ij} A_{ji} + \frac{1}{T_1} \sum_i (\rho_{ii}^e - \rho_{ii}) A_{ii} - i \sum_{i,j} \omega_{ij} \rho_{ij} A_{ji} - \frac{1}{T_2} \sum_{i \neq j} \rho_{ij} A_{ji} \quad (1.43)$$

Equation 1.43 may be simplified by noting that

$$\sum_{i,j} [\mathscr{H}', \rho]_{ij} A_{ji} = \text{Tr}\,([\mathscr{H}', \rho] A)$$
$$= \text{Tr}\,(\rho [A, \mathscr{H}']) \quad (1.44)$$
$$= \langle [A, \mathscr{H}'] \rangle$$

Similarly

$$\sum_{i,j} \hbar \omega_{ij} \rho_{ij} A_{ji} = \langle [A, \mathscr{H}_0] \rangle \quad (1.45)$$

Since $\rho_{ij}^e = 0$ for $i \neq j$, the summation $\sum_i \rho_{ii}^e A_{ii}$ may be written as

$$\sum_i \rho_{ii}^e A_{ii} = \text{Tr}\,(\rho A)^e = \langle A \rangle^e \quad (1.46)$$

where $\langle A \rangle^e$ is the equilibrium value for $\langle A \rangle$ when $\mathscr{H}' = 0$. The substitution of (1.44)–(1.46) into (1.43) and a rearrangement of some of the terms yield

$$\langle \dot{A} \rangle + \frac{\langle A \rangle}{T_2} - \frac{\langle A \rangle^e}{T_1} = \frac{1}{i\hbar} \langle [A, \mathscr{H}] \rangle + \left(\frac{1}{T_2} - \frac{1}{T_1} \right) \sum_i \rho_{ii} A_{ii} \quad (1.47)$$

where $\mathscr{H} = \mathscr{H}_0 + \mathscr{H}'$.

There are two cases of special interest:

1. All diagonal elements of A are zero. In this case, (1.47) reduces to

$$\langle \dot{A} \rangle + \frac{\langle A \rangle}{T_2} = \frac{1}{i\hbar} \langle [A, \mathscr{H}] \rangle \quad (1.48)$$

2. Only the diagonal elements of A are nonzero. Equation 1.47 then becomes

$$\langle \dot{A} \rangle + \frac{\langle A \rangle - \langle A \rangle^e}{T_1} = \frac{1}{i\hbar} \langle [A, \mathscr{H}] \rangle \quad (1.49)$$

Equations 1.48 and 1.49 will be applied to a variety of problems in later chapters.

We shall also require the second time derivative of operators whose diagonal elements are zero. If we differentiate Equation 1.48 with respect to time, we have

$$\langle \ddot{A} \rangle + \frac{\langle \dot{A} \rangle}{T_2} = \frac{1}{i\hbar} \langle [A, \dot{\mathscr{H}}] \rangle \quad (1.50)$$

The right-hand side of (1.50) is evaluated according to (1.47), whereupon we obtain

$$\langle \ddot{A} \rangle + \frac{2}{T_2}\langle \dot{A} \rangle + \frac{1}{T_2^2}\langle A \rangle = -\frac{1}{\hbar^2}\langle [[A, \mathcal{H}], \mathcal{H}]\rangle + \frac{1}{i\hbar}\left[\frac{1}{T_2} - \frac{1}{T_1}\right]\sum_i \rho_{ii}[A, \mathcal{H}']_{ii} \quad (1.51)$$

In general, we shall be interested in the case where $[A, \mathcal{H}'] = 0$, so that the last term on the right-hand side of (1.51) is zero.

The equations for the time derivatives of operators will be applied in Chapter 2 to the study of the dipole transition. We shall derive equations of motion in terms of observable quantities, thereby going from the formalism of quantum mechanics to phenomena that are investigated in the laboratory. Chapter 3 will be devoted to some of these experimentally observed phenomena, such as resonance absorption, saturation, and dispersion.

REFERENCES

1. P. Dirac, *The Principles of Quantum Mechanics*, 4th ed., Chaps. 1–5, Clarendon Press, Oxford, 1958.
2. V. F. Weisskopf and E. Wigner, *Z. Physik*, **63**, 54, 1930.
3. See, for example, W. Louisell, *Radiation and Noise in Quantum Electronics*, Sec. 6.6, McGraw-Hill, New York, 1964.
4. For additional discussion of continuous spectra and the Dirac delta function, see F. Mandl, *Quantum Mechanics*, Butterworths Scientific Publications, London, 1957, pp. 14–19.

A further discussion of the density operator is presented in U. Fano, "Description of States in Quantum Mechanics by Density Matrix and Operator Techniques," *Rev. Mod. Physics*, **29**, 74–93, January 1957. Additional texts on quantum mechanics:

(a) C. W. Sherwin, *Introduction to Quantum Mechanics*, Henry Holt and Co., New York, 1959.
(b) P. A. Lindsay, *Introduction to Quantum Mechanics for Electrical Engineers*, McGraw-Hill, New York, 1967.
(c) L. Pauling and E. B. Wilson, *Introduction to Quantum Mechanics*, McGraw-Hill, New York, 1935.
(d) R. L. White, *Basic Quantum Mechanics*, McGraw-Hill, New York, 1966.
(e) A. Messiah, *Quantum Mechanics*, Vols. I and II, North-Holland Publishing Co., Amsterdam, 1961.
(f) R. H. Dicke and J. P. Wittke, *Introduction to Quantum Mechanics*, Addison-Wesley, Reading, Mass., 1961.
(g) F. Mandl, *Quantum Mechanics*, Butterworths Scientific Publications, London, 1957.
(h) L. I. Schiff, *Quantum Mechanics*, McGraw-Hill, New York, 1955.

PROBLEMS

1.1 Prove or disprove the following statements:
(a) An operator with real eigenvalues is always Hermitian.

(b) The eigenvalues of an Hermitian operator are always real.
(c) If the expectation value of an operator is real for any arbitrary state vector, then the operator is always Hermitian.

1.2 If $|\psi(t_0)\rangle$ is a solution to the wave equation at $t = t_0$, and

$$|\psi(t)\rangle = U(t, t_0) |\psi(t_0)\rangle$$

is a general solution to the wave equation, determine the conditions that must be satisfied by the function $U(t, t_0)$.

1.3 Show that $\text{Tr}(\rho^2) \leq 1$. What is $\text{Tr}(\rho^2)$ for an eigenstate; a pure state; a mixed state?

1.4 The form of the density operator for a system in thermal equilibrium with its environment is given by the expression†

$$\rho = \frac{\exp(-\mathcal{H}/\kappa T)}{\text{Tr} \exp(-\mathcal{H}/\kappa T)}$$

where

$$\exp\left(\frac{-\mathcal{H}}{\kappa T}\right) |u_i\rangle = \exp\left(\frac{-E_i}{\kappa T}\right) |u_i\rangle$$

if $\mathcal{H} |u_i\rangle = E_i |u_i\rangle$. The constant κ is Boltzmann's constant. (In general, $f(A) |u_i\rangle = f(A_i) |u_i\rangle$ where A is any operator, $|u_i\rangle$ is an eigenfunction of A, A_i is the corresponding eigenvalue, and $f(A)$ is any function of A.) Show that the diagonal elements of ρ are given by the Boltzmann distribution and the off-diagonal elements are zero. Compare this result with the equilibrium expressions for the matrix elements of ρ given by (1.30) and (1.36).

1.5 The entropy of a system is a measure of how much is unknown about the the system. If the entropy is increased we know less about the state of the system, and conversely we know more if the entropy is reduced. In terms of the density operator the entropy $\mathcal{S}$ is given by the expression

$$\mathcal{S} = -\kappa \, \text{Tr}(\rho \ln \rho)$$

where κ is Boltzmann's constant. For a basis in which ρ is diagonal, perform the following:
(a) Show that the entropy is zero when the system is in a pure state.
(b) Obtain an expression for the entropy when there is equal probability for the system to be in any of its eigenstates (subject to the constraint that $\text{Tr}\,\rho = 1$).
(c) Show that the expression given for the entropy in part b is the maximum value the entropy can have.

† See W. Louisell, *Radiation and Noise in Quantum Electronics*, McGraw-Hill, New York, 1964, pp. 228–233.

1.6 The text gives a probabilistic interpretation for the diagonal elements of the density matrix. Show that $|\rho_{ij}|$ for $i \neq j$ is the geometric mean of the probabilities that $|u_i\rangle$ and $|u_j\rangle$ are occupied when the system is in a pure state.

1.7(a) In the derivation of (1.44) the relationship

$$\text{Tr}\,([\mathcal{H}', \rho]A) = \text{Tr}\,(\rho[A, \mathcal{H}'])$$

was used. Prove this relationship.

(b) Derive (1.45).

1.8 (a) Consider a two-level system with

$$\mathcal{H}_0 = \begin{pmatrix} E_1 & 0 \\ 0 & E_2 \end{pmatrix}$$

where E_1 and E_2 are the energy eigenvalues. If $\mathcal{H}' = bA$, where b is a constant and

$$A = \begin{pmatrix} 0 & a \\ a^* & 0 \end{pmatrix}$$

obtain a second-order differential equation in time for $\langle A \rangle$. Describe a physical system from which the operator A may have been obtained. In this system what is the observable corresponding to A?

(b) For the system specified in part a, consider the operator

$$B = \begin{pmatrix} 1 & 0 \\ 0 & 0 \end{pmatrix}$$

What is the observable that corresponds to the operator B? Obtain a first-order differential equation in time for $\langle B \rangle$. What is the physical significance of this equation?

1.9 For Hermitian operators it is generally assumed that you can always write

$$\int_{x_1}^{x_2} dx\, f^* A g = \int_{x_1}^{x_2} dx\, (Af)^* g$$

where A is the operator and f and g are functions of x. However, the validity of the above equation depends upon the boundary conditions. Show, for example, that if A is the momentum operator the equality holds only if

$$f^*(x_1)\, g(x_1) = f^*(x_2)\, g(x_2)$$

DIPOLE TRANSITIONS

2

2.1 INTRODUCTION

In this chapter we shall study in some detail the interaction between radiation and a two-level quantized medium. This interaction accounts for many of the phenomena associated with the properties of materials, such as the absorption and dispersion of light, the saturation or bleaching of transitions, and the generation of coherent electromagnetic radiation as in the maser or laser. Both the electric dipole and the magnetic dipole (spin $\frac{1}{2}$) transitions are considered.

These phenomena will be investigated using the semiclassical approach in which the fields are treated classically and the medium is treated quantum mechanically. Unless one is interested in noise phenomena, this approach is justified because of the high occupation numbers of macroscopically observable fields. A high state of excitation implies that large quantum numbers are involved, and this in turn, by the correspondence principle of quantum mechanics, implies that the fields can be expected to behave classically. A detailed comparison between the semiclassical and quantum electrodynamic approaches is deferred until Chapter 6, where the phenomenon of spontaneous emission, which is neglected in the semiclassical approach, will be explored.

2.2 HAMILTONIAN OF AN ATOM IN AN ELECTROMAGNETIC FIELD

We begin by considering a simplified model of the atom. Assume that the atom consists of a single electron of charge $-e$ and mass m moving about the nucleus in some effective potential $\mathscr{V}(\mathbf{r})$, where $\mathbf{r}$ is the position of the electron with respect to the nucleus, assumed infinitely heavy. The Hamiltonian for the electron, neglecting spin and relativistic effects, is usually written in the form[1]

$$\mathscr{H} = \frac{1}{2m}(\mathbf{p} + e\mathbf{A})^2 + \mathscr{V} - e\Phi \qquad (2.1)$$

The quantity **p** is the canonical momentum of the electron and is related to the ordinary linear momentum $m\dot{\mathbf{r}}$ by

$$\mathbf{p} = m\dot{\mathbf{r}} - e\mathbf{A}$$

where, in section 2.2, a dot over a variable indicates a total derivative with respect to time. The symbols Φ and $\mathbf{A}$ are the scalar and vector potentials, respectively, of an external electromagnetic field, and $e = |e|$ is the magnitude of the electronic charge.

It is important for what follows to recognize that the form of the Hamiltonian (2.1) is not unique, but is just one of many that we may obtain by starting with a suitably chosen Lagrangian $\mathscr{L}(q, \dot{q}, t)$ and applying the transformation

$$\mathscr{H}(q, p, t) = \sum_i \dot{q}_i p_i - \mathscr{L}(q, \dot{q}, t) \tag{2.2}$$

where q is the coordinate and the canonical momentum p_i is defined by $p_i = \partial\mathscr{L}/\partial\dot{q}_i$. The subscript i refers to a coordinate direction, so that in rectangular coordinates, for example, we have

$$\sum_i \dot{q}_i p_i = \dot{q}_x p_x + \dot{q}_y p_y + \dot{q}_z p_z$$

By a suitably chosen Lagrangian we mean, simply, one that yields the correct equations of motion when substituted into Lagrange's equations,[2]

$$\frac{d}{dt}\left(\frac{\partial\mathscr{L}}{\partial\dot{q}_i}\right) - \frac{\partial\mathscr{L}}{\partial q_i} = 0$$

The Lagrangian on which (2.1) is based is given by

$$\mathscr{L} = \tfrac{1}{2}m\dot{r}^2 - \mathscr{V} + e\Phi - e\dot{\mathbf{r}} \cdot \mathbf{A} \tag{2.3}$$

2.2.1 Multipole Expansion

The vector potential $\mathbf{A}$ seen by the electron varies, in general, from point to point in space. Given the relatively small size of the atom, however, the spatial variation of $\mathbf{A}$ over the dimensions of the atom is slight for the externally applied fields that will be of interest to us.

For frequencies below the ultraviolet portion of the spectrum, wavelengths are greater than hundreds of Angstroms or more (1 Angstrom = 10^{-8} cm). By comparison, the Bohr radius of the ground state of the hydrogen atom is $a_0 = 0.529$ Å. Therefore it is convenient to expand the vector potential $\mathbf{A}(x, y, z, t)$ in a Taylor's series about the nuclear position $\mathbf{R}$, as shown in Figure 2.1,

$$\mathbf{A}(\mathbf{R} + \mathbf{r}, t) = \mathbf{A}(\mathbf{R}, t) + (\mathbf{r} \cdot \nabla_\mathbf{R})\mathbf{A}(\mathbf{R}, t) + \cdots \tag{2.4}$$

In the above expression $\nabla_\mathbf{R}$ is in rectangular coordinates given by

$$\nabla_\mathbf{R} = \mathbf{1}_x \frac{\partial}{\partial R_x} + \mathbf{1}_y \frac{\partial}{\partial R_y} + \mathbf{1}_z \frac{\partial}{\partial R_z}$$

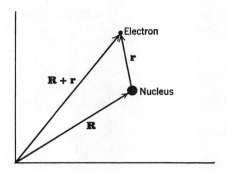

figure 2.1 Geometry used in multipole expansion.

where $\mathbf{1}_x$, $\mathbf{1}_y$, and $\mathbf{1}_z$ are unit vectors in the x, y, and z directions, respectively, and $\partial/\partial R_x$ denotes a partial derivative with respect to x evaluated at the position of the nucleus. An expansion of the form (2.4) leads to interaction terms in the Hamiltonian that correspond to electric dipole, magnetic dipole, electric quadrupole, and other interactions with an external electromagnetic field.

Rather than substitute the above expansion directly into the Hamiltonian (2.1), we find it useful to pursue an extension of a technique due originally to Goeppert-Mayer,[3,4] in which the expansion is instead substituted into the Lagrangian (2.3) where certain mathematical manipulations can be performed more easily.

2.2.2 Electric Dipole Interaction

The first term in the expansion (2.4), when substituted into the Lagrangian (2.3), yields

$$\mathscr{L} = \tfrac{1}{2}m\dot{\mathbf{r}}^2 - \mathscr{V} + e\Phi - e\dot{\mathbf{r}} \cdot \mathbf{A}(\mathbf{R}, t) \tag{2.5}$$

This approximation neglects the spatial variation of $\mathbf{A}$ over atomic dimensions altogether. Since a total time derivative can be added to a Lagrangian without changing the equations of motion of a system,[5] a term of the form

$$e\frac{d}{dt}[\mathbf{r} \cdot \mathbf{A}(\mathbf{R}, t)]$$

may be added in order to transform the Lagrangian into

$$\mathscr{L} = \tfrac{1}{2}m\dot{\mathbf{r}}^2 - \mathscr{V} + e\Phi + e\mathbf{r} \cdot \dot{\mathbf{A}}(\mathbf{R}, t) \tag{2.6}$$

This procedure is followed to shift the time derivative from the electronic coordinate to the vector potential because of the resulting simplification. It follows that in this case the canonical momentum $p_i = \partial \mathscr{L}/\partial \dot{q}_i$ is given by

$$\mathbf{p} = m\dot{\mathbf{r}}$$

and the corresponding Hamiltonian, found by applying (2.2), is

$$\mathscr{H} = \frac{\mathbf{p}^2}{2m} + \mathscr{V} - e\Phi - e\mathbf{r} \cdot \dot{\mathbf{A}}(\mathbf{R}, t) \tag{2.7}$$

In a charge-free region it is possible to choose the potential functions so that the scalar potential Φ vanishes and the fields can be derived from the vector potential alone by means of the relations[6]

$$\mathbf{B} = \nabla \times \mathbf{A}, \qquad \mathbf{E} = -\frac{\partial \mathbf{A}}{\partial t} \tag{2.8}$$

Therefore, (2.7) can be written in the form

$$\mathscr{H} = \frac{\mathbf{p}^2}{2m} + \mathscr{V} - \boldsymbol{\mu} \cdot \mathbf{E}(\mathbf{R}, t) \tag{2.9}$$

where the definition of the electric dipole moment $\boldsymbol{\mu} = -e\mathbf{r}$ has been introduced, and it has been noted that, at the nucleus,

$$\dot{\mathbf{A}}(\mathbf{R}, t) \equiv \frac{d\mathbf{A}(\mathbf{R}, t)}{dt} = \frac{\partial \mathbf{A}(\mathbf{R}, t)}{\partial t} = -\mathbf{E}(\mathbf{R}, t)$$

The derivation above, based on the first term in the expansion of $\mathbf{A}(\mathbf{R} + \mathbf{r}, t)$ about the nuclear position $\mathbf{R}$, is called the electric dipole approximation, since the interaction between the electron and the external field is in the form characteristic of an electric dipole in a quasistatic field. The electric dipole approximation is exceptionally good even for optical fields extending well into the far ultraviolet. Such fields change significantly only over distances on the order of one wavelength, typically several thousand Angstroms, as compared with atomic dimensions of a few Angstroms. Therefore, to a first approximation the spatial variation in $\mathbf{A}$ can be neglected and the Hamiltonian (2.9) employed.

2.2.3 Electric Quadrupole, Magnetic Dipole, and Diamagnetic Interactions

If the second term in the expansion (2.4) is included, an additional term

$$-e\dot{\mathbf{r}} \cdot \mathbf{r} \cdot \nabla_R \mathbf{A}(\mathbf{R}, t)$$

appears in the Lagrangian. Following the procedure established in the dipole case, we add a total time derivative of the form

$$\frac{e}{2} \frac{d}{dt} [\mathbf{r} \cdot \mathbf{r} \cdot \nabla_R \mathbf{A}(\mathbf{R}, t)]$$

Differentiation of the above expression followed by use of the vector identity

$$\mathbf{r} \cdot \dot{\mathbf{r}} \cdot \nabla_R \mathbf{A}(\mathbf{R}, t) = \dot{\mathbf{r}} \cdot \mathbf{r} \cdot \nabla_R \mathbf{A} + \dot{\mathbf{r}} \cdot \mathbf{r} \times [\nabla_R \times \mathbf{A}(\mathbf{R}, t)]$$

then leads to the Lagrangian

$$\mathscr{L} = \tfrac{1}{2} m\dot{\mathbf{r}}^2 - \mathscr{V} + e\Phi + e\mathbf{r} \cdot \dot{\mathbf{A}}(\mathbf{R}, t) + \frac{e}{2} \mathbf{r} \cdot \mathbf{r} \cdot \nabla_R \dot{\mathbf{A}}(\mathbf{R}, t)$$

$$+ \frac{e}{2} \dot{\mathbf{r}} \cdot \mathbf{r} \times [\nabla_R \times \mathbf{A}(\mathbf{R}, t)] \quad (2.10)$$

The canonical momentum $p_i = \partial \mathscr{L} / \partial \dot{q}_i$ is now given by

$$\mathbf{p} = m\dot{\mathbf{r}} + \frac{e}{2} \mathbf{r} \times [\nabla_R \times \mathbf{A}(\mathbf{R}, t)]$$

and the corresponding Hamiltonian is

$$\mathscr{H} = \frac{1}{2m} \left[\mathbf{p} - \frac{e}{2} \mathbf{r} \times \mathbf{B}(\mathbf{R}, t) \right]^2 + \mathscr{V} - \mathbf{\mu} \cdot \mathbf{E}(\mathbf{R}, t) + \frac{e}{2} \mathbf{r} \cdot \mathbf{r} \cdot \nabla_R \mathbf{E}(\mathbf{R}, t) \quad (2.11)$$

where once again the potential functions are chosen so that the scalar potential Φ vanishes and the electric and magnetic fields are determined by the vector potential $\mathbf{A}$ alone in accordance with (2.8).

If the bracketed term in (2.11) is squared and the resulting terms rearranged, the Hamiltonian can be cast in a form highly descriptive of the physical processes involved. In carrying out the square, cross-product terms of the form $\mathbf{p} \cdot \mathbf{r} \times \mathbf{B}$ can be written as $-\mathbf{r} \times \mathbf{p} \cdot \mathbf{B}$ by the usual triple scalar product rule. Although $\mathbf{r}$ and $\mathbf{p}$, when treated as quantum mechanical operators, do not commute, the care usually required in handling noncommutative operators is not required in this case; the vector $\mathbf{r} \times \mathbf{B}$ is perpendicular to $\mathbf{r}$ and therefore, when dotted into $\mathbf{p}$, does not yield noncommuting products. To obtain the final form of the Hamiltonian, we use the fact that a moving particle of charge $-e$ and mass m gives rise to a magnetic moment $\mathbf{m}$ that is related to its mechanical angular momentum $\mathbf{L} = \mathbf{r} \times \mathbf{p}$ by $\mathbf{m} = -(e/2m)\mathbf{L}$. The result is

$$\mathscr{H} = \frac{\mathbf{p}^2}{2m} + \mathscr{V} - \mathbf{\mu} \cdot \mathbf{E}(\mathbf{R}, t) - \mathbf{m} \cdot \mathbf{B}(\mathbf{R}, t) + \frac{e}{2} \mathbf{r} \cdot \mathbf{r} \cdot \nabla_R \mathbf{E}(\mathbf{R}, t)$$

$$+ \frac{e^2}{8m} [\mathbf{r} \times \mathbf{B}(\mathbf{R}, t)]^2 \quad (2.12)$$

The first two terms constitute the unperturbed Hamiltonian in the absence of an applied electromagnetic field; the third and fourth terms correspond to the electric and magnetic dipole interaction terms, respectively; the fifth term represents the electric quadrupole interaction that involves the gradient of the electric field; and the sixth term accounts, for example, for the diamagnetic property of the atom.

The procedure followed above can be extended to higher-order processes as desired, although the need to do so is rare because of the decreasing strength of the higher-order interactions. An elegant derivation for the general case in terms of a canonical transformation of the basic Hamiltonian (2.1) has been presented by Fiutak.[7]

For those cases in which the electric dipole term vanishes, the higher-order moments become significant. In the remainder of the text, however, we shall be interested primarily in those cases in which the electric dipole term predominates. The procedure that we shall employ in detail for the electric dipole case can, if necessary, be applied to the other cases as well.

2.2.4 Multiparticle Systems

Thus far we have considered the case of a single electron in the vicinity of a nucleus. However, problems of interest involve multiparticle systems, such as a multielectron atom or a crystal. Therefore it is appropriate to consider the electric dipole interaction for this case. The Hamiltonian (2.9) generalizes in a straightforward way to a sum over all the election energies

$$\mathcal{H} = \frac{1}{2m} \sum_k \mathbf{p}_k^2 + \mathcal{V}(\mathbf{r}_k) - \boldsymbol{\mu} \cdot \mathbf{E} \tag{2.13}$$

where now the electric dipole moment of the system is given by $\boldsymbol{\mu} = -e \sum_k \mathbf{r}_k$, and the potential energy term $\mathcal{V}(\mathbf{r}_k)$ includes Coulomb interactions between individual electrons located at coordinates $\mathbf{r}_k$ as well as between the electrons and nuclei.[8] It has also been assumed that we are considering in this case a region small enough with respect to spatial variations in the electric field that all of the electrons experience the same field.

The above Hamiltonian can be written in the form

$$\mathcal{H} = \mathcal{H}_0 + \mathcal{H}'$$

where $\mathcal{H}_0$ is the unperturbed Hamiltonian in the absence of an applied electromagnetic field

$$\mathcal{H}_0 = \frac{1}{2m} \sum_k \mathbf{p}_k^2 + \mathcal{V}(\mathbf{r}_k) \tag{2.14}$$

and $\mathcal{H}'$ is the interaction Hamiltonian

$$\mathcal{H}' = -\boldsymbol{\mu} \cdot \mathbf{E} \tag{2.15}$$

It is apparent that the generalization to the multiparticle system does not conceptually alter the previous considerations in any significant way as far as the interaction of the atom with an applied electromagnetic field through the electric dipole term is concerned, and the same argument applies to the other terms in the multipole expansion. The complexity of multiparticle systems lies, instead, in the determination of the eigenfunctions and eigenvalues of the unperturbed Hamiltonian (2.14), which in our work we can assume to be tabulated.

2.3 MATRIX ELEMENTS AND THE CONCEPT OF PARITY

The interaction of an atomic or molecular system with an applied electromagnetic field results from the presence of an $\mathcal{H}'$ term in the Hamiltonian. The strength of the interaction depends on the matrix elements of $\mathcal{H}'$ which in principle can be determined by a quantum mechanical calculation if the wave functions are known. Even if the wave functions are not known explicitly, however, whether a strong interaction occurs can often be determined by arguments concerning the symmetry of the wave functions alone, which, in turn, are determined by the symmetry of the system. Such considerations are described by the concept of parity, which we shall now discuss.

2.3.1 Electric Dipole Transition

For the electric dipole interaction, the matrix elements of $\mathcal{H}'$ are of the form

$$\mathcal{H}'_{mn} = \langle u_m | \mathcal{H}' | u_n \rangle = -\langle u_m | \boldsymbol{\mu} \cdot \mathbf{E} | u_n \rangle \tag{2.16}$$

where $|u_n\rangle$ are taken to be eigenvectors of the unperturbed Hamiltonian $\mathcal{H}_0$ that satisfy the equation

$$\mathcal{H}_0 |u_n\rangle = E_n |u_n\rangle \tag{2.17}$$

As discussed in the previous section, in the dipole approximation the spatial variation of the electric field $\mathbf{E}$ is assumed negligible over atomic dimensions. Therefore, at a given atomic site the electric field is a function of time only and can be taken outside the quantum mechanical scalar product in (2.16). As a result, the matrix elements of $\mathcal{H}'$ are simply the matrix elements of the electric dipole operator $\boldsymbol{\mu}$ with electric field $\mathbf{E}$ as a multiplier,

$$\mathcal{H}'_{mn} = -\boldsymbol{\mu}_{mn} \cdot \mathbf{E} \tag{2.18}$$

It will now be shown that it is possible to determine from rather elementary considerations just which of the matrix elements $\boldsymbol{\mu}_{mn}$ exist and which vanish for a particular system. Such knowledge allows us to determine, for example, whether an electric dipole transition can take place between a certain pair of states under the influence of an applied electromagnetic field, or whether a permanent electric dipole moment exists. Both of these questions are important in determining, for example, the absorption spectra of a material.

For a large class of systems the operation $\mathbf{r} \rightarrow -\mathbf{r}$ applied to the unperturbed Hamiltonian $\mathcal{H}_0(\mathbf{r})$ leaves the form of the Hamiltonian unchanged; that is,

$$\mathcal{H}_0(\mathbf{r}) = \mathcal{H}_0(-\mathbf{r}) \tag{2.19}$$

If the Hamiltonian is a function of several coordinates $r_1, r_2, \ldots$, this operation is to be applied simultaneously to all the coordinates. The invariance of the Hamiltonian with respect to this operation occurs, for example, in a system that possesses a center of symmetry. Another important example is the isolated multielectron atom where the potential energy $\mathscr{V}(r_k)$ is a function only of the distance between particles,

$$\mathscr{V}(r_k) = \mathscr{V}(r_{ij}), \qquad r_{ij} = |r_i - r_j|$$

With the symmetry condition of (2.19) assumed, consider first the case where the eigenfunctions of the Hamiltonian are nondegenerate, and let the Hamiltonian operate on an eigenfunction in accordance with (2.17),

$$\mathscr{H}_0(r) u_n(r) = E_n u_n(r) \tag{2.20}$$

Applying the operation $r \to -r$ to the above and using (2.19), we obtain

$$\mathscr{H}_0(r) u_n(-r) = E_n u_n(-r) \tag{2.21}$$

We see from (2.20) and (2.21) that $u_n(r)$ and $u_n(-r)$ are both eigenfunctions belonging to E_n. For a nondegenerate energy level there is only one such eigenfunction; therefore, in order for (2.20) and (2.21) to be true simultaneously,[9,10]

$$u_n(-r) = \pm u_n(r)$$

That is, the eigenfunction is either an even or an odd function of r. When the eigenfunction is an even function of r, it is said to be a function of even parity and the corresponding state is said to be a state of even parity. If the eigenfunction is an odd function of r, it is said to be a function of odd parity. Therefore the eigenstates are either of even or of odd parity. Such states are said to have a definite parity. It can be shown that even in the case of degeneracy the eigenfunctions can still be chosen so that they exhibit a definite parity.[9]

To summarize, then, if the Hamiltonian is invariant under the transformation $r \to -r$, the eigenstates have a definite parity, which may be either odd or even. The usefulness of the parity concept as a means of classification of states will be apparent in the following discussion:

1. If two eigenstates $|u_m\rangle$ and $|u_n\rangle$ belonging to different energy eigenvalues E_m and E_n have the same parity, then the matrix element μ_{mn} vanishes, since the calculation involves integrals over all space of the form

$$\int u_m^*(r) \, r \, u_n(r) \, dV$$

and the integrand is of odd parity. If the matrix element is zero, transitions between the two states due to an applied electromagnetic field are forbidden, as we shall show in Section 2.4. By "forbidden" we mean forbidden in the electric dipole approximation. Transitions may still be possible in higher-order processes, such as the electric quadrupole transition, but these are generally several orders of magnitude less probable. *Therefore, electric dipole transitions can occur only between states of opposite parity.* This type of constraint is known as a selection rule.

2. If the eigenstates have a definite parity *and* are nondegenerate, the diagonal elements of the dipole operator $\boldsymbol{\mu}$, and hence of the interaction Hamiltonian (2.18), vanish, since the calculation involves integrals of the form

$$\int u_n^*(\mathbf{r})\,\mathbf{r}\,u_n(\mathbf{r})\,dV$$

Under this condition, a permanent electric dipole moment, that is, a permanent or static expectation value of the operator $\boldsymbol{\mu} = -e\mathbf{r}$, cannot exist. This can be seen by the following calculation, which, at the same time, illustrates the basic procedure for making the transition from a quantum mechanical operator to an observable. The observable, the dipole moment in this case, is obtained simply by forming the expectation value of the corresponding operator. Thus

$$\langle \boldsymbol{\mu} \rangle = \text{Tr}\,(\rho\boldsymbol{\mu}) = \text{Tr} \begin{pmatrix} \rho_{11} & \rho_{12} & \cdots \\ \rho_{21} & \rho_{22} & \cdots \\ \vdots & \vdots & \ddots \end{pmatrix} \begin{pmatrix} 0 & \mu_{12} & \cdots \\ \mu_{21} & 0 & \cdots \\ \vdots & \vdots & \ddots \end{pmatrix}$$

Note that the result contains only off-diagonal terms in ρ_{ij}. By (1.30), ρ_{ij} goes to zero in the steady state in the absence of coupling to an interaction Hamiltonian, and therefore the system lacks a permanent or static electric dipole moment. This is true, for example, for the ground states of all atoms.

3. If, on the other hand, the eigenstates, although possessing a definite parity, are degenerate, or nearly so with respect to thermal excitation as in the case of certain complex molecules, then it is possible for a permanent dipole moment to exist.[10-12] In this case, in constructing a set of orthogonal basis vectors with which to calculate matrix elements, we must mix states of different parity, and therefore the basis vectors do not have a definite parity. Diagonal elements of the electric dipole operator that give rise to a permanent dipole moment can then exist. The rotational spectra of molecules are due entirely to interactions with permanent electric dipole moments.[13]

4. In those systems for which the invariance condition (2.19) does not hold, the eigenstates do not possess a definite parity. In this case, all matrix elements of the dipole operator $\boldsymbol{\mu}$ may exist. Thus, dipole transitions between all states are in general possible, and a permanent electric dipole moment may appear. Placement of an atom in an environment where an electrostatic field exists (for example, in a crystalline field or in an externally applied field) provides an example of such a

system. The field contributes a term to $\mathscr{V}$ in (2.14), which upsets the symmetry and prevents the eigenstates from having a definite parity. The permanent dipole moment that is induced corresponds to the appearance of diagonal terms in the dipole operator μ. Therefore the application of an electrostatic field in this manner can allow certain processes to take place in a given material that would otherwise be prohibited by symmetry.

An important application of the parity concept is its use in conjunction with symmetry considerations to determine selection rules in complex systems even though no information exists as to the precise form of the wave functions.

2.3.2 Electric Quadrupole and Magnetic Dipole Transitions

As we have seen, electric dipole transitions between certain states are forbidden because of parity considerations. For such cases it is useful to consider whether transitions between the states in question can take place by means of higher-order multipole terms such as the magnetic dipole or electric quadrupole term. As an example of the application of parity concepts to higher-order transitions we shall consider the electric quadrupole case.

According to the expansion (2.12), the appropriate interaction Hamiltonian for the electric quadrupole interaction has matrix elements of the form

$$\mathscr{H}'_{mn} = \langle u_m | \mathscr{H}' | u_n \rangle = \frac{e}{2} \langle u_m | \mathbf{r} \cdot \mathbf{r} \cdot \nabla_\mathbf{R} \mathbf{E}(\mathbf{R}, t) | u_n \rangle$$

$$= \frac{e}{2} \langle u_m | \mathbf{r} \cdot \mathbf{r} \cdot | u_n \rangle \nabla_\mathbf{R} \mathbf{E}(\mathbf{R}, t) \quad (2.22)$$

Removal of the function $\nabla_\mathbf{R} \mathbf{E}$ from the quantum mechanical scalar product is possible because it is not a function of the electronic coordinate $\mathbf{r}$. In the evaluation of the above vector products, $\mathbf{r} \cdot \nabla_\mathbf{R}$ is taken first, which yields a scalar. This scalar then operates on $\nabla_\mathbf{R} \mathbf{E}$ to yield a vector. A scalar product of the first $\mathbf{r}$ with this result is then formed. Any other possible sequence of operations will be found to violate vector algebra.

It is seen for this case that if two eigenstates $|u_m\rangle$ and $|u_n\rangle$ corresponding to different energy eigenvalues E_m and E_n have opposite parity, then the matrix element in (2.22) vanishes, since the calculation involves integrals over all space of the form

$$\int u_m^*(\mathbf{r}) \, xy \, u_n(\mathbf{r}) \, dV$$

and the integrand is of odd parity. *Therefore, electric quadrupole transitions can occur only between states of the same parity.* Since this condition for the electric quadrupole case is the opposite of the one that applies in the electric dipole case, when a transition is ruled out for one case because of parity considerations, it is allowed for the other.

By similar arguments it can be shown that, in common with the electric quadrupole case, *magnetic dipole transitions can occur only between states of the same parity*. Such parity considerations can be similarly applied to higher-order multipole interactions.

2.4 EQUATIONS OF MOTION FOR THE ELECTRIC DIPOLE TRANSITION

2.4.1 Hamiltonian

The interaction that takes place between a classical electromagnetic field and a quantized medium by means of the electric dipole transition can be described by a set of coupled nonlinear differential equations, which we shall derive in this section. Electric dipole transitions are generally responsible for the absorption and dispersion properties of materials in the ultraviolet, visible, and near infrared portions of the optical spectrum.

It is assumed that the medium consists of a collection of atoms or molecules possessing a pair of nondegenerate eigenstates of opposite parity, $|u_1\rangle$ and $|u_2\rangle$, separated in energy by $E_2 - E_1 = \hbar\Omega$, as shown in Figure 2.2.

The Hamiltonian for the system is given by

$$\mathcal{H} = \mathcal{H}_0 + \mathcal{H}' \tag{2.23}$$

where $\mathcal{H}_0$ is the Hamiltonian for the undisturbed medium and $\mathcal{H}'$ is the interaction Hamiltonian corresponding to the electric dipole transition

$$\mathcal{H}' = -\boldsymbol{\mu} \cdot \mathbf{E} = -\mu_\alpha E_\alpha \tag{2.24}$$

The subscript $\alpha = x, y, z$ denotes the coordinate direction. Summation over repeated subscripts is understood; that is,

$$\mu_\alpha E_\alpha \equiv \mu_x E_x + \mu_y E_y + \mu_z E_z$$

The basis Hamiltonian $\mathcal{H}_0$ that satisfies the eigenvalue equation $\mathcal{H}_0 |u_n\rangle = E_n |u_n\rangle$ has the matrix elements $(\mathcal{H}_0)_{mn} \equiv \langle u_m| \mathcal{H}_0 |u_n\rangle = E_n \delta_{mn}$, and thus it

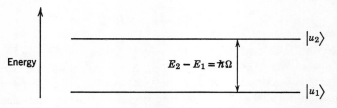

figure 2.2 Energy-level diagram for electric dipole transition.

can be written in matrix form,

$$\mathcal{H}_0 = \begin{pmatrix} E_1 & 0 \\ 0 & E_2 \end{pmatrix} \quad (2.25)$$

In light of the parity discussion of the previous section, the electric dipole operator of interest for a two-level system possessing nondegenerate eigenstates of opposite parity is given by†

$$\mu_\alpha = \begin{pmatrix} 0 & \mu_\alpha \\ \mu_\alpha^* & 0 \end{pmatrix} \quad (2.26)$$

In the above matrix the matrix element μ_x, for example, is simply the x component of the matrix element of the dipole operator $\boldsymbol{\mu}$ taken between states $|u_1\rangle$ and $|u_2\rangle$,

$$\mu_x = \langle u_1 | \boldsymbol{\mu} | u_2 \rangle \cdot \mathbf{1}_x$$

where $\mathbf{1}_x$ is a unit vector in the x direction. The complex conjugate that appears in (2.26) satisfies the requirement that $\boldsymbol{\mu}$, which corresponds to a physical observable, must be represented by a Hermitian operator.

With the electric dipole operator given by (2.26), the corresponding interaction Hamiltonian (2.24) is therefore given by

$$\mathcal{H}' = \begin{pmatrix} 0 & -\mu_\alpha E_\alpha \\ -\mu_\alpha^* E_\alpha & 0 \end{pmatrix} \quad (2.27)$$

Again, summation over repeated subscripts is understood.

We now derive the equations of motion for the classical observables of interest. The technique that we use here is of general applicability and will be used to handle numerous problems throughout the text.

2.4.2 Electric Dipole Moment

An important observable of interest in the electric dipole case is the expectation value of the dipole moment, $\langle \mu_\alpha \rangle$, given by the trace prescription,

$$\langle \mu_\alpha \rangle = \text{Tr}(\rho \mu_\alpha) \quad (2.28)$$

† Note that the symbol μ_α is being used interchangeably for the dipole operator and for its matrix elements. However, it is always clear from the context which is meant, and a considerable simplification in notation results. Otherwise, superfluous notation such as

$$\mu_\alpha = \begin{pmatrix} 0 & (\mu_\alpha)_{12} \\ (\mu_\alpha)_{12}^* & 0 \end{pmatrix}$$

would have to be used.

According to (2.26), all the diagonal elements of μ_α are zero. Therefore the appropriate equation of motion for the first derivative of $\langle \mu_\alpha \rangle$ is given by (1.48),

$$\langle \dot\mu_\alpha \rangle + \frac{\langle \mu_\alpha \rangle}{T_2} = \frac{1}{i\hbar} \langle [\mu_\alpha, \mathcal{H}] \rangle \tag{2.29}$$

where the dot denotes $\partial/\partial t$.

Evaluation of the commutator $[\mu_\alpha, \mathcal{H}]$ with the aid of (2.23)–(2.27) yields

$$[\mu_\alpha, \mathcal{H}] = [\mu_\alpha, \mathcal{H}_0 + \mathcal{H}'] = [\mu_\alpha, \mathcal{H}_0] + [\mu_\alpha, \mathcal{H}'] = [\mu_\alpha, \mathcal{H}_0] = \hbar\Omega \begin{pmatrix} 0 & \mu_\alpha \\ -\mu_\alpha^* & 0 \end{pmatrix} \tag{2.30}$$

where $\hbar\Omega = E_2 - E_1$. The resulting matrix (2.30) is not recognizable as one of the operators that have been previously defined. However, we wish to obtain an expression entirely in terms of the expectation values of variables of interest. Therefore we progress to the expression for the second derivative of $\langle \mu_\alpha \rangle$, given by (1.51),

$$\langle \ddot\mu_\alpha \rangle + \frac{2}{T_2} \langle \dot\mu_\alpha \rangle + \frac{1}{T_2^2} \langle \mu_\alpha \rangle = -\frac{1}{\hbar^2} \langle [[\mu_\alpha, \mathcal{H}], \mathcal{H}] \rangle \tag{2.31}$$

where the second term on the right-hand side of (1.51) vanishes, since $[\mu_\alpha, \mathcal{H}'] = 0$.

The inside commutator on the right-hand side of (2.31) is given by (2.30). Evaluation of the outside commutator proceeds similarly, yielding

$$[[\mu_\alpha, \mathcal{H}], \mathcal{H}] = \begin{pmatrix} -2\hbar\Omega\mu_\alpha\mu_\beta^* E_\beta & \hbar^2\Omega^2 \mu_\alpha \\ \hbar^2\Omega^2 \mu_\alpha^* & 2\hbar\Omega\mu_\alpha\mu_\beta^* E_\beta \end{pmatrix} \tag{2.32}$$

In deriving this equation, we use the fact that $\mu_\alpha\mu_\beta^*$ is real, as proved in Appendix 3.

The matrix (2.32) can be separated into two matrices, one of which corresponds to the dipole moment operator μ_α given by (2.26), the other to a diagonal matrix;

$$[[\mu_\alpha, \mathcal{H}], \mathcal{H}] = \hbar^2\Omega^2 \mu_\alpha - 2\hbar\Omega\mu_\alpha\mu_\beta^* E_\beta D$$

where D is the diagonal matrix:

$$D = \begin{pmatrix} 1 & 0 \\ 0 & -1 \end{pmatrix} \tag{2.33}$$

To determine which observable the diagonal matrix D represents, we note from (2.28) that

$$\langle D \rangle = \mathrm{Tr}\left[\begin{pmatrix} \rho_{11} & \rho_{12} \\ \rho_{21} & \rho_{22} \end{pmatrix} \begin{pmatrix} 1 & 0 \\ 0 & -1 \end{pmatrix} \right] = (\rho_{11} - \rho_{22}) \tag{2.34}$$

That is, we observe that the diagonal matrix corresponds to the difference in occupation probabilities of the lower and upper energy eigenstates. Equation 2.31 therefore yields the following equation of motion for the electric dipole moment:

$$\langle \ddot{\mu}_\alpha \rangle + \frac{2}{T_2} \langle \dot{\mu}_\alpha \rangle + \Omega^2 \langle \mu_\alpha \rangle = \frac{2\Omega}{\hbar} (\mu_\alpha \mu_\beta^*)(\rho_{11} - \rho_{22}) E_\beta^{\text{loc}} \qquad (2.35)$$

In this final form we have taken $\Omega^2 \gg 1/T_2^2$, which holds for most cases of interest. This corresponds physically to the fact that the frequency linewidth of the transition is small compared to the resonant frequency. In addition, we have written the field as E_β^{loc} in order to emphasize that this is the local field acting on an individual atom or molecule. In a dense medium this field is not the same as the macroscopic field that appears in Maxwell's equations. The local field seen by a molecule differs from the macroscopic field because of the influence of nearby polarizable matter. This subject is discussed in detail in Section 2.4.6.

2.4.3 Molecular Averaging

Assuming that there are $N = \mathcal{N}/V$ atoms or molecules per unit volume, each contributing a dipole moment in accordance with (2.35), we determine the total macroscopic polarization of the medium by summing the contributions from each molecule,

$$P_\alpha = \frac{1}{V} \sum_{i=1}^{\mathcal{N}} \langle \mu_\alpha \rangle^i = N_V \overline{\langle \mu_\alpha \rangle} \qquad (2.36)$$

where the overbar indicates an average over all the molecules. The value for $\langle \mu_\alpha \rangle$ may differ from one molecule to another because each molecule may have a different orientation.

In summing and taking the orientational average of both sides of (2.35), we encounter on the right-hand side the term

$$N_V \overline{(\mu_\alpha \mu_\beta^*)(\rho_{11} - \rho_{22})} \qquad (2.37)$$

where it has been assumed that the field E_β^{loc} is constant over the volume of interest. If the spatial variation of E_β^{loc} is important, the field must be included in the averaging process. Since ρ_{ii} is the probability of occupation of the ith state and $N_V \overline{\rho_{ii}}$ is the orientationally averaged expectation value of the number of molecules per unit volume in the ith state, we shall use the definition

$$(N_1 - N_2) \equiv N_V \overline{\rho_{11}} - N_V \overline{\rho_{22}} = N_V \langle D \rangle \qquad (2.38)$$

The quantity $(N_1 - N_2)$ will be referred to as the population difference per unit volume.

Let us first assume that $(\rho_{11} - \rho_{22})$ is approximately the same for all molecules regardless of orientation, that is, the value of $(\rho_{11} - \rho_{22})$ is independent of the molecular orientation. This condition often holds since $(\rho_{11} - \rho_{22})$ is generally established by equilibrium with an isotropic bath, for example, thermal equilibrium with the surroundings or equilibrium with an isotropic radiation source. Under this condition the spatial average (2.37) may be written

$$\overline{(\mu_\alpha \mu_\beta^*) N_V (\rho_{11} - \rho_{22})} = \overline{(\mu_\alpha \mu_\beta^*)} (N_1 - N_2) \tag{2.39}$$

As a second case, if the value of $(\rho_{11} - \rho_{22})$ depends on the orientation of the molecule but all molecules have the same orientation as in a single domain crystal, then (2.37) again reduces to (2.39). If a strong polarized field acts on a collection of molecules, $(\rho_{11} - \rho_{22})$ may change because a large number of transitions take place. If the molecules have arbitrary orientation, then $(\rho_{11} - \rho_{22})$ depends on the orientation, as does $\mu_\alpha \mu_\beta^*$. For this case, (2.37) becomes

$$b \overline{(\mu_\alpha \mu_\beta^*)} (N_1 - N_2) \tag{2.40}$$

where

$$b \equiv \frac{\overline{(\mu_\alpha \mu_\beta^*)(\rho_{11} - \rho_{22})}}{\overline{(\mu_\alpha \mu_\beta^*)} \overline{(\rho_{11} - \rho_{22})}}$$

and $0 \leq b \leq 1$. In the work that follows we shall assume that $b \approx 1$, so that (2.39) holds. In this case the orientational average of the product $(\mu_\alpha \mu_\beta^*)(\rho_{11} - \rho_{22})$ equals the product of the averages, and the two functions in the product are said to be independent as far as their orientation is concerned. If $b \neq 1$, the same equations apply if $(\mu_\alpha \mu_\beta^*)$ is multiplied by b. In Section 3.4.1 and Appendix 5 an estimate is made of the error introduced when there is a strong field and (2.39) is assumed to apply. It is found that, in general, this error is not significant. The equation of motion for the macroscopic polarization $P_\alpha = N_V \langle \mu_\alpha \rangle$, found by summing contributions on both sides of (2.35) and taking into account the above considerations, is

$$\ddot{P}_\alpha + \frac{2}{T_2} \dot{P}_\alpha + \Omega^2 P_\alpha = \frac{2\Omega}{\hbar} \overline{(\mu_\alpha \mu_\beta^*)} (N_1 - N_2) E_\beta^{\text{loc}} \tag{2.41}$$

The required averaging of the quantity $(\mu_\alpha \mu_\beta^*)$ consists of an average over all possible orientations, since this quantity will, in general, vary from molecule to molecule depending on its orientation with respect to the fixed coordinate system used to calculate the matrix elements of the electric dipole moment operator $\boldsymbol{\mu}$. For example, in a gas of anisotropic molecules oriented at random, any given molecule may, according to (2.35), possess a y component of dipole moment resulting from an x component of applied field because of a particular orientation that yields a finite value for $(\mu_y \mu_x^*)$. On the average, however, the y components of dipole moment cancel, since the overall symmetry of a gas is isotropic, which implies that the x component of field is effective only in producing an x component of polarization. In certain crystalline structures, on the other hand, there may be only one orientation possible, in which case the average of the matrix element product is the same as the value for the individual atom.

As an example of orientational averaging, consider the case of the isotropic gas of anisotropic molecules mentioned above. As discussed, $\overline{(\mu_\alpha \mu_\beta^*)} = 0$ for $\alpha \neq \beta$, since for an isotropic medium the induced dipole moment averaged over many molecules must lie in the same direction as the applied field. Therefore, only the terms $\overline{|\mu_x|^2}$, $\overline{|\mu_y|^2}$, and $\overline{|\mu_z|^2}$ need be considered. In addition, in an isotropic medium there is no distinction among the x, y, and z directions and thus $\overline{|\mu_x|^2} = \overline{|\mu_y|^2} = \overline{|\mu_z|^2}$. Let us introduce the quantity

$$|\mathbf{\mu}_{12}|^2 = \overline{|\mu_x|^2} + \overline{|\mu_y|^2} + \overline{|\mu_z|^2} = 3\overline{|\mu_x|^2} \tag{2.42}$$

Then, for the isotropic case, with the aid of (2.42) we can write (2.41) as

$$\ddot{\mathbf{P}} + \frac{2}{T_2}\dot{\mathbf{P}} + \Omega^2 \mathbf{P} = \frac{2\Omega\,|\mathbf{\mu}_{12}|^2}{\hbar}\,\frac{(N_1 - N_2)}{3}\mathbf{E}^{\mathrm{loc}} \tag{2.43}$$

Media with more complicated symmetry properties are considered in Chapter 3.

Equation 2.41 is the basic polarization equation for the quantized medium. We see that the polarization of the medium acts as a harmonic oscillator of frequency Ω driven by an electric field through a coupling coefficient proportional to the population difference. One conclusion that can be drawn from (2.41) is that in the absence of an applied field an initial polarization will decay with a transverse relaxation time constant T_2 because of internal dephasing of the individual dipoles through mutual interaction. Another important conclusion is that the ability of the electric field to drive the polarization is weakened as the populations of the two levels become more nearly equal. In particular, if the populations are equal, the coupling to the electric field is zero, which means that the polarization oscillator does not "see" the driving field at all. We shall see later that this effect constitutes the basis of the phenomenon of saturation.

2.4.4 Population Difference

Another observable of interest in the electric dipole transition problem is the population difference per unit volume defined in (2.38), which derives from the diagonal matrix operator of (2.33) and (2.34):

$$(N_1 - N_2) = N_V \overline{(\rho_{11} - \rho_{22})} = N_V \langle D \rangle \tag{2.44}$$

where

$$D = \begin{pmatrix} 1 & 0 \\ 0 & -1 \end{pmatrix}$$

N_V is the number of atoms or molecules per unit volume, and the overbar indicates an orientational average.

The population difference is a measure of the energy stored in the quantized medium. The power per unit volume delivered to the medium, $\mathscr{P}$, is given by

$$\mathscr{P} = -\frac{\partial}{\partial t}\left[\frac{\hbar\Omega}{2}(N_1 - N_2)\right] \tag{2.45}$$

That is, if an atom within a unit volume makes a transition from level 1 to level 2, N_1 decreases by 1 and N_2 increases by 1. The difference decreases by 2, and the energy stored in the unit volume increases by $\hbar\Omega$.

We begin our consideration of the population difference variable defined in (2.44) by obtaining an equation of motion for the first derivative of the observable $\langle D \rangle = (\rho_{11} - \rho_{22})$. From (1.49), which applies to the case of an operator whose matrix is diagonal, we have

$$\frac{\partial}{\partial t}(\rho_{11} - \rho_{22}) + \frac{(\rho_{11} - \rho_{22}) - (\rho_{11} - \rho_{22})^e}{T_1} = \frac{1}{i\hbar}\langle[D, \mathscr{H}]\rangle \tag{2.46}$$

Evaluation of the commutator $[D, \mathscr{H}]$ with the aid of (2.23)–(2.27) yields

$$[D, \mathscr{H}] = [D, \mathscr{H}_0 + \mathscr{H}'] = [D, \mathscr{H}_0] + [D, \mathscr{H}']$$

$$= [D, \mathscr{H}'] = -2E_\alpha \begin{pmatrix} 0 & \mu_\alpha \\ -\mu_\alpha^* & 0 \end{pmatrix} \tag{2.47}$$

which is recognized to be the matrix that appears in (2.30). Combination of (2.29), (2.30), and (2.47) followed by substitution into (2.46) leads to

$$\frac{\partial}{\partial t}(N_1 - N_2) + \frac{(N_1 - N_2) - (N_1 - N_2)^e}{T_1} = -\frac{2}{\hbar\Omega}\dot{P}_\alpha E_\alpha^{\text{loc}} \tag{2.48}$$

where the spatial averages indicated in (2.36) and (2.44) have been performed, the electric field has been labeled to emphasize that it is the local field seen by the molecules, and P_α/T_2 has been dropped in comparison with $\dot{P}_\alpha$, since $\Omega T_2 \gg 1$ for cases of interest.

Equation 2.48, the basic population difference equation for the quantized medium, is an alternative form of the original density matrix equations for the diagonal elements. This equation is a power balance equation in which changes in the stored energy per unit volume in the quantized medium are driven by a power term of the form $\dot{\mathbf{P}} \cdot \mathbf{E}$. A term of this form should be familiar from Maxwell's equations, where in Poynting's theorem it represents energy delivered to polarizable matter from the fields.

As a result of the second term on the left-hand side of (2.48), an initial population difference $(N_1 - N_2)$ will, in the absence of a driving field, decay to an equilibrium value with a longitudinal or thermalizing relaxation time constant T_1. Therefore the second term corresponds to energy exchange with the surrounding heat bath. The equilibrium value to which the population difference relaxes is typically that established by a Boltzmann distribution, although, as we shall see later, it is possible to establish other equilibrium population differences by certain excitation processes. This use of excitation processes is of significance in the operation of masers and lasers.

Equations 2.41–2.48 for the polarization and population difference per unit volume have been derived on the basis of the density matrix approach of Section 1.4.5. This approach leads to differential equations in terms of classical variables of interest. Such equations lend themselves to physical interpretation, which helps to understand the basic processes at work and guides the approximation procedures to be used.

2.4.5 Field Equations

The equations of motion for the polarization and population difference describe the behavior of the medium in the presence of an electromagnetic field. In order to close the system of equations, it is necessary to include a field equation that takes into account the effects of the dynamical properties of the medium back on the field.

The field equations for an isotropic polarizable medium in the absence of free charge are

$$\nabla \cdot \mathbf{B} = 0 \qquad \nabla \cdot \mathbf{D} = 0$$

$$\nabla \times \mathbf{H} = \mathbf{J} + \frac{\partial \mathbf{D}}{\partial t} \qquad \nabla \times \mathbf{E} = -\frac{\partial \mathbf{B}}{\partial t} \qquad (2.49)$$

$$\mathbf{B} = \mu_0 \mathbf{H} \qquad \mathbf{D} = \epsilon_0 \mathbf{E} + \mathbf{P}^{tot}$$

$$\mathbf{J} = \sigma \mathbf{E}$$

where a conductivity term $\mathbf{J} = \sigma \mathbf{E}$ has been included to account phenomenologically for losses resulting from absorption and scattering caused by transitions other than the one of interest.

We shall show in Section 2.4.7 that the total polarization $\mathbf{P}^{tot}$ can be separated into two parts. One part is a source term resulting from the transition of interest $\mathbf{P}^s$, which is related to the polarization $\mathbf{P}$ appearing in the equations of motion for the transition, (2.41) and (2.48). The other is a source term resulting from all other transitions. As a result of this separation, it is found that the expression for $\mathbf{D}$ can be rewritten

$$\mathbf{D} = \epsilon \mathbf{E} + \mathbf{P}^s$$

When this expression is written in this manner, the dielectric effects of transitions other than the one of interest are taken into account by the use of ϵ in place of ϵ_0, and the effects of the transition of interest are explicitly accounted for by the polarization $\mathbf{P}^s$. The appropriate value to be used for ϵ is that which is measured on the high-frequency side of the transition of interest.[14]

A wave equation in the variable $\mathbf{E}$ can be obtained by taking the curl of the $\nabla \times \mathbf{E}$ equation and substituting from the other equations as appropriate. The result is

$$\nabla \times (\nabla \times \mathbf{E}) + \frac{\eta \mathscr{A}}{c} \frac{\partial \mathbf{E}}{\partial t} + \frac{\eta^2}{c^2} \frac{\partial^2 \mathbf{E}}{\partial t^2} = -\mu_0 \frac{\partial^2 \mathbf{P}^s}{\partial t^2} \qquad (2.50)$$

where $c^2 = 1/\mu_0 \epsilon_0$, $\eta = \sqrt{\epsilon/\epsilon_0}$ is the refractive index of the medium excluding the effect of the transition of interest, and we have defined a damping coefficient, $\mathscr{A} = \mu_0 \sigma c/\eta$ so that in the absence of polarization the energy of a plane wave which is proportional to $|\mathbf{E}|^2$, decays as $e^{-\mathscr{A} z}$ for propagation in the z direction.

The field equation in the form given above readily lends itself to analyses in terms of traveling plane waves of the form

$$\mathbf{E}(\mathbf{r}, t) = \sum_l \frac{\mathbf{E}_0^l}{2} e^{i(\omega_l t - \mathbf{k}_l \cdot \mathbf{r})} + \text{c.c.}$$

where c.c. is the complex conjugate. This is the approach generally taken in discussing topics such as absorption, gain, and harmonic generation.

An alternative form of the field equation that is useful for resonators is obtained by expanding the fields in terms of the normal mode patterns of the resonator as described by Slater.[15] The normal mode patterns are simply the familiar standing wave patterns characteristic of, for example, microwave cavities.

The normal mode expansion is obtained by considering the problem of the distribution of fields inside a cavity with perfectly conducting walls enclosing a dielectric medium of permittivity ϵ. The boundary conditions imposed by such a cavity are that the tangential components of $\mathbf{E}$ and normal component of $\mathbf{B}$ vanish on the surface. It is found that the solutions to the field equations that satisfy the boundary conditions may be written as a sum of an infinite discrete set of normal mode solutions $\mathbf{E}_a(\mathbf{r})$, $\mathbf{H}_a(\mathbf{r})$, which satisfy the equations

$$\nabla \times \mathbf{E}_a = \frac{\omega_a \eta}{c} \mathbf{H}_a, \qquad \nabla \times \mathbf{H}_a = \frac{\omega_a \eta}{c} \mathbf{E}_a \qquad (2.51)$$

From the above equations we note that $\mathbf{E}_a(\mathbf{r})$ and $\mathbf{H}_a(\mathbf{r})$ are the eigenfunctions of the operator $(\nabla \times \nabla \times)$ corresponding to the eigenvalues $(\omega_a \eta/c)^2$. These functions are orthogonal and may be normalized so that

$$\int \mathbf{E}_a \cdot \mathbf{E}_b \, dV = \delta_{ab}, \qquad \int \mathbf{H}_a \cdot \mathbf{H}_b \, dV = \delta_{ab} \qquad (2.52)$$

An arbitrary field can be expressed as a sum of normal mode solutions by the expansions

$$\mathbf{E} = -\frac{1}{\sqrt{\epsilon}} \sum_a p_a(t) \mathbf{E}_a(\mathbf{r}), \qquad \mathbf{H} = \frac{1}{\sqrt{\mu}} \sum_a \omega_a q_a(t) \mathbf{H}_a(\mathbf{r})$$

where the time dependence of the modes is expressed in $q_a(t)$, $p_a(t)$.

2 DIPOLE TRANSITIONS

Substituting the normal mode expansions into the wave equation (2.50) and applying (2.51) and (2.52), we find that the differential equation for $p_a(t)$ is

$$\ddot{p}_a + \frac{\mathscr{A}c}{\eta}\dot{p}_a + \omega_a^2 p_a = \frac{1}{\sqrt{\epsilon}}\int \ddot{\mathbf{P}}^s \cdot \mathbf{E}_a(\mathbf{r}) \, dV \tag{2.53}$$

If we assume that a single normal mode $\mathbf{E}_c(\mathbf{r})$ predominates, then the total field is given by $\mathbf{E} = -p_c(t)\mathbf{E}_c(\mathbf{r})/\sqrt{\epsilon}$. Therefore, (2.53) can be written in the form

$$\ddot{\mathbf{E}} + \frac{1}{\tau_c}\dot{\mathbf{E}} + \omega_c^2 \mathbf{E} = -\frac{1}{\epsilon}\mathbf{E}_c(\mathbf{r})\int \ddot{\mathbf{P}}^s \cdot \mathbf{E}_c(\mathbf{r}) \, dV \tag{2.54}$$

where $\tau_c = \eta/\mathscr{A}c$ is the cavity lifetime, which is the length of time required for the energy in the cavity to decay to $1/e$ in the absence of a driving term.†

In many cases the polarization $\mathbf{P}$ is itself driven by the field and therefore has the same spatial distribution as the normal mode. In this case (2.54) can be simplified to

$$\ddot{\mathbf{E}} + \frac{1}{\tau_c}\dot{\mathbf{E}} + \omega_c^2 \mathbf{E} = -\frac{1}{\epsilon}\ddot{\mathbf{P}}^s \tag{2.55}$$

In the derivation of (2.55) it has been assumed that the polarization source fills the cavity. If this is not the case then the right hand side of (2.55) must be multiplied by a filling factor F given by $F = \int |\mathbf{E}_c(\mathbf{r})|^2 \, dV$ where the integral is over the volume of the polarization source. Since $\mathbf{E}_c(\mathbf{r})$ is normalized over the volume of the resonator, $F = 1$ when the polarization source fills the cavity. In most cases a good approximation for F is the ratio of the volume of the source to the volume of the resonator. Unless specified otherwise we shall take $F = 1$.

The field equation, in the form of either (2.50) or (2.54)–(2.55), when added to the equations of the preceding section completes the set necessary to describe in full the electric dipole interaction of a classical electromagnetic field and a two-level quantized medium.

† The cavity lifetime, although derived here on the basis of conductivity losses, is meant to account for all the loss mechanisms within the cavity. In an optical resonator, for example,

$$\frac{1}{\tau_c} = \frac{1}{\tau_m} + \frac{1}{\tau_s} + \frac{1}{\tau_d}$$

where τ_m is the contribution to the cavity lifetime resulting from the loss of radiation through the end mirrors alone, τ_s corresponds to the scattering and absorption losses excluding the effects of the transition of interest, and τ_d represents the diffraction losses. An expression for τ_m is derived in Section 4.4 in terms of the reflectivity of the mirrors forming the resonator; $\tau_s = \eta/\mathscr{A}c$, where $\mathscr{A}$ is the power attenuation constant for a plane wave propagating through the medium; τ_d depends on geometrical factors such as the radii of the mirrors and mirror separation.[16]

2.4.6 Electric Dipole Transition—Summary

We summarize here the development used to obtain the set of equations for the electric dipole transition.

1. An interaction Hamiltonian of interest is chosen.
2. Expectation values of pertinent observables are formed by the trace prescription.
3. By use of the expressions of Section 1.4.5 for derivatives of expectation values, a set of differential equations in terms of the variables of interest is obtained.
4. A field equation is added to close the system.

In the case of the electric dipole transition an assumed interaction Hamiltonian of the form

$$\mathcal{H}' \rightarrow \begin{pmatrix} 0 & -\mu_\alpha E_\alpha \\ -\mu_\alpha^* E_\alpha & 0 \end{pmatrix}$$

leads to equations of motion, which for an isotropic medium are

$$\ddot{\mathbf{P}} + \frac{2}{T_2}\dot{\mathbf{P}} + \Omega^2 \mathbf{P} = \frac{2\Omega}{\hbar} \frac{|\mu_{12}|^2}{3}(N_1 - N_2)\mathbf{E}^{\text{loc}}$$

$$\frac{\partial}{\partial t}(N_1 - N_2) + \frac{(N_1 - N_2) - (N_1 - N_2)^e}{T_1} = -\frac{2}{\hbar\Omega}\dot{\mathbf{P}}\cdot\mathbf{E}^{\text{loc}}$$
(2.56)

The field equation is

$$\nabla \times (\nabla \times \mathbf{E}) + \frac{\eta \mathscr{A}}{c}\frac{\partial \mathbf{E}}{\partial t} + \frac{\eta^2}{c^2}\frac{\partial^2 \mathbf{E}}{\partial t^2} = -\mu_0 \frac{\partial^2 \mathbf{P}^s}{\partial t^2} \quad \text{(traveling-wave case)}$$

or
(2.57)

$$\ddot{\mathbf{E}} + \frac{1}{\tau_c}\dot{\mathbf{E}} + \omega_c^2 \mathbf{E} = -\frac{1}{\epsilon}\mathbf{E}_c \int \ddot{\mathbf{P}}^s \cdot \mathbf{E}_c(\mathbf{r})\, dV \quad \text{(cavity case)}$$

2.4.7 Local Field Correction Factors

One step yet remains in the progression that has brought us from the microscopic to the macroscopic domain, namely, the determination of the relationships that exist, on the one hand, between the local and macroscopic fields and, on the other, between the polarization source term $\mathbf{P}^s$ that drives the macroscopic field equation and the actual polarization $\mathbf{P}$ that appears in the equations for the medium. These differences are due to the influence of nearly polarizable matter on the local electric field seen by a given atom or molecule.

For two media that are, oddly enough, extreme opposites in density, polarization effects can be ignored. In the first, a dilute gas, there are no nearby neighbors to influence the local field, which is then identical with the macroscopic field. In the second medium, conduction electrons in semiconductors and metals, the electronic wavefunctions are spread throughout the crystal rather than localized at a given atomic site and, as a result, sample the macroscopic field.[17] For these two cases $\mathbf{E}^{loc}$ can be taken equal to $\mathbf{E}$ and $\mathbf{P}^s$ can be taken equal to $\mathbf{P}$.

For other dense media a distinction between the local and macroscopic fields is made by introducing what is known as a Lorentz local field correction factor, which takes into account the modification of the local field by polarization of the surroundings. The pertinent calculation is carried out in Appendix 4 for the case of an isotropic medium. The results are the following:

1. The source term $\mathbf{P}^s$ to be used in the field equation to represent the transition of interest is related to the actual polarization $\mathbf{P}$ by

$$\mathbf{P}^s = \left(\frac{\eta^2 + 2}{3}\right)\mathbf{P} \tag{2.58}$$

where $\eta = \sqrt{\epsilon/\epsilon_0}$ is the refractive index of the medium excluding the effects of the transition of interest.

2. The local field is related to the macroscopic field by the relationship

$$\mathbf{E}^{loc} = \left(\frac{\eta^2 + 2}{3}\right)\mathbf{E} \tag{2.59}$$

With the aid of (2.58) and (2.59), the equations of motion for an isotropic medium are replaced by

$$\ddot{\mathbf{P}}^s + \frac{2}{T_2}\dot{\mathbf{P}}^s + \Omega^2 \mathbf{P}^s = \frac{2\Omega}{\hbar} \frac{L\,|\mathbf{\mu}_{12}|^2}{3}(N_1 - N_2)\mathbf{E} \tag{2.60a}$$

$$\frac{\partial}{\partial t}(N_1 - N_2) + \frac{(N_1 - N_2) - (N_1 - N_2)^e}{T_1} = -\frac{2}{\hbar\Omega}\dot{\mathbf{P}}^s \cdot \mathbf{E} \tag{2.60b}$$

where $L = [(\eta^2 + 2)/3]^2$ is the Lorentz correction factor for the isotropic case. The field equations remain:

$$\nabla \times (\nabla \times \mathbf{E}) + \frac{\eta \mathscr{A}}{c}\frac{\partial \mathbf{E}}{\partial t} + \frac{\eta^2}{c^2}\frac{\partial^2 \mathbf{E}}{\partial t^2} = -\mu_0 \frac{\partial^2 \mathbf{P}^s}{\partial t^2} \quad \text{(traveling-wave case)} \tag{2.60c}$$

$$\ddot{\mathbf{E}} + \frac{1}{\tau_c}\dot{\mathbf{E}} + \omega_c^2 \mathbf{E} = -\frac{1}{\epsilon}\mathbf{E}_c(\mathbf{r})\int \ddot{\mathbf{P}}^s \cdot \mathbf{E}_c(r)\,dV \quad \text{(cavity case)} \tag{2.60d}$$

Thus we see that the effect of the polarization of the surroundings on the form of the equations for the electric dipole case is to introduce the Lorentz correction factor L.

Equation 2.60a indicates that the polarization acts as a harmonic oscillator driven by an electric field through a coupling coefficient proportional to the population difference. The second equation for the population difference, (2.60b), is a power balance equation that relates changes in the energy stored per unit volume in the medium to a driving term of the form $\dot{\mathbf{P}}^s \cdot \mathbf{E}$. Equations 2.60c and 2.60d describe the behavior of the macroscopic electromagnetic field as influenced by the medium through the driving source term $\ddot{\mathbf{P}}^s$. This set of coupled equations will be taken in Chapter 3 as the basis for discussion of various phenomena such as absorption, dispersion, saturation, and maser action.

Equations 2.60 apply to an isotropic medium. For an anisotropic medium such as a complex crystal, equations of the same form apply when the polarization and field are polarized along principal axes of the system. (Principal axes are those directions for which the polarization component depends on the field only through that component lying in the same direction, that is, the tensor relating field to polarization is diagonal.) Therefore the results derived in this chapter on the basis of (2.60) apply also to the principal axis case under the substitutions

$$\epsilon \to \epsilon_\alpha$$
$$\eta \to \eta_\alpha$$
$$L \to L_\alpha \quad (2.61)$$
$$\frac{|\boldsymbol{\mu}_{12}|^2}{3} \to |(\mu_\alpha)_{12}|^2$$

The substituted values are those appropriate to the polarization direction α and will, in general, be different for different directions.

For polarizations in arbitrary directions in an anisotropic medium, we must return to (2.41) and (2.48) and carry through for the general case the development that is presented for the isotropic case.

2.5 EQUATIONS OF MOTION FOR THE MAGNETIC DIPOLE SPIN ½ SYSTEM

A magnetic case of interest that closely parallels the electric dipole transition is found in the magnetic dipole spin ½ system. One example of a spin ½ system is the electron. Other examples include protons, neutrons, and certain "one-electron" atoms such as silver that consist of closed shells and subshells whose spins cancel apart from a single valence electron in an s state.[18] In the discussion that follows we shall confine our attention to the electron. The magnetic dipole term considered in this section is not the induced dipole considered in the multipole expansion of Section 2.2.3, but corresponds instead to a permanent magnetic dipole.

The electron has associated with it a measurable spin angular momentum $\hbar/2$ and a magnetic dipole moment $e\hbar/2m$ in MKS units.[19] These properties, although suggestive of a spinning charge distribution, have no satisfactory classical

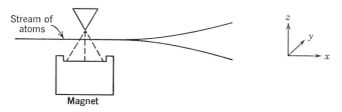

figure 2.3 Stern-Gerlach experiment.

explanation. They are therefore considered as purely quantum mechanical in origin and are taken simply to represent certain internal degrees of freedom of the particle.

Experimentally it is found that the spin can assume only two possible orientations with respect to some fixed direction in space, usually taken as the z axis. One orientation is parallel, the other antiparallel, to the reference axis. We account for this behavior quantum mechanically by postulating the existence of two original spin eigenstates represented by $|+\rangle$ and $|-\rangle$ that satisfy an eigenvalue equation

$$s_z |\pm\rangle = \pm \frac{\hbar}{2} |\pm\rangle \tag{2.62}$$

where s_z is a spin operator corresponding to a measurement of the z component of spin. The eigenvalues of this operator are $\pm \hbar/2$. A measurement of this type can be performed by means of the Stern-Gerlach experiment (Figure 2.3), in which a stream of particles passing through an inhomogeneous magnetic field is divided into two well-defined spatially separated beams according to their individual spins. The deflection is caused by the force experienced by the magnetic dipole moment associated with the spin.[20]

In the study of spin it is convenient to introduce the notation characteristic of matrix mechanics, where we let the orthogonal eigenstates $|\pm\rangle$ be represented by the matrices

$$|+\rangle \rightarrow \begin{pmatrix} 1 \\ 0 \end{pmatrix}, \quad |-\rangle \rightarrow \begin{pmatrix} 0 \\ 1 \end{pmatrix} \tag{2.63}$$

In this notation an arbitrary spin state

$$|\psi_s\rangle = a_+ |+\rangle + a_- |-\rangle$$

is represented by the column matrix

$$|\psi_s\rangle = a_+ \begin{pmatrix} 1 \\ 0 \end{pmatrix} + a_- \begin{pmatrix} 0 \\ 1 \end{pmatrix} = \begin{pmatrix} a_+ \\ a_- \end{pmatrix}$$

The form assumed by the operator s_z must be

$$s_z = \frac{\hbar}{2}\begin{pmatrix} 1 & 0 \\ 0 & -1 \end{pmatrix} \tag{2.64}$$

in order that (2.62) and (2.63) be satisfied simultaneously.

If, in addition to postulating the existence of two spin eigenstates, we also postulate that the spin operators satisfy the usual angular momentum commutation relations

$$[s_i, s_j] = i\hbar s_k \tag{2.65}$$

where i, j, k form a cyclic permutation of x, y, z, then the operators s_x and s_y are given by

$$s_x = \frac{\hbar}{2}\begin{pmatrix} 0 & 1 \\ 1 & 0 \end{pmatrix}, \quad s_y = \frac{\hbar}{2}\begin{pmatrix} 0 & -i \\ i & 0 \end{pmatrix}$$

The postulate (2.65) is justified in the end by comparison of predicted results with experimental observations. The spin operators introduced above are often written in the form $\mathbf{s} = (\hbar/2)\boldsymbol{\sigma}$, where the components of $\boldsymbol{\sigma}$ are the Pauli spin operators

$$\sigma_x = \begin{pmatrix} 0 & 1 \\ 1 & 0 \end{pmatrix}, \quad \sigma_y = \begin{pmatrix} 0 & -i \\ i & 0 \end{pmatrix}, \quad \sigma_z = \begin{pmatrix} 1 & 0 \\ 0 & -1 \end{pmatrix} \tag{2.66}$$

The electron has a magnetic moment $\mathbf{m}$, which is related to the spin $\mathbf{s}$ by

$$\mathbf{m} = -\frac{e}{m}\mathbf{s} = -\frac{\gamma\hbar}{2}\boldsymbol{\sigma} \tag{2.67}$$

where $e = |e|$ is the magnitude of the electronic charge and γ is defined as the gyromagnetic ratio.† The magnetic moment operators are, from (2.66) and (2.67),

$$m_x = -\frac{\gamma\hbar}{2}\begin{pmatrix} 0 & 1 \\ 1 & 0 \end{pmatrix}, \quad m_y = -\frac{\gamma\hbar}{2}\begin{pmatrix} 0 & -i \\ i & 0 \end{pmatrix}, \quad m_z = -\frac{\gamma\hbar}{2}\begin{pmatrix} 1 & 0 \\ 0 & -1 \end{pmatrix} \tag{2.68}$$

Thus far in our discussion we have been treating the spin eigenstates of the electron as if they existed independently of the rest of the electronic wavefunction. Strictly speaking, we have been able to make this approximation only by neglecting the interaction between the orbital and spin coupling, that is, by neglecting the interaction between the orbital and spin magnetic moments of the atom. This

---

† Equation 2.67 is a special case of a general theorem of quantum mechanics that states that the magnetic dipole moment $\mathbf{m}$ and angular momentum $\hbar\mathcal{J}$ of an isolated system are related by

$$\mathbf{m} = -g\beta\mathcal{J} = -\gamma\hbar\mathcal{J}$$

where β is the Bohr magneton $\beta = e\hbar/2m$, and γ is the gyromagnetic ratio $ge/2m$, with g the spectroscopic splitting factor, equal to 2 for the electron spin.

approximation is, in fact, good for all but the heavier elements because only a small amount of interaction energy is ordinarily involved in spin-orbit coupling. When this approximation holds we are allowed to write the eigenvector of the electron as a product of space and spin eigenvectors.[21-24] Therefore, associated with a particular electronic spatial eigenstate $|u_k\rangle$ there are two eigenvectors,

and
$$|\varphi_1\rangle = |u_k\rangle |+\rangle$$
$$|\varphi_2\rangle = |u_k\rangle |-\rangle \qquad (2.69)$$

Spatial and spin operators operate separately on the spatial and spin parts, respectively, of the total eigenvector $|\varphi\rangle$.

2.5.1 Hamiltonian

In order to account for the interaction of the magnetic moment of the electron with magnetic fields, a term of the form $-\mathbf{m} \cdot \mathbf{B}$ must be added to the Hamiltonian (2.1) for the electron, where $\mathbf{B}$ is the magnetic flux density. The total Hamiltonian $\mathcal{H}$ can be written in the form

$$\begin{aligned} \mathcal{H} &= \mathcal{H}_0 + \mathcal{H}' \\ &= (\mathcal{H}_{01} + \mathcal{H}_{02}) + \mathcal{H}' \end{aligned} \qquad (2.70)$$

The first term $\mathcal{H}_{01}$ is taken to be that part of the Hamiltonian associated with the electron-electron and electron-nuclear interactions that give rise to the spatial eigenstates $|u_k\rangle$ that satisfy the eigenvalue equation

$$\mathcal{H}_{01} |u_k\rangle = E_k |u_k\rangle \qquad (2.71)$$

The term $\mathcal{H}_{02}$ is that part of the $-\mathbf{m} \cdot \mathbf{B}$ term that is due to a static component of magnetic field $\mathbf{B}_0$, chosen to lie in the z direction,

$$\mathcal{H}_{02} = -\mathbf{m} \cdot \mathbf{B}_0 = -m_z B_{0z} \qquad (2.72)$$

The symbol $\mathcal{H}'$ is an interaction term that accounts for the effects of a time-dependent magnetic field $\mathbf{B}'$,

$$\mathcal{H}' = -\mathbf{m} \cdot \mathbf{B}' \qquad (2.73)$$

The matrix elements of the above Hamiltonians are calculated in terms of the basis vectors (2.69); $\mathcal{H}_{01}$ operates only on the spatial part of the eigenvector $|u_k\rangle$, whereas $\mathcal{H}_{02}$ and $\mathcal{H}'$ operate only on the spin part of the eigenvector. As in the electric dipole case discussed in the preceding section, the field is treated classically in the spirit of the semiclassical approach. Therefore the matrix elements of $\mathcal{H}_{02}$ and $\mathcal{H}'$ in (2.72) and (2.73) are simply the matrix elements of $\mathbf{m}$ with magnetic flux density as a multiplier.

Consider first the basis Hamiltonian $\mathcal{H}_0 = \mathcal{H}_{01} + \mathcal{H}_{02}$. A typical matrix element is determined as follows:

$$(\mathcal{H}_0)_{11} = (\mathcal{H}_{01})_{11} + (\mathcal{H}_{02})_{11} = \langle \varphi_1 | \mathcal{H}_{01} | \varphi_1 \rangle + \langle \varphi_1 | \mathcal{H}_{02} | \varphi_1 \rangle$$

With $|\varphi_1\rangle$ given by (2.69), we have

$$\begin{aligned}(\mathcal{H}_0)_{11} &= \langle u_k | \langle + | \mathcal{H}_{01} | u_k \rangle | + \rangle + \langle u_k | \langle + | \mathcal{H}_{02} | u_k \rangle | + \rangle \\ &= \langle + | + \rangle \langle u_k | \mathcal{H}_{01} | u_k \rangle + \langle u_k | u_k \rangle \langle + | \mathcal{H}_{02} | + \rangle \end{aligned} \quad (2.74)$$

From (2.63) we have $\langle + | + \rangle = 1$; from (2.71) we see that $\langle u_k | \mathcal{H}_{01} | u_k \rangle = E_k \langle u_k | u_k \rangle = E_k$, assuming that the eigenvectors $|u_k\rangle$ are orthonormal; from (2.67) and (2.72), we see that $\langle + | \mathcal{H}_{02} | + \rangle = (\gamma \hbar B_{0z})/2$. Therefore, (2.74) reduces to

$$(\mathcal{H}_0)_{11} = E_k + \frac{\gamma \hbar B_{0z}}{2}$$

The other matrix elements are determined similarly, with the result that in matrix form the basis Hamiltonian $\mathcal{H}_0 = \mathcal{H}_{01} + \mathcal{H}_{02}$ is written

$$\mathcal{H}_0 = \begin{pmatrix} E_k + \dfrac{\gamma \hbar B_{0z}}{2} & 0 \\ 0 & E_k - \dfrac{\gamma \hbar B_{0z}}{2} \end{pmatrix} \quad (2.75)$$

We see from (2.17) that the eigenvalue equations

$$\mathcal{H}_0 | \varphi_1 \rangle = E_1 | \varphi_1 \rangle$$
$$\mathcal{H}_0 | \varphi_2 \rangle = E_2 | \varphi_2 \rangle$$

are satisfied, provided that we take

$$E_1 = E_k + \frac{\gamma \hbar B_{0z}}{2}$$

$$E_2 = E_k - \frac{\gamma \hbar B_{0z}}{2}$$

Therefore, in the presence of a static magnetic field the electronic eigenstate $|u_k\rangle$ splits into a pair of energy levels separated in energy by an amount $\hbar \Omega = \hbar \gamma B_{0z}$, where the higher energy corresponds to the spin eigenstate $|+\rangle$ (see Figure 2.4). Note that the numbering scheme that labels level 1 as the higher energy state is reversed from the electric dipole case shown in Figure 2.2.

By a procedure similar to the one that was followed to determine the matrix elements of $\mathcal{H}_0$, the interaction Hamiltonian (2.73) is found to be

$$\mathcal{H}' = \frac{\gamma \hbar}{2} \begin{pmatrix} B'_z & B'_x - i B'_y \\ B'_x + i B'_y & -B'_z \end{pmatrix} \quad (2.76)$$

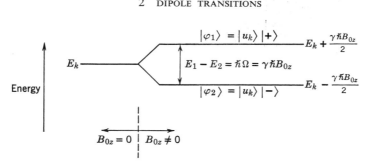

figure 2.4 Energy level splitting scheme for magnetic dipole spin $\frac{1}{2}$ transition.

2.5.2 Equations of Motion

We are now in a position to derive the equations of motion for the spin $\frac{1}{2}$ system. For the dipole moment operator m_x given by (2.68) we apply (1.48) for the first derivative of the expectation value $\langle m_x \rangle = \text{Tr}(\rho m_x)$,

$$\langle \dot{m}_x \rangle + \frac{\langle m_x \rangle}{T_2} = \frac{1}{i\hbar} \langle [m_x, \mathcal{H}] \rangle \tag{2.77}$$

Evaluation of the commutator $[m_x, \mathcal{H}]$ with the aid of (2.75) and (2.76) yields

$$[m_x, \mathcal{H}] = [m_x, \mathcal{H}_0 + \mathcal{H}'] = [m_x, \mathcal{H}_0] + [m_x, \mathcal{H}']$$
$$= -i\hbar\Omega m_y + i\hbar\gamma(B'_y m_z - B'_z m_y) \tag{2.78}$$

Therefore, (2.77) reduces to

$$\langle \dot{m}_x \rangle + \frac{\langle m_x \rangle}{T_2} = \gamma(\mathbf{B'} \times \langle \mathbf{m} \rangle)_x - \Omega \langle m_y \rangle \tag{2.79}$$

Similarly, the expectation value for the dipole moment operator m_y is found to satisfy

$$\langle \dot{m}_y \rangle + \frac{\langle m_y \rangle}{T_2} = \gamma(\mathbf{B'} \times \langle \mathbf{m} \rangle)_y + \Omega \langle m_x \rangle \tag{2.80}$$

For the dipole moment operator m_z given by (2.68), we apply (1.49), which is applicable to operators with diagonal matrices, to obtain the equation of motion for $\langle m_z \rangle = \text{Tr}(\rho m_z)$,

$$\langle \dot{m}_z \rangle + \frac{\langle m_z \rangle - \langle m_z \rangle^e}{T_1} = \gamma(\mathbf{B'} \times \langle \mathbf{m} \rangle)_z \tag{2.81}$$

Equations 2.79–2.81 can be rewritten in the form known as the Bloch equations:[25]

$$\dot{M}_x + \frac{M_x}{T_2} = \gamma(\mathbf{B'} \times \mathbf{M})_x - \Omega M_y$$

$$\dot{M}_y + \frac{M_y}{T_2} = \gamma(\mathbf{B'} \times \mathbf{M})_y + \Omega M_x \quad (2.82)$$

$$\dot{M}_z + \frac{M_z - M_z^e}{T_1} = \gamma(\mathbf{B'} \times \mathbf{M})_z$$

In the above equations $\mathbf{M}$ is the macroscopic magnetization $\mathbf{M} = N_V \overline{\langle \mathbf{m} \rangle}$, where N_V is the number of electrons per unit volume contributing to the total magnetization, and the overbar indicates an orientational average.

We also note at this point that

$$M_z = N_V \overline{\langle m_z \rangle} = N_V \, \overline{\mathrm{Tr}\,(\rho m_z)} = -N_V \frac{\gamma \hbar}{2} \overline{\mathrm{Tr} \left[\begin{pmatrix} \rho_{11} & \rho_{12} \\ \rho_{21} & \rho_{22} \end{pmatrix} \begin{pmatrix} 1 & 0 \\ 0 & -1 \end{pmatrix} \right]}$$

$$= N_V \frac{\gamma \hbar}{2} \overline{(\rho_{22} - \rho_{21})} = \frac{\gamma \hbar}{2} (N_2 - N_1)$$

Thus we see that the magnetization parallel to the applied dc magnetic field is expressed in terms of diagonal density matrix elements only and corresponds to the difference in occupation probabilities of the two energy levels. The z component of magnetization is therefore a measure of the energy stored in the spin system. The constant M_z^e is the equilibrium value to which M_z decays in the absence of an applied field $\mathbf{B'}$. Since Ω is related to the static magnetic field by $\Omega = \gamma B_{0z}$ and the total magnetic field is given by

$$\mathbf{B} = \hat{1}_z B_{0z} + \mathbf{B'}$$

(2.82) can be simplified to

$$\dot{M}_x + \frac{M_x}{T_2} = \gamma(\mathbf{B} \times \mathbf{M})_x$$

$$\dot{M}_y + \frac{M_y}{T_2} = \gamma(\mathbf{B} \times \mathbf{M})_y \quad (2.83)$$

$$\dot{M}_z + \frac{M_z - M_z^e}{T_1} = \gamma(\mathbf{B} \times \mathbf{M})_z$$

Although the Bloch equations may appear complex, they readily permit a straightforward physical interpretation. Without relaxation terms, (2.83) reduces to

$$\frac{d\mathbf{M}}{dt} = \gamma(\mathbf{B} \times \mathbf{M}) \quad (2.84)$$

2 DIPOLE TRANSITIONS

which is of the same form as the classical expression for a charged spinning top placed in a magnetic field. Rotation of such a top results in a current distribution, and the current, in turn, generates a magnetic moment that can interact with the magnetic field.

Let us pursue the analogy with the spinning top in more detail. A classical top made up of negative charged particles with a charge-to-mass ratio $-e/m$ possesses a magnetic dipole moment **m** and angular momentum **L** as shown in Figure 2.5. If ρ_m is the mass density and **v** is the particle velocity at any given point, then **m** and **L** are given by[26]

$$\mathbf{m} = \tfrac{1}{2}\int \mathbf{r} \times \left(-\frac{e}{m}\rho_m\mathbf{v}\right) dV$$

and

$$\mathbf{L} = \int \mathbf{r} \times (\rho_m\mathbf{v})\, dV$$

figure 2.5 Classical charged spinning top.

and are therefore related by

$$\mathbf{m} = -\frac{e}{2m}\mathbf{L} \tag{2.85}$$

In a uniform magnetic field **B** the top experiences a net torque

$$\mathbf{T} = \mathbf{m} \times \mathbf{B}$$

The equation of motion under such a torque is

$$\frac{d\mathbf{L}}{dt} = \mathbf{m} \times \mathbf{B}$$

which can be rewritten with the aid of (2.85) and the definition of γ as

$$\frac{d\mathbf{m}}{dt} = \frac{\gamma}{2}(\mathbf{B} \times \mathbf{m})$$

Thus the Bloch equations without relaxation and the equation of motion for a charged spinning top are formally identical except for a factor of 2. This factor of 2 is characteristic of the difference between the classical and quantum pictures of a charged spinning body.

In a static z-directed magnetic field, the magnetization precesses about the z axis, in analogy with the spinning top. This is evident from (2.84), where the magnetization is driven by a torque that is perpendicular to the z-directed magnetic field and therefore lies in the xy plane. The effect of the torque in the xy plane is to

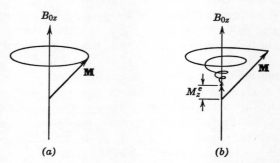

figure 2.6 Precession of magnetization **M** in a static magnetic field B_{0z}: (a) loss neglected; (b) loss included.

cause the magnetization to precess, as shown in Figure 2.6a, at the frequency $\Omega = \gamma B_{0z}$, where $\hbar\Omega$ is the energy difference between the quantized levels of the spin system.

When the effects of the loss terms in (2.82) are taken into account, it is found that the precessional motion described above decays as in Figure 2.6b. To see this, let $\mathbf{B}' = 0$ in (2.82). Then

$$\dot{M}_x + \frac{M_x}{T_2} = -\Omega M_y$$

$$\dot{M}_y + \frac{M_y}{T_2} = \Omega M_x$$

$$\dot{M}_z + \frac{M_z - M_z^e}{T_1} = 0$$

Thus, T_2 accounts for the decay of M_x and M_y, which are transverse to the dc field B_{0z}, and T_1 accounts for the decay of M_z, which is parallel to B_{0z}. The constant T_2 is the transverse or spin-spin relaxation time constant that accounts for the spin-spin interactions that, by their mutual perturbing influence, dephase the motion of the individual spins so that the expectation value of the transverse component of the total magnetization goes to zero; T_1 is the longitudinal or spin-lattice relaxation time constant that accounts for the exchange of energy between the spin system and its surrounding environment. This exchange results in the decay of an initial M_z, which represents the energy $-\mathbf{M} \cdot \mathbf{B} = -M_z B_{0z}$ stored per unit volume in the spin system. The final value to which M_z decays, M_z^e, is the value at which the spin system establishes equilibrium with its environment.

A field $\mathbf{B}'$ polarized in the xy plane that alternates near the precessional frequency Ω will, in accordance with the Bloch equations 2.83, induce precessional motion of the magnetization. This results in the transfer of energy between the field and the spin system. Absorption of energy from the field is known as electron spin paramagnetic resonance absorption. This phenomenon is examined in more detail in Chapter 3.

2.5.3 Field Equations for Magnetic Dipole Spin $\frac{1}{2}$ Case

The Bloch equations describe the behavior of the spin $\frac{1}{2}$ system under the influence of an applied magnetic field. The system of equations is closed by the addition of field equations that describe the behavior of the fields in the presence of matter.

The field equations in the absence of free charge are

$$\nabla \cdot \mathbf{B} = 0 \qquad \nabla \cdot \mathbf{D} = 0$$
$$\nabla \times \mathbf{H} = \mathbf{J} + \frac{\partial \mathbf{D}}{\partial t} \qquad \nabla \times \mathbf{E} = -\frac{\partial \mathbf{B}}{\partial t} \qquad (2.86)$$
$$\mathbf{B} = \mu_0(\mathbf{H} + \mathbf{M}) \qquad \mathbf{D} = \epsilon \mathbf{E}$$
$$\mathbf{J} = \sigma \mathbf{E}$$

where the conductivity term $\mathbf{J} = \sigma \mathbf{E}$ has been included to account phenomenologically for losses as in the electric dipole case. A wave equation in the variable $\mathbf{B}$ is obtained by taking the curl of the $\nabla \times \mathbf{H}$ equation after substituting for $\mathbf{H}$ in terms of $\mathbf{B}$. The result is

$$\nabla \times (\nabla \times \mathbf{B}) + \frac{\eta \mathscr{A}}{c} \frac{\partial \mathbf{B}}{\partial t} + \frac{n^2}{c^2} \frac{\partial^2 \mathbf{B}}{\partial t^2} = \mu_0 \nabla \times (\nabla \times \mathbf{M}) \qquad (2.87)$$

where $c^2 = 1/\mu_0 \epsilon_0$, $\eta = \sqrt{\epsilon/\epsilon_0}$ is the refractive index of the medium, and $\mathscr{A} = \mu_0 \sigma c/\eta$ is a damping coefficient so that in the absence of a magnetization $\mathbf{M}$ the energy of a plane wave, proportional to $|\mathbf{H}|^2$, decays as $e^{-\mathscr{A} z}$ for propagation in the z direction. As in the electric dipole case, it is possible to express the fields either in terms of traveling waves or in terms of a cavity normal mode expansion, although we shall not pursue the subject further here.

The field equation (2.87) in conjunction with the Bloch equations (2.83) provides the necessary set of coupled equations from which we may obtain self-consistent solutions for the interaction of an electromagnetic field with a spin system. These equations will be taken as the starting point for the analysis of electron spin paramagnetic resonance absorption presented in Chapter 3.

REFERENCES

1. H. Goldstein, *Classical Mechanics*, Addison-Wesley, Reading, Mass., 1950, pp. 222ff.
2. *Ibid.*, p. 18.
3. M. Goeppert-Mayer, "Uber Elementarakte mit zwei Quantensprungen," *Ann. Physik*, **9**, 273, 1931.

4. N. Bloembergen, *Nonlinear Optics*, W. A. Benjamin, New York, 1965, pp. 36, 174.
5. Reference 1, pp. 30ff.
6. M. Born and E. Wolf, *Principles of Optics*, Pergamon Press, New York, 1964, pp. 72ff.
7. J. Fiutak, "The Multipole Expansion in Quantum Theory," *Can. J. Phys.*, **41**, 12, 1963.
8. S. Raimes, *The Wave Mechanics of Electrons in Metals*, North-Holland Publishing Company, Amsterdam, 1961, pp. 135ff.
9. A. Messiah, *Quantum Mechanics*, Vol. I, North-Holland Publishing Company, Amsterdam, 1961, pp. 112–113.
10. D. Bohm, *Quantum Theory*, Prentice-Hall, Englewood Cliffs, N.J., 1951, pp. 431ff.
11. L. I. Schiff, *Quantum Mechanics*, McGraw-Hill, New York, 1955, p. 160.
12. C. H. Townes and A. L. Schawlow, *Microwave Spectroscopy*, McGraw-Hill, New York, 1955, p. 132.
13. G. Herzberg, *Infrared and Raman Spectra*, Van Nostrand, New York, 1945, p. 55.
14. C. Kittel, *Introduction to Solid-State Physics*, 3rd ed., Wiley, New York, 1966, pp. 384ff.
15. J. C. Slater, *Microwave Electronics*, Van Nostrand, New York, 1950, Chap. 4.
16. B. A. Lengyel, *Introduction to Laser Physics*, Wiley, New York, 1966, pp. 71ff.
17. See Ref. 4, p. 68.
18. See Ref. 8, p. 110.
19. See Ref. 8, pp. 106ff.
20. See Ref. 9, pp. 24–27.
21. R. H. Dicke and J. P. Wittke, *Introduction to Quantum Mechanics*, Addison-Wesley Publishing Company, Reading, Mass., 1960, p. 194.
22. See Ref. 8, p. 112.
23. See Ref. 16, pp. 32ff.
24. See Ref. 21, p. 324.
25. G. E. Pake, *Paramagnetic Resonance*, W. A. Benjamin, New York, 1962, p. 23.
26. See Ref. 1, pp. 176ff.

PROBLEMS

2.1 Show that if $\mathcal{H}(-\mathbf{r}) = \mathcal{H}(\mathbf{r})$ it is always possible to choose degenerate eigenfunctions with a definite parity. Show that in general these functions are not orthogonal.

2.2 Obtain the equations of motion for the matrix elements of the density operator for a two-level system where $\mathcal{H}'$ results from electric quadrupole interaction. If both states have opposite parity, show that the populations are unaltered by the interaction.

2.3 Show how terms of the form

$$\int u_m^*(\mathbf{r})\, xy\, u_n(\mathbf{r})\, dV$$

are obtained from (2.22).

2.4 In (2.48), $(\hbar\Omega/2)(\partial/\partial t)(N_2 - N_1)$ is the rate of change of energy in the quantized medium. From Maxwell's equations show that $\dot{P}_\alpha E_\alpha$ is the rate of exchange of energy between a field and a polarizable medium.

2.5 From (2.43) and (2.48) show that if relaxation is not considered (that is, $T_1, T_2 \to \infty$) in the presence of an applied field E, $(N_2 - N_1)$ behaves as shown:

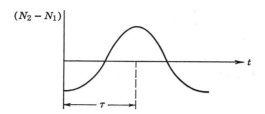

If the field is turned off at $t = \tau$, a population inversion has been achieved. Determine τ. If it is valid to neglect relaxation, it is necessary that $\tau \ll T_1, T_2$. Determine a minimum value for $|E|^2$ so that this condition is satisfied.

The method that has been described for population inversion is known as 180° pulse inversion. It provides for a transient inversion as opposed to the steady-state inversion techniques discussed in Chapter 4.

2.6 In equation 2.12 a magnetic moment **m** was defined as $\mathbf{m} = (-e/2m)(\mathbf{r} \times \mathbf{p})$.

For an electron in a circular orbit show that this definition is consistent with the energy of a magnetic dipole

$$\mathcal{W} = -\mathbf{m} \cdot \mathbf{B}$$

and the torque exerted on the dipole

$$\mathbf{T} = \mathbf{m} \times \mathbf{B}$$

2.7 In (2.54) a cavity lifetime τ_c was introduced. Calculate the value for τ_c for an interferometer resonator using planar mirrors that are 2 cm in diameter and 1 m apart. The light beam $\lambda = 6000$ Å may be considered to be a TEM_{00n} mode (a plane wave) with a diameter of 0.5 cm. The medium filling the resonator attenuates the light at 0.1% per cm and each mirror reflects 95% of the incident light. Determine τ_c for the resonator and show that diffraction losses are negligible.

2.8 From

$$S_z = \frac{\hbar}{2}\begin{pmatrix} 1 & 0 \\ 0 & -1 \end{pmatrix}$$

and $[s_i, s_j] = i\hbar s_k$, derive the two other spin matrices.

2.9 In going from (2.54) to (2.55) it was stated that

$$\mathbf{E}_c(\mathbf{r}) \int \mathbf{\ddot{P}}^s \cdot \mathbf{E}_c(\mathbf{r}) \, dV = \mathbf{\ddot{P}}^s$$

if $\mathbf{P}^s$ has the same spatial dependence as $\mathbf{E}_c$, where $\mathbf{E}_c$ is the normal mode field. Show that the above equality does hold under the stated condition.

2.10 We have considered an electric dipole interaction between two states of opposite parity. Show that if the interaction occurs between states of mixed parity, the equations of motion are not significantly altered. Diagonal terms will be added to both $\mathcal{H}'$ and μ. The diagonal terms in $\mathcal{H}'$ introduce a small change in the transition frequency Ω. Obtain an expression for this frequency shift. The diagonal terms in μ provide a much smaller contribution to $\langle \mu \rangle$ than do the off-diagonal terms if $\omega \simeq \Omega$. Estimate the ratio of the contribution of diagonal to off-diagonal terms for comparable matrix element magnitudes, that is

$$|\mu_{ii}| \simeq |\mu_{ij}|$$

2.11 In nuclear magnetic resonance (NMR), interaction occurs between the spin states of the proton and electromagnetic radiation. If a material is subjected to a dc magnetic of 1 weber/m² in the z direction, what is the resonant absorption frequency resulting from splitting of the spin states of the protons? (Caution: the spin operator for the proton is not the same as for the electron.)

2.12 If ω_0 is the precession frequency, sketch the time behavior of the magnetization **M** for the following cases:

$$(i) \quad \omega_0 \gg \frac{1}{T_2} \gg \frac{1}{T_1}$$

$$(ii) \quad \omega_0 \gg \frac{1}{T_2} \simeq \frac{1}{T_1}$$

$$(iii) \quad \omega_0 \simeq \frac{1}{T_2} \simeq \frac{1}{T_1}$$

$$(iv) \quad \omega_0 < \frac{1}{T_2}$$

2.13 In the presence of a static field B_{oz}, an rf, circularly polarized wave is applied in the xy plane at the precession frequency ($\omega = \gamma B_{oz}$) for a time interval Δt. If **M** is initially in the z direction, determine the rf field amplitude and Δt so that at the end of the pulse **M** is in the xy plane. (Take $T_1 = 10^{-3}$ sec, $T_2 = 10^{-5}$ sec.) Sketch the evolution of **M** after such a pulse is turned off. Would you expect a signal to be radiated after the pulse, and if so how would it vary in time? (This effect is known as a $\pi/2$ pulse.)

RESONANT PROCESSES

3

3.1 INTRODUCTION

When an electromagnetic field of a given frequency interacts with a medium that has a transition near the same frequency, the interaction is a resonant one, and several important phenomena associated with resonance occur. (We are using the term "resonant process" to include transitions between any pair of energy levels, rather than the more restrictive spectroscopic convention that "resonant lines" refer only to those transitions terminating on the ground state.) For example, the radiation may be absorbed as it passes through the medium, resulting in the decay of the incident wave. This is known as resonance absorption. Or an incident wave passing through the medium may experience growth instead of decay because of a process known as negative absorption or stimulated emission. This occurs when the atomic populations of the levels involved in the transition are inverted, that is, when the medium is excited by some additional source in such a way that there are more atoms in the higher energy state than in the lower. The maser and laser depend on such a process for their operation. Another important phenomenon associated with resonance is the saturation of the transition by a strong signal so that no further absorption takes place. Radiation then passes through the medium unattenuated. This effect is often termed *bleaching* of the transition and has important applications in the control of lasers to obtain large peak output powers.

All of these phenomena can be investigated with the aid of the coupled equations derived in the preceding chapter for the electric and magnetic dipole transitions. It is the purpose of this chapter to demonstrate how to apply these equations to such problems.

In this chapter only first-order or resonant processes will be considered, that is, processes for which the frequencies of the electromagnetic field and transition are approximately the same. In a representation in which the field as well as the medium is quantized, first-order processes correspond to the absorption or emission of a single photon and are therefore called single-photon processes. Higher-order processes, which are generally much smaller in magnitude than the resonance effects described here, will be considered in Chapter 5.

3.2 STEADY-STATE BEHAVIOR OF THE ELECTRIC DIPOLE TRANSITION; ABSORPTION, DISPERSION, AND SATURATION

Large-scale macroscopic phenomena such as the absorption and dispersion of light, or the generation of coherent radiation as in a laser, generally take place by means of electric dipole transitions between the various atomic or molecular energy levels. In order to study such phenomena, we begin with the appropriate equations derived in Chapter 2 for an isotropic medium, Equations 2.60;

$$\ddot{\mathbf{P}} + \frac{2}{T_2}\dot{\mathbf{P}} + \Omega^2 \mathbf{P} = \frac{2\Omega}{\hbar} L \frac{|\boldsymbol{\mu}_{12}|^2}{3}(N_1 - N_2)\mathbf{E} \tag{3.1}$$

$$\frac{\partial}{\partial t}(N_1 - N_2) + \frac{(N_1 - N_2) - (N_1 - N_2)^e}{T_1} = -\frac{2}{\hbar\Omega}\dot{\mathbf{P}} \cdot \mathbf{E} \tag{3.2}$$

and

$$\nabla \times (\nabla \times \mathbf{E}) + \frac{\eta \mathscr{A}}{c}\frac{\partial \mathbf{E}}{\partial t} + \frac{\eta^2}{c^2}\frac{\partial^2 \mathbf{E}}{\partial t^2} = -\mu_0 \frac{\partial^2 \mathbf{P}}{\partial t^2} \quad \text{(traveling wave case)} \tag{3.3a}$$

$$\ddot{\mathbf{E}} + \frac{1}{\tau_c}\dot{\mathbf{E}} + \omega_c^2 \mathbf{E} = -\frac{1}{\epsilon}\mathbf{E}_c(\mathbf{r})\int \dot{\mathbf{P}} \cdot \mathbf{E}_c(\mathbf{r})\, dV \quad \text{(cavity case)} \tag{3.3b}$$

The superscript s, which appears in the original equations, has been dropped with the understanding that the polarization that appears in the above equations represents the polarization source term $\mathbf{P}^s$, which is $(\eta^2 + 2)/3$ times the actual polarization $\mathbf{P}$ associated with the transition of interest. [See (2.58).]

As discussed in Section 2.4.7, Equations 3.1 through 3.3 apply also to an anisotropic medium when the polarization and field are polarized along principal axes of the system, provided that the substitutions (2.61) are made. Therefore the results derived on the basis of (3.1)–(3.3) apply also to the principal axis case in anisotropic media. For arbitrary directions in an anisotropic medium, we must return to (2.41) and (2.48) and carry through for the general case the development that is presented for the isotropic case.

In our initial examination of these equations, we shall find that the polarization equation is primarily responsible for the absorption and dispersion properties of the medium, whereas the population difference equation accounts for saturation effects. Such phenomena are essentially steady-state phenomena and can therefore be studied by means of a relatively simple steady-state approach.

3.2.1 Susceptibility

At optical or microwave frequencies we are generally interested in the polarization at frequency ω induced by a field at the same frequency, where ω does not differ greatly from the transition frequency Ω. Therefore, for plane wave propagation in the z direction, we assume a solution to (3.1) of the form

$$\mathbf{P} = \tfrac{1}{2}\tilde{\mathbf{P}}e^{i(\omega t - kz)} + \text{c.c.} \tag{3.4}$$

and similarly for $\mathbf{E}$, where c.c. denotes the complex conjugate and the tilde ($\sim$) indicates a complex magnitude. The solution to (3.1) is given by

$$\tilde{\mathbf{P}} = \frac{1}{\hbar} L \frac{|\mu_{12}|^2}{3}(N_1 - N_2)\frac{1}{(\Omega - \omega) + i(1/T_2)}\tilde{\mathbf{E}}$$

where we have made use of the near-resonance condition $\omega \approx \Omega$. In addition, we have assumed that the population difference $(N_1 - N_2)$ is time-independent. It is apparent from (3.2) that in the steady-state $(N_1 - N_2)$ has high-frequency components, but these are generally small compared to the dc term because of the fact that $\omega T_1 \gg 1$.

It is convenient to express the above relationship in terms of a linear susceptibility $\chi(\omega)$, defined by

$$\tilde{\mathbf{P}} = \epsilon_0 \chi(\omega)\tilde{\mathbf{E}} \tag{3.5}$$

where ϵ_0 is the permittivity of free space. With the above definition, $\chi(\omega)$ is given by

$$\chi(\omega) = \frac{\pi}{\hbar \epsilon_0} L \frac{|\mu_{12}|^2}{3}(N_1 - N_2)\tilde{g}_L(\omega, \Omega) \tag{3.6}$$

where $\tilde{g}_L(\omega, \Omega)$ is a frequency-dependent lineshape factor referred to as a complex Lorentzian function, plotted in Figure 3.1:

$$\begin{aligned}\tilde{g}_L(\omega, \Omega) &= \frac{1}{\pi}\frac{1}{(\Omega - \omega) + i(1/T_2)}\\ &= \frac{1}{\pi}\frac{(\Omega - \omega)}{(\Omega - \omega)^2 + (1/T_2)^2} - i\frac{1}{\pi}\frac{1/T_2}{(\Omega - \omega)^2 + (1/T_2)^2}\end{aligned} \tag{3.7}$$

The linear susceptibility is therefore a frequency-dependent property of the medium that relates the polarization to the electric field. It is termed a *linear* susceptibility because the polarization is linear in the field that produces it.

For the isotropic case under consideration, $\chi(\omega)$ is a scalar that relates like components of polarization and field, as shown by (3.6). For the general anisotropic case, $\chi(\omega)$ is a tensor that relates a given component of polarization to all field components on which it depends. This more general case is taken up in Section 3.3.

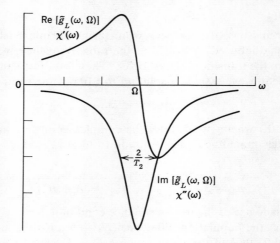

figure 3.1 Real and imaginary parts of the complex Lorentzian lineshape factor

$$\tilde{g}_L(\omega, \Omega) = \frac{1}{\pi} \frac{1}{[\Omega - \omega + i(1/T_2)]}$$

to which is proportional the real and imaginary parts, respectively, of the susceptibility of the electric dipole transition, $\chi = \chi' + i\chi''$.

3.2.2 Absorption and Dispersion

The susceptibility $\chi(\omega)$ has both real and imaginary parts, as indicated by (3.6) and (3.7), and can therefore be expressed in the form

$$\chi(\omega) = \chi'(\omega) + i\chi''(\omega) \tag{3.8}$$

where

$$\chi'(\omega) = \frac{\pi}{\hbar\epsilon_0} L \frac{|\mu_{12}|^2}{3} (N_1 - N_2) \left[\frac{1}{\pi} \frac{(\Omega - \omega)}{(\Omega - \omega)^2 + (1/T_2)^2} \right] \tag{3.9}$$

and

$$\chi''(\omega) = -\frac{\pi}{\hbar\epsilon_0} L \frac{|\mu_{12}|^2}{3} (N_1 - N_2) \left[\frac{1}{\pi} \frac{1/T_2}{(\Omega - \omega)^2 + (1/T_2)^2} \right] \tag{3.10}$$

The frequency dependence of the real and imaginary parts of the susceptibility follows the real and imaginary parts, respectively, of the complex Lorentzian line shown in Figure 3.1. We shall now show that $\chi'(\omega)$ and $\chi''(\omega)$ account, separately, for the dispersion and absorption properties of the medium.

The propagation of an electromagnetic wave through the medium is governed by the wave equation (3.3a). If (3.6) is substituted into (3.3a) and plane wave propagation of the form (3.4) is assumed, the wave equation reduces to an expression for the propagation constant k, given by

$$k^2 = \frac{\eta^2 \omega^2}{c^2} \left[1 + \frac{\chi(\omega)}{\eta^2} \right] \tag{3.11a}$$

where we have assumed that the background loss $\mathscr{A}$ caused by transitions other than the one of interest is negligible. Upon taking the square root and assuming $\chi(\omega)/\eta^2 \ll 1$ so that the expansion $\sqrt{1 + \chi/\eta^2} = 1 + \chi/2\eta^2$ holds (which is the case for other than very strong transitions), we obtain

$$k = k' + ik'' \approx \frac{\eta \omega}{c} \left[1 + \frac{\chi'(\omega)}{2\eta^2} \right] + i \frac{\omega \chi''(\omega)}{2\eta c} \tag{3.11b}$$

The real and imaginary parts of k are plotted in Figure 3.2 with the aid of Equations 3.9–3.11. The real part of k depends on $\chi'(\omega)$ and accounts for the dispersive properties of the medium since the phase velocity of the wave is given by $v_p = \omega/k'$. We see that the phase velocity is approximately c/η with a superimposed rapid variation with frequency in the immediate vicinity of the transition. This is often termed the anomalous dispersion characteristic, since as we approach the transition frequency from the lower side, the gentle increase of k' with frequency experiences a sudden reversal in direction.

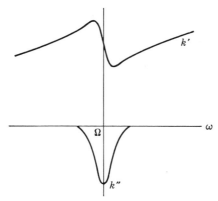

figure 3.2 Complex propagation constant $k = k' + ik''$ in the vicinity of an electric dipole transition located at frequency Ω.

The imaginary part of k depends on $\chi''(\omega)$ and accounts for the absorption properties of the medium. In particular, since the field propagates as e^{-ikz}, the time-averaged power per unit area $I = \eta \epsilon_0 c \, |\tilde{E} e^{-ikz}|^2/2$ carried by the wave decays as

$$I = I_0 e^{-\Gamma z} \tag{3.12}$$

where Γ is the absorption constant, defined by $\Gamma = -2k''$. From (3.10) and (3.11) we find that for $\omega \approx \Omega$, the absorption constant is given by

$$\Gamma = \frac{\Omega \pi}{\hbar \epsilon_0 c \eta} L \frac{|\mu_{12}|^2}{3} (N_1 - N_2) g_L(\omega, \Omega) \tag{3.13}$$

where $g_L(\omega, \Omega)$ is the Lorentzian lineshape factor (see Figure 3.5)

$$g_L(\omega, \Omega) = \frac{1}{\pi} \frac{1/T_2}{(\Omega - \omega)^2 + (1/T_2)^2} \tag{3.14}$$

discussed in detail in Section 3.2.4. Comparison of (3.7) and (3.14) shows that $g_L(\omega, \Omega)$ is the negative of the imaginary part of the complex lineshape factor $\tilde{g}_L(\omega, \Omega)$.

The characteristic linewidth for the Lorentzian line, $\Delta \omega_L$, is given by

$$\Delta \omega_L = \frac{2}{T_2}$$

where $\Delta \omega_L$ is the frequency difference between the points on either side of the central maximum where $g_L(\omega, \Omega)$ drops to one-half its line-center value. The quantity $\Delta \omega_L$ is called the full width at half maximum.

We observe that the absorption constant Γ is proportional to the population difference per unit volume, $(N_1 - N_2)$. The constant Γ has its maximum value when all of the atoms are in the lower state, that is, $N_2 = 0$. Under this condition the only process taking place is the absorption of energy from the electromagnetic field by atoms in the lower state that then make transitions to the upper state. For values of N_2 other than zero, the absorption constant Γ is less. It is apparent from the form of (3.13) that the N_2 term contributes a negative absorption, or emission process, that counteracts the absorption by atoms in the lower state. This

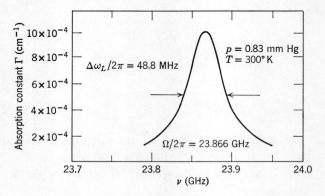

figure 3.3 Absorption line of NH_3 molecule for inversion transition.

counteracting process, which is proportional to N_2, is an induced emission process that corresponds to the transfer of energy from the medium to the field by atoms initially in the upper state. Where N_2 is allowed to become greater than N_1, Γ becomes negative, and an electromagnetic wave experiences growth while traveling through the medium. This latter condition constitutes the principle upon which the maser and laser are based.

It is evident from (3.12) that a measurement of the decrease in power per unit area as a wave traverses a lightly absorbing sample

$$dI = -\Gamma I\, dz$$

provides a direct measurement of the absorption constant Γ. Such a measurement is shown in Figure 3.3 for the "flip-flop" or "inversion" transition of the NH_3 ammonia molecule that is located in the microwave region of the spectrum.[1,2] The transition is so named because it corresponds to a symmetrical vibration of the nitrogen atom through a plane containing the three hydrogen atoms, thus producing an inversion of the structure as shown in Figure 3.4.

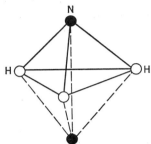

figure 3.4 Inversion vibration of the NH_3 molecule. The single nitrogen molecule vibrates through the plane containing the three hydrogen molecules in a symmetrical fashion and thus resembles a one-dimensional harmonic oscillator.

The dipole matrix element $|\mu_{12}|$ can be obtained from the absorption data by means of (3.13), and the ammonia transition will be used as an example of this calculation. The number of molecules per unit volume, N_V, at a pressure $p = 0.83$ mm Hg and a temperature $T = 298°K$ is, according to the ideal gas law,†

$$N_V = \frac{p}{\kappa T} = \frac{0.83 \times 1.013 \times 10^5}{760 \times 1.38 \times 10^{-23} \times 298} = 2.69 \times 10^{22} \text{ molecules/m}^3$$

where κ is Boltzmann's constant, $\kappa = 1.38 \times 10^{-23}$ joules/deg. For a Boltzmann distribution, it is estimated that approximately 6.4% of the molecules are in the lower of the two states[3] or $N_1 = 1.71 \times 10^{21}$ molecules/m³. The population difference, evaluated in terms of a ratio, is

$$\frac{N_2 - N_1}{N_1} = \left(1 - \frac{N_2}{N_1}\right) = \left((1 - e^{-\hbar\Omega/\kappa T}\right) \approx \frac{\hbar\Omega}{\kappa T} = 3.83 \times 10^{-3}$$

For a dilute gas $\eta \approx 1$ and with $\Omega/2\pi = 23.86$ GHz, $\Gamma = 10^{-3}$ cm⁻¹, and $\Delta\omega_L = 2/T_2 = 2\pi \times 48.8 \times 10^6$ rad/sec, we obtain from (3.13)

$$|\mu_{12}| = 1.14 \text{ debye}$$

† The conversion for pressure is given by

$$p(\text{Newtons/m}^2) = \frac{p(\text{mm Hg})}{760(\text{mm Hg/atm})} \times 1.013 \times 10^5 \left(\frac{\text{Newtons/m}^2}{\text{atm}}\right)$$

The debye is a convenient unit for measuring dipole moments:

$$1 \text{ debye} = 3.33 \times 10^{-30} \text{ coul m} = 10^{-18} \text{ esu}$$

In a comparison of the above value for $|\mu_{12}|$ with that listed in handbooks, care must be taken to note that the handbook value is an effective "permanent" dipole moment μ associated with the pyramidal configuration of the NH$_3$ molecule that would exist if the configuration shown in Figure 3.3 were static.[4] For a symmetric top molecule such as ammonia, the two values are related by

$$|\mu_{12}|^2 = \mu^2 \frac{K^2}{\mathcal{J}(\mathcal{J}+1)}$$

where $\mathcal{J}$ and K are the angular momentum and magnetic quantum numbers of the top as it undergoes inversion vibration. With $\mathcal{J} = K = 3$ for the case under consideration and a handbook value $\mu = 1.46$ debye based on high-accuracy Stark measurements, we obtain an infrared value for $|\mu_{12}|$ of 1.26 debye, in reasonable agreement with the value obtained using the data from the experimental absorption curve discussed above.

As an alternative to expressing the absorption properties of a transition in terms of the absorption constant Γ, a per-atom absorption cross-section σ_c is sometimes used. The cross-section σ_c is defined in the following manner. Consider the case where all the atoms are in the lower energy state, that is, $N_1 = N_V$, $N_2 = 0$, so that only absorption and no emission takes place. The cross-section σ_c is then defined under this condition as the ratio of the power absorbed per atom to the incident power per unit area,

$$\sigma_c = \frac{\mathcal{P}/N_V}{I} \tag{3.15}$$

where $\mathcal{P}$ is the power absorbed per unit volume, N_V is the number of atoms per unit volume, and I is the incident power per unit area. This definition can be shown to be consistent with the concept that each atom has an effective area σ_c as far as its ability to remove energy from an incident wave is concerned (see Problem 3.1). We can think of σ_c as an effective cross-sectional area of the absorbing atom.

In terms of the absorption constant Γ, $\mathcal{P}$ for an ensemble of atoms or molecules is given by

$$\mathcal{P} = -\frac{dI}{dz} = \Gamma I$$

Therefore, under the condition of N_V atoms per unit volume, all in the lower level, we can obtain from the above two expressions an expression for σ_c,

$$\sigma_c = \frac{\Gamma}{N_V} = \frac{\Omega \pi}{\hbar \epsilon_0 c \eta} L \frac{|\mu_{12}|^2}{3} g_L(\omega, \Omega) \tag{3.16}$$

For the ammonia transition at 23.9 GHz just discussed, the cross-section at resonance is $\sigma_c(\Omega) = 1.68 \times 10^{-16}$ cm². The per-atom cross-section σ_c is useful because it is a measure of the strength of the transition that is independent of the level populations N_1 and N_2, which vary with temperature and density.

3.2.3 Oscillator Strength and Sum Rules

The strength of a transition is often specified by yet another measure, the *oscillator strength*. The procedure that led to the expression (3.16) for the cross-section, when applied to the corresponding classical case of a collection of oscillating dipoles, each with an electric dipole moment $\mathbf{p} = -e\mathbf{r}$ and charge-to-mass ratio $-e/m$, leads to the scattering cross-section (see Problem 3.2)

$$\sigma_c = \frac{\pi e^2}{2m\epsilon_0 c\eta} Lg_L(\omega, \Omega) \tag{3.17}$$

A comparison between (3.16) and (3.17) invites the definition of an *oscillator strength* as the ratio of the quantum mechanical to the classical cross-sections,

$$f_{ji} = \frac{2\Omega_{ji}}{\hbar} \frac{|\mathbf{\mu}_{ji}|^2}{3} \bigg/ \frac{e^2}{m} \tag{3.18}$$

where the subscripts i and j label the states involved; f_{ij} is the strength associated with the transition $i \to j$. Since $\Omega_{ji} \equiv (E_j - E_i)/\hbar$ may be either positive or negative, oscillator strengths may also be either positive or negative. Some typical oscillator strengths are listed in Table 3.1.[5]

TABLE 3.1 OSCILLATOR STRENGTH FOR SELECTED TRANSITIONS OF THE HYDROGEN ATOM

Wavelength (Å)	Transition	$|f_{ij}|$
1216	$1s - 2p$	0.416
6563	$2p - 3d$	0.694
18751	$3d - 4f$	1.016
40532	$4f - 5g$	1.345

Where experimental results are lacking, the magnitude of a given matrix element may be estimated from a convenient sum rule involving the oscillator strengths. For example, assume that we desire information about the electric dipole matrix elements associated with electronic transitions between a given level i and several levels j. A straightforward calculation using commutator relationships for transitions in which a single electron is involved yields ($\mu\mu = \mathbf{\mu} \cdot \mathbf{\mu}$)

$$[[\mathbf{\mu}, \mathcal{H}_0], \mathbf{\mu}] = \frac{3e^2\hbar^2}{m} \tag{3.19}$$

where $\mu = -er$, $\mathcal{H}_0 = \dfrac{\mathbf{p}^2}{2m} + \mathcal{V}$. An expansion of the commutator yields

$$[[\mu, \mathcal{H}_0], \mu] = 2\mu \cdot \mathcal{H}_0 \mu - \mathcal{H}_0 \mu \cdot \mu - \mu \cdot \mu \mathcal{H}_0$$

By repeated insertion of the identity operator $I = \sum_j |u_j\rangle\langle u_j|$ into the above expression, we find that

$$\langle u_i| [[\mu, \mathcal{H}_0], \mu] |u_i\rangle = 2\hbar \sum_j \Omega_{ji} |\mu_{ji}|^2 \tag{3.20}$$

where $|u_i\rangle$, $|u_j\rangle$ are taken to be eigenvectors of $\mathcal{H}_0$. Combination of (3.18) and (3.19) then results in the expression

$$\sum_j f_{ji} = 1 \tag{3.21}$$

If more than one electron partakes in a particular transition, then the sum of oscillator strengths to a given level i is equal to the number of electrons involved. This is known as the Kuhn-Thomas sum rule.[6]

For transitions involving a single electron only, the sum of oscillator strengths to a given level is unity. For this case an individual oscillator strength may, of course, have a magnitude greater than unity, since, in general, both positive and negative values are contained in the sum. As an example of how the sum rule (3.21) provides an estimate of a matrix element, consider the one-electron case where i refers to the ground state 0. Then all terms in the sum (3.21) are positive, and as a result the largest single oscillator strength possible is unity. In this case any given matrix element must satisfy the inequality

$$|\mu_{j0}|^2 \leq \frac{3e^2 \hbar}{2m\Omega_{j0}}$$

An additional sum rule sometimes useful for estimating the magnitude of a matrix element can be obtained directly by a procedure involving the removal of the identity factor $I = \sum_j |u_j\rangle\langle u_j|$; thus

$$\sum_j |\mu_{ji}|^2 = \sum_j |\langle u_j| \mu |u_i\rangle|^2 = \sum_j \langle u_i| \mu |u_j\rangle \cdot \langle u_j| \mu |u_i\rangle = \langle u_i| \mu^2 |u_i\rangle$$

Since all terms in the sum are positive, any given matrix element must satisfy the inequality

$$|\mu_{ji}|^2 \leq (\mu^2)_{ii}$$

Before leaving the discussion on oscillator strengths, we should point out that the close similarity in form between the classical and quantum mechanical cross-sections is representative of a more general correspondence principle for the electric dipole case. This principle states that the classical and quantum mechanical expressions for absorption and dispersion are identical in form provided that we make the associations, in the isotropic case,

$$\frac{e^2}{m} \leftrightarrow \frac{2\Omega}{\hbar} \frac{|\mu_{12}|^2}{3}, \quad N_V \leftrightarrow N_1 - N_2$$

which is equivalent to an oscillator strength of unity. Correspondences such as these, which are relatively easy to derive, are often useful in checking quantum mechanical derivations.

3.2.4 Lorentzian Lineshape—Homogeneous Broadening

The frequency dependence of the absorption line given in Equations 3.13 and 3.14 corresponds to what is known as the Lorentzian lineshape. This lineshape appears often in studies of the frequency characteristics of physical systems and applies, for example, to the response of high-Q RLC electric circuits. H. A. Lorentz originally derived this lineshape in somewhat different form as the statistical distribution of the frequencies emitted by colliding gas molecules when the collision frequency is small compared to the undisturbed emission frequency.

The lineshape factor, as given by (3.14),

$$g_L(\omega, \Omega) = \frac{1}{\pi} \frac{1/T_2}{(\Omega - \omega)^2 + (1/T_2)^2} \tag{3.14}$$

shown in Figure 3.5, is normalized so that the area under the curve is unity,

$$\int_{-\infty}^{\infty} g_L(\omega, \Omega) \, d\omega = 1$$

As mentioned earlier, the characteristic linewidth for the Lorentzian line, $\Delta\omega_L$, is given by $\Delta\omega_L = 2/T_2$, where $\Delta\omega_L$ is the frequency difference between the points on either side of the central maximum where $g_L(\omega, \Omega)$ drops to one-half its line-center value. In terms of $\Delta\omega_L$, (3.14) can be written

$$g_L(\omega, \Omega) = \frac{1}{\pi} \frac{\Delta\omega_L/2}{(\Omega - \omega)^2 + (\Delta\omega_L/2)^2} \tag{3.22}$$

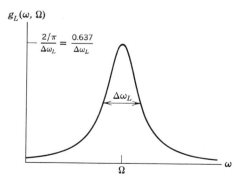

figure 3.5 Lorentzian lineshape.

The term "homogeneous" is applied to Lorentzian line broadening for the following reasons. In deriving the polarization equation, (3.1), which led to the Lorentzian lineshape factor (3.14), we began with an equation for the dipole moment of an individual atom, Equation 2.35, which was characterized by a transition frequency Ω and a relaxation time constant T_2. The polarization equation was then obtained by a spatial average followed by multiplication by N_V, the number of atoms per unit volume. As a result, the frequency characteristics of the total polarization is the same as that of the individual atoms. Thus the characteristics of the total macroscopic polarization are due simply to the sum of a homogeneous set of contributions from individual atoms, all of which have the same transition frequency Ω and linewidth $\Delta\omega_L = 2/T_2$. Therefore it can be said in summary that the frequency characteristics of the absorption line as presented earlier corresponds to a homogeneously broadened Lorentzian line; a line that results whenever the broadening is due to relaxation processes that act equally on all atoms, each of which has the same transition frequency. Coupling to lattice vibrations, collisions, and coupling between atoms (for example, spin-spin coupling) are such relaxation processes.

Another type of broadening, known as *inhomogeneous* broadening, also occurs under certain circumstances. Therefore, before we leave the subject of line broadening, we shall consider this case.

3.2.5 Gaussian Lineshape—Inhomogeneous Broadening

If the linewidth of a collection of atoms or molecules arises primarily from the fact that each atom or molecule has a different transition frequency Ω_i, the line is said to be inhomogeneously broadened. An important example is provided by the case of Doppler broadening in a gas. Here different molecules have different transition frequencies because of their motion, which results in Doppler shifts. Other sources of inhomogeneous broadening include any other mechanisms that result in a shift of the basic transition frequency on an atom-to-atom basis, such as crystalline imperfections or, in the case of spin systems, inhomogeneities in the magnetic field that determines the transition frequency. If such variations have a Gaussian random distribution, which is often the case, then the resulting inhomogeneous line will have a Gaussian lineshape.

In an inhomogeneously broadened line, individual atoms within the collection typically possess homogeneous lineshapes that are significantly narrower than the overall lineshape of the collection of atoms as a whole. These narrow homogeneous lines, which each correspond to the absorption line of those atoms centered at a given resonant transition frequency Ω_i, are referred to as spin packets, a term based on studies of spin paramagnetism (Sections 2.5 and 3.5). The overall inhomogeneous lineshape then results from an ensemble of spin packets whose span of transition frequencies Ω_i determines the overall linewidth, as indicated in Figure 3.6.

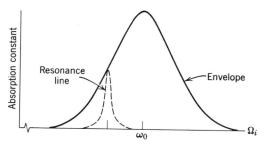

figure 3.6 Diagram illustrating the distinction between the resonance line for an individual atomic response or spin packet and the envelope caused by inhomogeneous broadening.

If a field is applied to such a collection of atoms, the polarization that results is a summation of independent responses from individual spin packets. An expression for the susceptibility in this case is obtained in terms of an integral over the various overlapping contributions from each of the spin packets.[7] Following is an example of such a calculation applied specifically to the case of Doppler broadening in a gas. The results of the calculation are, however, applicable to any case where the inhomogeneous lineshape is Gaussian.

In the case of Doppler broadening in a gas, the molecules in each spin packet have different transition frequencies Ω_i because of the motion of the gas molecules. For example, if a given molecule is moving with a velocity component v_i opposite to the direction of a propagating electromagnetic wave, then the interaction frequency is (nonrelativistic)

$$\Omega_i = \omega_0\left(1 + \frac{v_i}{c}\right)$$

where ω_0 is the transition frequency of the stationary molecule. This frequency-shift effect, the Doppler shift, gives rise to a line that consists of an ensemble of homogeneously broadened Lorentzian lines centered at different transition frequencies Ω_i.

The number of molecules per unit volume $d\mathcal{M}$ in the transition frequency range $d\Omega_i$ is determined from the Maxwellian thermal equilibrium velocity distribution in a gas, and is given by[8]

$$d\mathcal{M} = N_V g_G(\Omega_i, \omega_0) \, d\Omega_i \tag{3.23}$$

where N_V is the total number of molecules per unit volume and $g_G(\Omega_i, \omega_0)$ is the Gaussian lineshape factor plotted in Figure 3.7,

$$\begin{aligned} g_G(\Omega_i, \omega_0) &= \frac{[(4/\pi) \ln 2]^{1/2}}{\Delta\omega_G} \exp\left[-4(\ln 2)\frac{(\Omega_i - \omega_0)^2}{(\Delta\omega_G)^2}\right] \\ &= \frac{0.939}{\Delta\omega_G} \exp\left[-4(\ln 2)\frac{(\Omega_i - \omega_0)^2}{(\Delta\omega_G)^2}\right] \end{aligned} \tag{3.24}$$

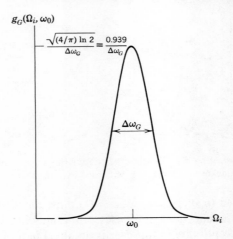

figure 3.7 Gaussian lineshape.

The quantity $\Delta\omega_G$ is the characteristic linewidth in radians per second for the Gaussian line and corresponds to the frequency difference between the points on either side of the central maximum where $g_G(\Omega_i, \omega_0)$ drops to one-half its line-center value. The Gaussian lineshape factor is normalized so that the area under the curve is unity,

$$\int_{-\infty}^{\infty} g_G(\Omega_i, \omega_0)\, d\Omega_i = 1$$

A comparison between Gaussian and Lorentzian lines of the same linewidth, shown in Figure 3.8, indicates that the Gaussian line has a higher maximum and less pronounced skirts than the Lorentzian line.

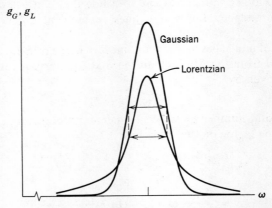

figure 3.8 Comparison of Gaussian and Lorentzian lines of same linewidth.

We shall now proceed to determine the susceptibility of the inhomogeneously broadened Doppler line. When a field $\tilde{\mathbf{E}}$ is applied to the gas, assumed isotropic, any given homogeneously broadened spin packet contributes an incremental polarization $d\tilde{\mathbf{P}}$ at frequency ω in accordance with (3.5)–(3.7) for the homogeneously broadened case,

$$d\tilde{\mathbf{P}} = \epsilon_0 \tilde{\mathbf{E}}\, d\chi(\omega, \Omega_i) \tag{3.25}$$

The appropriate expression for $d\chi(\omega, \Omega_i)$ is obtained from (3.6) by noting that instead of N_V molecules per unit volume contributing, as in the case of homogeneous broadening, we have in accordance with (3.23) a contribution from $d\mathcal{M} = N_V g_G(\Omega_i, \omega_0)\, d\Omega_i$ molecules per unit volume in a frequency range $d\Omega_i$. Therefore we have in place of (3.6)

$$d\chi(\omega, \Omega_i) = \frac{\pi}{\hbar \epsilon_0} L \frac{|\mu_{12}|^2}{3} (N_1 - N_2) \tilde{g}_L(\omega, \Omega_i) g_G(\Omega_i, \omega_0)\, d\Omega_i \tag{3.26}$$

where we have used the relationship $N_V \overline{(\rho_{11} - \rho_{22})} = N_1 - N_2$. The function $\tilde{g}_L(\omega, \Omega_i)$ is the complex Lorentzian given by (3.7),

$$\tilde{g}_L(\omega, \Omega_i) = \frac{1}{\pi} \frac{1}{(\Omega_i - \omega) + i(1/T_2)} \tag{3.27}$$

Although the Lorentz local field correction factor L is unity for a gas, we retain it so that the results will also be applicable to a dense medium in which the inhomogeneous broadening is Gaussian.

The polarization at frequency ω that results from the overlapping contributions of all of the spin packets is given by a summation of components of the form (3.25), which, with the aid of (3.26) and (3.27), results in an integral over the span of transition frequencies Ω_i. The total susceptibility is therefore given by

$$\chi(\omega) = \frac{\pi}{\hbar \epsilon_0} L \frac{|\mu_{12}|^2}{3} (N_1 - N_2) \int_{-\infty}^{\infty} \tilde{g}_L(\omega, \Omega_i) g_G(\Omega_i, \omega_0)\, d\Omega_i \tag{3.28}$$

With $\tilde{g}_L(\omega, \Omega_i)$ and $g_G(\Omega_i, \omega_0)$ given by (3.27) and (3.24), respectively, we find that (3.28) can be evaluated in terms of the tabulated integral,[9]

$$\omega(z) = \frac{i}{\pi} \int_{-\infty}^{\infty} \frac{e^{-t^2}\, dt}{z - t}, \quad \text{Im } z > 0$$

where z is complex.

It is often the case that the homogeneous linewidth $2/T_2$ associated with the Lorentzian spin packet is much narrower than the overall width $\Delta\omega_G$ of the Gaussian line. Under this condition, evaluation of (3.28) can be simplified by the assumption that the width of the spin packet becomes vanishingly small. Therefore, in the evaluation of the integral (3.28) we may take[10]

$$\lim_{T_2 \to \infty} [\tilde{g}_L(\omega, \Omega_i)] = \lim_{T_2 \to \infty} \left[\frac{1}{\pi} \frac{\Omega_i - \omega}{(\Omega_i - \omega)^2 + (1/T_2)^2} - i \frac{1}{\pi} \frac{(1/T_2)}{(\Omega_i - \omega)^2 + (1/T_2)^2} \right]$$

$$= \frac{1}{\pi} \frac{1}{\Omega_i - \omega} - i\delta(\Omega_i - \omega) \tag{3.29}$$

where $\delta(\Omega_i - \omega)$ is the Dirac delta function. The substitution of (3.29) into (3.28) then yields the result

$$\chi(\omega) = \chi'(\omega) + i\chi''(\omega)$$

where

$$\chi''(\omega) = -\frac{\pi}{\hbar\epsilon_0} L \frac{|\mu_{12}|^2}{3} (N_1 - N_2) g_G(\omega, \omega_0) \tag{3.30}$$

$$\chi'(\omega) = -\frac{1}{\pi} \text{PP} \int_{-\infty}^{\infty} \frac{\chi''(\Omega_i)}{\Omega_i - \omega} d\Omega_i \tag{3.31}$$

and PP stands for the principal part.†

We see from (3.30) that in the case where the individual spin packets are much narrower than the overall line, the absorption line for the inhomogeneously broadened case is of the same form as that for the homogeneously broadened line, (3.10), except that the Gaussian lineshape factor $g_G(\omega, \omega_0)$ replaces the Lorentzian factor $g_L(\omega, \Omega)$.

The dispersion given by (3.31) is found to be related to the absorption line by a convolution integral. This type of interrelationship between χ' and χ'' holds also for the homogeneously broadened case given by (3.9) and (3.10). These two cases are particular examples of what are known as the Kramers-Kronig relations,

$$\chi'(\omega) = -\frac{1}{\pi} \text{PP} \int_{-\infty}^{\infty} \frac{\chi''(\omega')}{\omega' - \omega} d\omega'$$

$$\chi''(\omega) = \frac{1}{\pi} \text{PP} \int_{-\infty}^{\infty} \frac{\chi'(\omega')}{\omega' - \omega} d\omega' \tag{3.32}$$

Proof of the above rests on a fundamental theorem of the theory of complex variables that requires that the real and imaginary parts of a complex function $\chi(z)$ be related in a certain way when $\chi(z)$ is free of poles in one half of the z plane (either upper or lower). The proof involves a simple contour integration, given in Appendix 5. We shall return to the case of the inhomogeneously broadened line in considering the helium-neon gas laser in Chapter 4.

† In an integral expression, PP indicates the following calculation:

$$\text{PP} \int_{-\infty}^{\infty} \frac{f(\Omega_i)}{\Omega_i - \omega} d\Omega_i = \lim_{\epsilon \to 0} \left[\int_{-\infty}^{\omega-\epsilon} \frac{f(\Omega_i)}{\Omega_i - \omega} d\Omega_i + \int_{\omega+\epsilon}^{\infty} \frac{f(\Omega_i)}{\Omega_i - \omega} d\Omega_i \right]$$

3.2.6 Saturation

When radiation is absorbed, the average energy delivered to the two-level dipole system is generally lost rather rapidly to the surrounding medium because of relaxation processes. As a result the characteristics of the dipole system as seen by the electromagnetic wave are essentially left unaltered by the absorption process. If the intensity is high enough, however, the relaxation processes may not be sufficiently rapid to dissipate the absorbed energy without a significant readjustment of the level populations in such a way that the absorption and dispersion properties of the medium begin to saturate. To study this phenomenon, we must take into account the consequences of the solution to the population difference equation (3.2), repeated here for easy reference:

$$\frac{\partial}{\partial t}(N_1 - N_2) + \frac{(N_1 - N_2) - (N_1 - N_2)^e}{T_1} = -\frac{2}{\hbar\Omega}\dot{\mathbf{P}}\cdot\mathbf{E} \tag{3.33}$$

The first term represents power being delivered to the dipole system, the second represents the net power interchange with the surrounding medium, and the right-hand side of (3.33) corresponds to the power delivered by the applied field.

In the absence of an applied field $\mathbf{E}$, the steady-state solution to the population difference equation (3.33) is simply

$$(N_1 - N_2) = (N_1 - N_2)^e$$

That is, the net interchange with the surrounding medium is zero, and the population difference assumes a constant equilibrium value. If the system is in thermal equilibrium with its surroundings then the equilibrium population is given by the Boltzmann distribution.

If we take into account the effects of the applied field resulting from the term in the right-hand side of (3.33), however, we find that the power delivered to the dipole system results in the establishment of a new equilibrium value. Assuming traveling waves of the form (3.4) for $\mathbf{P}$ and $\mathbf{E}$ and equating the time-independent components on either side of (3.33), we obtain

$$\frac{\hbar\Omega}{2}\frac{(N_1 - N_2) - (N_1 - N_2)^e}{T_1} = -\frac{i\omega}{4}(\tilde{\mathbf{P}}\cdot\tilde{\mathbf{E}}^* - \tilde{\mathbf{P}}^*\cdot\tilde{\mathbf{E}}) \tag{3.34}$$

where the asterisk denotes the complex conjugate. Since the polarization can be expressed in terms of the field in accordance with (3.5) as

$$\tilde{\mathbf{P}} = \epsilon_0\chi(\omega)\tilde{\mathbf{E}} = \epsilon_0(\chi'(\omega) + i\chi''(\omega))\tilde{\mathbf{E}}$$

Equation 3.34 can be written in the form

$$\frac{\hbar\Omega}{2}\frac{(N_1 - N_2) - (N_1 - N_2)^e}{T_1} = \tfrac{1}{2}\Omega\chi''(\omega)|\tilde{\mathbf{E}}|^2 \tag{3.35}$$

The physical significance of (3.35) is that in the steady state the time average of the population difference adjusts itself so that the net power lost to the surrounding medium corresponds to the power delivered to the dipole system during the absorption process.

The susceptibility can be eliminated from (3.35) with the aid of (3.10), yielding, for the homogeneously broadened line,

$$(N_1 - N_2) = \frac{(N_1 - N_2)^e}{1 + \dfrac{I}{I_{sat}} \dfrac{g_L(\omega, \Omega)}{T_2/\pi}} \tag{3.36}$$

where $I = \eta \epsilon_0 c |\tilde{\mathbf{E}}|^2/2$ is the power per unit area carried by the wave and η is the refractive index of the medium. The term I_{sat} is a saturation parameter defined by

$$I_{sat} \equiv \frac{\eta \epsilon_0 c}{(2T_1 T_2/\hbar^2) L(|\mathbf{\mu}_{12}|^2/3)} \tag{3.37}$$

where I_{sat} is the power per unit area that a wave on resonance must carry in order to reduce the population difference to one-half its unsaturated value.

Equation 3.36 is plotted in Figure 3.9. At high field intensities it is seen that the population difference approaches zero, that is, the populations of the upper and lower levels become nearly equal. Regardless of the intensity of the applied field, it is impossible by the resonant absorption type of process described here to "invert" the population difference, that is, to place more atoms in the upper level than in the lower. Population inversion, which is required in the operation of lasers and masers, is accomplished by processes involving other energy levels. The means for

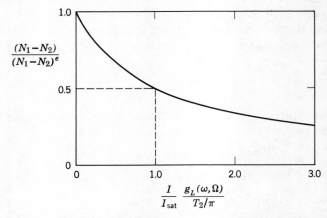

figure 3.9 Saturation of population difference per unit volume with increasing field intensity.

obtaining population inversion are discussed in Chapter 4. The factor that multiplies I/I_{sat} in (3.36) and in Figure 3.9 is unity for $\omega = \Omega$ and less elsewhere, indicating that the effectiveness of the field in saturating the transition is highest at resonance.

A combination of (3.13), (3.14), and (3.36) yields an expression for the absorption constant when saturation becomes important,

$$\Gamma = \frac{\Omega \pi}{\hbar \epsilon_0 c \eta} L \frac{|\mu_{12}|^2}{3} (N_1 - N_2)^e \left[\frac{1}{\pi} \frac{1/T_2}{(\Omega - \omega)^2 + (1/T_2)^2(1 + I/I_{\text{sat}})} \right] \quad (3.38)$$

The unsaturated value for Γ may be obtained from (3.38) by letting $I = 0$. The effects of saturation are the flattening and broadening of the lineshape factor so that the peak absorption is decreased by the amount

$$\Gamma(\Omega) = \Gamma_{\text{unsat}}(\Omega) \frac{1}{1 + I/I_{\text{sat}}} \quad (3.39)$$

and the linewidth increased by the amount

$$\Delta \omega_L = \Delta \omega_{L_{\text{unsat}}} \sqrt{1 + I/I_{\text{sat}}} \quad (3.40)$$

These effects are illustrated in Figure 3.10.

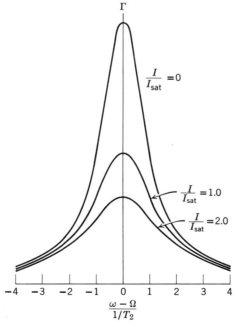

figure 3.10 Absorption constant under saturating conditions.

As saturation power levels are approached, the combination of a decreasing absorption constant with increasing intensity leads to the result that the power absorbed per unit volume, $\mathscr{P} = \Gamma I$, saturates at a constant level $\mathscr{P}_{sat}$ shown in Figure 3.11. Combining Equations 3.37, 3.38, and 3.39, for $I/I_{sat} \gg 1$ we obtain on resonance

$$\mathscr{P}_{sat}(\Omega) = \Gamma_{unsat}(\Omega) I_{sat} = \frac{\hbar\Omega}{2} \frac{(N_1 - N_2)^e}{T_1} \qquad (3.41)$$

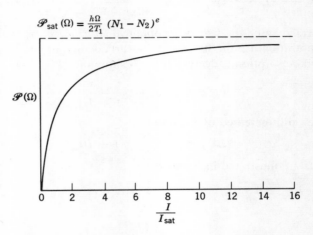

figure 3.11 Power absorbed per unit volume at resonance under saturating conditions.

From the first equality in (3.41) we see that at complete saturation the power absorbed per unit volume approaches a value corresponding to that which would be absorbed at an intensity level $I = I_{sat}$ if the absorption constant maintained its unsaturated value. Since no further power can be absorbed by the dipole system under this condition, the transition is said to be *bleached*, and any additional power added to the wave propagates through the medium unattenuated.

Measurement of saturation effects is sometimes useful in determining the relaxation time constant T_1. All that is required is a measurement of the unsaturated absorption constant and linewidth (to determine the matrix element $|\mu_{12}|^2$ and linewidth $\Delta\omega_{L unsat} = 2/T_2$) followed by a measurement of either the absorption constant or the linewidth under saturating conditions to determine the parameter I_{sat} from either (3.39) or (3.40). The time constant T_1 can then be obtained by solving (3.37).

Experimentally, saturation effects are sometimes observable at the relatively modest power level of 1 mW/cm². For the ammonia case treated earlier, at a pressure of 4×10^{-2} mm Hg, $T_1 \sim 1/3 \times 10^6$ sec, $T_2 \sim 1/6.4 \times 10^6$ sec, $|\mu_{12}| = 1.14$ debye, yielding, by Equation 3.37, $I_{sat} \sim 6$ mW/cm².

3.2.7 Degeneracy

In the development thus far we have assumed that the levels involved were nondegenerate and thus left unanswered a question of considerable importance in practical applications, namely, how are the equations modified in the case of degeneracy?

To answer this question, we consider the two-level system where the degeneracy of the upper and lower levels are g_2 and g_1, respectively. That is, there are g_2 eigenstates with an energy E_2, and g_1 corresponding to an energy E_1. We include here the possibility that the levels are not precisely degenerate, but are spaced closely both with respect to the bandwidth of the exciting radiation and with respect to κT. Figure 3.12 illustrates the type of system under consideration, where for clarity the individual states in each level are shown separated.

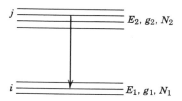

figure 3.12 Two-level system with upper and lower level degeneracies g_2 and g_1, respectively.

The polarization $\mathbf{P}_{ij}$ generated as a result of transitions between the ith and jth state is, for the isotropic case, given by (3.1):

$$\ddot{\mathbf{P}}_{ij} + \frac{2}{T_2}\dot{\mathbf{P}}_{ij} + \Omega^2 \mathbf{P}_{ij} = \frac{2\Omega}{\hbar} L \frac{|\mu_{ij}|^2}{3}(N_i - N_j)\mathbf{E} \qquad (3.42)$$

In the usual case where no selective mechanism has been applied to separate out molecules of different states in a given degenerate energy level, energy considerations indicate that all states of the same level will be equally populated at the beginning of an interaction. Therefore, at least initially, $N_j = N_2/g_2$ and $N_i = N_1/g_1$, where N_2 and N_1 are the total number of atoms or molecules per unit volume in the upper and lower levels, respectively. Under this condition (3.42) may be written

$$\ddot{\mathbf{P}}_{ij} + \frac{2}{T_2}\dot{\mathbf{P}}_{ij} + \Omega^2 \mathbf{P}_{ij} = \frac{2\Omega}{\hbar} L \frac{|\mu_{ij}|^2}{3}\left(\frac{N_1}{g_1} - \frac{N_2}{g_2}\right)\mathbf{E} \qquad (3.43)$$

Summing over the multiplicity of upper and lower states to obtain the total polarization, we obtain

$$\ddot{\mathbf{P}} + \frac{2}{T_2}\dot{\mathbf{P}} + \Omega^2 \mathbf{P} = \frac{2\Omega}{\hbar} L \frac{|\mu_{12}|^2}{3}\left(\frac{N_1}{g_1} - \frac{N_2}{g_2}\right)\mathbf{E} \qquad (3.44)$$

where now $|\mu_{12}|^2$ is defined by

$$|\mu_{12}|^2 = \sum_{i=1}^{g_1}\sum_{j=1}^{g_2}|\mu_{ij}|^2 \qquad (3.45)$$

If we wish to follow changes in the level populations as transitions occur, we need an expression for the degenerate case that corresponds to (3.2). As transitions occur it is typically the case, at least in a solid or liquid, that fast internal cross-relaxation processes between degenerate states in a given level exist.[11] As a result, regardless of which states are involved in transitions, the internal relaxation processes maintain the distributions $N_j = N_2/g_2$ and $N_i = N_1/g_1$. For this case Equations 3.43–3.45 continue to apply as N_1 and N_2 change. (Otherwise (3.42) must be used.) The population difference equation for this case is derived as follows.

First, we note that for the nondegenerate case, (3.2) can be written as two equations by the application of the relationship $N_1 + N_2 = N_V$, where N_V is a constant. These two equations are

$$\frac{\partial N_1}{\partial t} + \frac{N_1 - N_1^e}{T_1} = -\frac{1}{\hbar\Omega}\dot{\mathbf{P}}\cdot\mathbf{E} \qquad (3.46)$$

$$\frac{\partial N_2}{\partial t} + \frac{N_2 - N_2^e}{T_1} = \frac{1}{\hbar\Omega}\dot{\mathbf{P}}\cdot\mathbf{E} \qquad (3.47)$$

Applying (3.46) to the degenerate case, we must write

$$\frac{\partial N_i}{\partial t} + \frac{N_i - N_i^e}{T_1} = -\frac{1}{\hbar\Omega}\frac{1}{g_1}\sum_{i=1}^{g_1}\sum_{j=1}^{g_2}\dot{\mathbf{P}}_{ij}\cdot\mathbf{E} = -\frac{1}{\hbar\Omega}\frac{1}{g_1}\dot{\mathbf{P}}\cdot\mathbf{E} \qquad (3.48)$$

The right-hand side of (3.48) takes this form because, with cross-relaxation between degenerate levels, the population of state i changes as a result of *all* transitions between levels 1 and 2 and, in fact, changes as the fraction $1/g_1$ of the total change in the population of level 1. (If no cross-relaxation exists, the right-hand side of (3.48) is simply $-(1/\hbar\Omega)\sum_{j=1}^{g_2}\dot{\mathbf{P}}_{ij}\cdot\mathbf{E}$, which involves transitions just between level i and all levels j.) A similar argument applies for level j. The result, with substitution of $N_j = N_2/g_2$ and $N_i = N_1/g_1$, is

$$\frac{\partial}{\partial t}\frac{N_1}{g_1} + \frac{N_1/g_1 - N_1^e/g_1}{T_1} = -\frac{1}{\hbar\Omega}\frac{1}{g_1}\dot{\mathbf{P}}\cdot\mathbf{E} \qquad (3.49)$$

$$\frac{\partial}{\partial t}\frac{N_2}{g_2} + \frac{N_2/g_2 - N_2^e/g_2}{T_2} = \frac{1}{\hbar\Omega}\frac{1}{g_2}\dot{\mathbf{P}}\cdot\mathbf{E} \qquad (3.50)$$

These two equations can be combined in two ways. First, we observe that multiplication of (3.49) by g_1 and (3.50) by g_2 followed by subtraction of the second from the first gives (3.2) back again. From this we conclude that the population difference equation that represents energy balance does not change form when applied to the degenerate case. However, since the population difference in the polarization equation, (3.44), appears in terms of the variable $(N_1/g_1 - N_2/g_2)$, it is sometimes convenient for simultaneous solutions to write (3.49) and

(3.50) in the form obtained by subtraction of the second from the first,

$$\frac{\partial}{\partial t}\left(\frac{N_1}{g_1} - \frac{N_2}{g_2}\right) + \frac{(N_1/g_1 - N_2/g_2) - (N_1/g_1 - N_2/g_2)^e}{T_1} = -\frac{1}{\hbar\Omega}\left(\frac{1}{g_1} + \frac{1}{g_2}\right)\dot{\mathbf{P}} \cdot \mathbf{E}$$

(3.51)

Therefore, for the degenerate case with cross-relaxation between degenerate states, Equation 3.44 replaces (3.1), and (3.2) may be used as is or replaced by an equivalent in the form of (3.51).

Finally, we derive the relationship that exists between the populations of different energy levels in thermal equilibrium when there is degeneracy. In thermal equilibrium at absolute temperature T the distribution of atoms among different states will follow Boltzmann's law. If there are several states in a given energy level, the population per unit volume of any given state is related to the population per unit volume of another state at another energy level by the relationship

$$N_j = N_i e^{-(E_j - E_i)/\kappa T}$$

Therefore, in terms of the level populations per unit volume $N_2 = g_2 N_j$ and $N_1 = g_1 N_i$, we have

$$\frac{N_2}{g_2} = \frac{N_1}{g_1} e^{-(E_2 - E_1)/\kappa T} \qquad (3.52)$$

3.3 TENSOR PROPERTIES OF THE SUSCEPTIBILITY

Our treatment of the absorption, dispersion, and saturation properties of an isotropic medium has been based on the use of a scalar susceptibility $\chi(\omega)$, which relates the polarization to the electric field according to (3.5),

$$\tilde{\mathbf{P}} = \epsilon_0 \chi(\omega) \tilde{\mathbf{E}} \qquad (3.53)$$

The susceptibility concept can be extended to the anisotropic case as well. Because of the significance of anisotropic media in practical applications, it is useful to indicate how this extension is carried out and to discuss some of the important consequences that result. If isotropic media alone are of interest, however, this section can be bypassed without loss of continuity.

For the anisotropic case the polarization equation is given by (2.41)

$$\ddot{P}_\alpha + \frac{2}{T_2}\dot{P}_\alpha + \Omega^2 P_\alpha = \frac{2\Omega}{\hbar}\overline{(\mu_\alpha \mu_\beta^*)}(N_1 - N_2)E_\beta^{\text{loc}} \qquad (2.41)$$

If the variables are written as

$$P_\alpha = \tfrac{1}{2}\tilde{P}_\alpha e^{i(\omega t - kz)} + \text{c.c.}$$

$$E_\beta = \tfrac{1}{2}\tilde{E}_\beta e^{i(\omega t - kz)} + \text{c.c.}$$

then from (2.41) we obtain

$$\tilde{P}_\alpha = \frac{\pi}{\hbar} \overline{(\mu_\alpha \mu_\beta^*)} (N_1 - N_2) \tilde{g}_L(\omega, \Omega) \tilde{E}_\beta$$
$$\equiv \epsilon_0 \chi_{\alpha\beta}(\omega) \tilde{E}_\beta \qquad (3.54)$$

where $\tilde{g}_L(\omega, \Omega)$ is the complex Lorentzian function defined by (3.7), and we have taken $\tilde{E}_\beta^{\text{loc}} = \tilde{E}_\beta$. For dense media, $\tilde{E}_\beta^{\text{loc}} \neq \tilde{E}_\beta$ and it is necessary to introduce a local field correction factor as in the isotropic case, although in anisotropic media the expressions are more complex.[12]

In Appendix 3 it is shown that for a two-level system $\mu_\alpha \mu_\beta^* = \mu_\beta \mu_\alpha^*$. Therefore, for a two-level system and unity Lorentz correction factor, we have

$$\chi_{\beta\alpha}(\omega) = \chi_{\alpha\beta}(\omega) \qquad (3.55)$$

When more than two levels are involved and a local field correction factor appears, which is itself a function of the coordinate subscripts, (3.55) still applies and may be proved from thermodynamical considerations.[13]

The susceptibility $\chi_{\alpha\beta}(\omega)$ relates two physical variables that have a significance independent of any coordinate system in which they happen to be represented. Therefore the susceptibility that connects the two variables must in some sense also be independent of the coordinate system. This implies that the susceptibility, which has a particular form in a given coordinate system, must transform in a certain prescribed manner when the coordinate system is changed, in order to maintain the required physical relationship. Such a property is known as a tensor property. In particular, since $\chi_{\alpha\beta}(\omega)$ relates the components of one vector to another and therefore requires two subscripts, it is termed a second-rank tensor. The number of coordinate subscripts indicates the rank of a tensor; a vector is a first-rank tensor, a scalar is a zero-rank tensor.

A second-rank tensor is specified by nine numbers. In an orthogonal Cartesian coordinate system, the components of $\chi_{\alpha\beta}(\omega)$ are written in array form as

$$\begin{pmatrix} \chi_{xx} & \chi_{xy} & \chi_{xz} \\ \chi_{yx} & \chi_{yy} & \chi_{yz} \\ \chi_{zx} & \chi_{zy} & \chi_{zz} \end{pmatrix}$$

Of the nine elements in the above array, only six are independent because of the restriction imposed by (3.55). A tensor for which (3.55) holds is said to be a symmetric tensor.

Of particular interest are the transformation properties of the susceptibility tensor under various coordinate rotations. These properties are intimately connected with the symmetry properties of the medium, which are in turn important in determining the form of the second-rank linear susceptibility tensor $\chi_{\alpha\beta}$. For

3 RESONANT PROCESSES

example, let us consider a medium whose symmetry properties are such that regardless of its orientation with respect to a given applied field, the induced polarization is always aligned with the field. Then the off-diagonal elements of the susceptibility tensor $\chi_{\alpha\beta}$ must obviously vanish, for they correspond to the polarization generated in one direction when a field is applied in another. (See Equation 3.54.) Furthermore, if the amplitude of the polarization generated by the given field is independent of the orientation of the medium, the values for the diagonal elements of the susceptibility tensor are clearly equal. We have, in fact, just described the symmetry properties of an isotropic medium whose susceptibility tensor $\chi_{\alpha\beta}$ is of the form

$$\chi_{\alpha\beta} = \begin{pmatrix} \chi_{xx} & 0 & 0 \\ 0 & \chi_{xx} & 0 \\ 0 & 0 & \chi_{xx} \end{pmatrix}$$

For the more general case it is necessary to consider the transformation properties of a second-rank tensor under various rotations of the coordinate system. By a rotation of coordinate system, we mean a change from one set of mutually perpendicular axes, denoted by x_1, x_2, x_3, to another set x_1', x_2', x_3' with the same origin (Figure 3.13). The transformation from the old unprimed set of axes to the new primed set is represented by the coordinate transformation equation

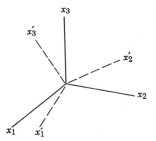

figure 3.13 Transformation of axes.

$$x_\alpha' = a_{\alpha\beta} x_\beta$$

where the $a_{\alpha\beta}$ are the cosines of the angles between the new x_α' and the old x_β axes. Summation over repeated subscripts is understood.

The above can be written alternatively in matrix notation as (the symbol $\approx$ indicates a matrix)

$$\underset{\approx}{x'} = \underset{\approx}{A} \underset{\approx}{x}$$

or, in full,

$$\begin{pmatrix} x' \\ y' \\ z' \end{pmatrix} = \begin{pmatrix} a_{xx} & a_{xy} & a_{xz} \\ a_{yx} & a_{yy} & a_{yz} \\ a_{zx} & a_{zy} & a_{zz} \end{pmatrix} \begin{pmatrix} x \\ y \\ z \end{pmatrix}$$

where the three-by-three array of coefficients is the transformation matrix $\underset{\approx}{A}$. Figure 3.13 is an illustration of the original and primed coordinate systems. The transformation of coordinates is a transformation of the components of a position vector. The components of any other vector, such as the polarization or electric field, transform in the same way as the position vector.[14] Thus the components of **P** in the new coordinate system are related to those in the old by

$$P'_\alpha = a_{\alpha\beta} P_\beta \tag{3.56}$$

and the inverse transformation is given by

$$P_\alpha = a_{\beta\alpha} P'_\beta \tag{3.57}$$

Note that in going from the old system to the new, as in (3.56), the repeated indices occur as close together as possible, whereas the opposite is true for the inverse transformation.

Now consider the transformation of a second-rank tensor that relates two vectors. In particular, we are interested in the transformation properties of the second-rank susceptibility tensor $\chi_{\alpha\beta}$. From (3.54) we have

$$\tilde{P}_\alpha = \epsilon_0 \chi_{\alpha\beta} \tilde{E}_\beta$$

The polarization in a new coordinate system is given by (3.56), which, with the aid of the above, can be written

$$\tilde{P}'_\alpha = \epsilon_0 a_{\alpha\beta} \chi_{\beta\gamma} \tilde{E}_\gamma$$

However, $\tilde{E}_\gamma$ transforms as in (3.57); therefore

$$\tilde{P}'_\alpha = \epsilon_0 a_{\alpha\beta} \chi_{\beta\gamma} a_{\delta\gamma} \tilde{E}'_\delta$$

or, finally,

$$\tilde{P}'_\alpha = \epsilon_0 \chi'_{\alpha\delta} \tilde{E}'_\delta$$

where the susceptibility $\chi'_{\alpha\delta}$ in the new coordinate system is related to that in the old by

$$\chi'_{\alpha\delta} = a_{\alpha\beta} \chi_{\beta\gamma} a_{\delta\gamma} \tag{3.58}$$

In matrix notation, (3.58) can be written

$$\underset{\sim}{\chi}' = \underset{\sim}{A}\underset{\sim}{\chi}\underset{\sim}{A}^t = \underset{\sim}{A}\underset{\sim}{\chi}\underset{\sim}{A}^{-1} \tag{3.59}$$

where $\underset{\sim}{\chi}$ and $\underset{\sim}{A}$ are both three-by-three matrices, $\underset{\sim}{A}^t$ designates the transpose of $\underset{\sim}{A}$ in which the rows and columns are interchanged, and $\underset{\sim}{A}^{-1}$ designates the inverse of $\underset{\sim}{A}$, which satisfies $\underset{\sim}{A}^{-1}\underset{\sim}{A} = \underset{\sim}{A}\underset{\sim}{A}^{-1} = I$, where I is the identity matrix. For orthogonal transformations, that is, transformations between two orthogonal Cartesian coordinate systems, the inverse and transpose matrices are identical. A transformation equation of the form (3.59) is known as a similarity transformation.

We are now ready to examine how the transformation properties of the susceptibility tensor under coordinate rotations are related to the physical properties of materials. First we measure the components of any vector that represents a physical quantity such as polarization or electric field in a given coordinate

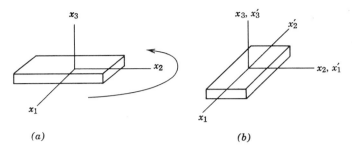

figure 3.14 Diagram illustrating operation on a medium to determine possible symmetry elements. A measurement to specify the elements of a tensor is made in the unprimed coordinate system shown in (a). The object is then rotated and the tensor elements are measured along the same directions in absolute space. That is, the tensor elements are referred to the unprimed coordinate system in (b). If the susceptibility tensor is the same for these two measurements, then the tensor referred to the primed and unprimed coordinates in (b) is the same. The medium is said to possess $\pi/2$ symmetry.

system for the medium in a specified position as illustrated in Figure 3.14a. The elements of the susceptibility tensor that relates these vectors are then determined.

We now alter the position of the medium such as by a $\pi/2$ rotation, and repeat the measurements in the same directions in absolute space. If the susceptibility tensor is unaltered this is equivalent to stating that the tensor referred to the primed and unprimed coordinates in Figure 3.14b is the same. This gives

$$\underset{\approx}{\chi}' = \underset{\approx}{\chi} \tag{3.60}$$

The susceptibility is then said to possess the symmetry element specified by the rotation operator.

If $\underset{\approx}{A}$ is the coordinate transformation matrix associated with a rotation that leads to an identity between the tensor elements as in (3.60), then (3.59) yields

$$\underset{\approx}{\chi} = \underset{\approx}{A} \underset{\approx}{\chi} \underset{\approx}{A}^{-1}$$

or

$$\underset{\approx}{\chi} \underset{\approx}{A} = \underset{\approx}{A} \underset{\approx}{\chi} \underset{\approx}{A}^{-1} \underset{\approx}{A} = \underset{\approx}{A} \underset{\approx}{\chi}$$

We therefore obtain the useful result that if $\underset{\approx}{A}$ corresponds to a symmetry operation, then

$$[\underset{\approx}{A}, \underset{\approx}{\chi}] = 0 \tag{3.61}$$

The above relationship is a statement of how the symmetry properties of the medium constrain the form of $\underset{\approx}{\chi}$. This is best illustrated by consideration of an example. First, however, some background is necessary.

In the study of the physical properties of crystals the possible macroscopic symmetry elements possessed by the crystal and corresponding operations can be listed as follows:[15,16]

1. Center of symmetry: operation known as inversion in which each point located at (x, y, z) is moved to the position $(-x, -y, -z)$.
2. Mirror plane: operation in which each point is moved to the position of its mirror image in a given plane.
3. n-fold rotation axis, $n = 1, 2, 3, 4, 6$: rotation of $2\pi/n$ about a given axis known, respectively, as a monad, diad, triad, tetrad, or hexad axis.
4. n-fold inversion axis, $n = 1, 2, 3, 4, 6$: rotation of $2\pi/n$ followed by an inversion.

Possible combinations of these macroscopic symmetry elements fall into 32 point groups, defining 32 crystal classes according to the point-group symmetry possessed by the crystal. The term "point group" arises from the fact that the operations listed above are with respect to a point without regard to translational symmetry. The physical structure of the crystal may, in fact, also possess two additional symmetry elements known as a glide plane and an n-fold screw axis, both of which are concerned with the translational properties of the crystal. If these additional symmetry elements are admitted into consideration, we find that there are 230 possible combinations of symmetry elements known as space groups. However, in the study of the macroscopic properties of crystals we need not be concerned with this additional refinement.

The 32 crystal classes, conveniently grouped into seven crystal systems according to certain similarities in the combination of symmetry elements, are listed in Table 3.2 along with several properties of interest. Each of the 32 classes is designated by a notation such as $\bar{3}m$, as shown in column 1. The numbers refer to an n-fold rotation axis, the overbar indicates an inversion axis, and m designates a mirror plane. The rules for constructing the full symbol are presented in texts on crystal properties. A brief summary is as follows:

1. Rotation axis: X.
2. Rotation axis followed by inversion: $\bar{X}$.
3. Rotation axis, mirror plane normal to it: X/m.
4. Rotation axis, diad axis (axes) normal to it: $X2$.
5. Rotation axis, mirror plane (planes) parallel to it: Xm.
6. Rotation axis, mirror plane normal to it and mirror planes parallel to it: X/mm.
7. Rotation axis, mirror plane normal to it, mirror plane parallel to it, mirror plane parallel to it and at 45° to the latter plane: X/mmm.

As an example, $\bar{3}m$ indicates a three-fold inversion axis with a mirror plane or planes parallel to it.

The nomenclature for the seven crystal systems according to the geometry of the crystal structure is listed in column 2. Column 3 indicates the breakdown of the crystals according to their optical properties. An anaxial crystal is one for which the susceptibility is the same for propagation of fields polarized parallel to

TABLE 3.2 CRYSTAL CLASSES

Class symbol[a]	Crystal system	Optical properties	Linear susceptibility tensor
$1, \bar{1}$	Triclinic	Biaxial	$\begin{pmatrix} \chi_{xx} & \chi_{xy} & \chi_{xz} \\ \chi_{xy} & \chi_{yy} & \chi_{yz} \\ \chi_{xz} & \chi_{yz} & \chi_{zz} \end{pmatrix}$
$2, m, 2/m$	Monoclinic	Biaxial	$\begin{pmatrix} \chi_{xx} & 0 & \chi_{xz} \\ 0 & \chi_{yy} & 0 \\ \chi_{xz} & 0 & \chi_{zz} \end{pmatrix}$
$222, mm2, mmm$	Orthorhombic	Biaxial	$\begin{pmatrix} \chi_{xx} & 0 & 0 \\ 0 & \chi_{yy} & 0 \\ 0 & 0 & \chi_{zz} \end{pmatrix}$
$4, \bar{4}, 4/m, 422, 4mm, \bar{4}2m, 4/mmm$	Tetragonal	Uniaxial	$\begin{pmatrix} \chi_{xx} & 0 & 0 \\ 0 & \chi_{xx} & 0 \\ 0 & 0 & \chi_{zz} \end{pmatrix}$
$3, \bar{3}, 32, 3m, \bar{3}m$	Trigonal	Uniaxial	$\begin{pmatrix} \chi_{xx} & 0 & 0 \\ 0 & \chi_{xx} & 0 \\ 0 & 0 & \chi_{zz} \end{pmatrix}$
$6, \bar{6}, 6/m, 622, 6mm, \bar{6}m2, 6/mmm$	Hexagonal	Uniaxial	$\begin{pmatrix} \chi_{xx} & 0 & 0 \\ 0 & \chi_{xx} & 0 \\ 0 & 0 & \chi_{zz} \end{pmatrix}$
$23, m3, 432, \bar{4}3, m3m$	Cubic	Anaxial	$\begin{pmatrix} \chi_{xx} & 0 & 0 \\ 0 & \chi_{xx} & 0 \\ 0 & 0 & \chi_{xx} \end{pmatrix}$

[a] Drawings illustrating the symmetry operations for the crystal classes are presented in Table 21, pp. 284–288, of Ref. 13.

any of the three x, y, or z axes. Such a crystal is optically isotropic and possesses a single index of refraction. In a uniaxial crystal the susceptibility for the propagation of field components polarized parallel to one of the axes, known as the optic axis, differs from propagation of components polarized parallel to the other axes. This produces two indices of refraction, one for those components parallel to the optic axis (extraordinary) and one for those perpendicular to the optic axis (ordinary). For a biaxial crystal the susceptibility differs for propagation of components polarized parallel to each of the three axes, resulting in three refractive indices.

The form of the susceptibility tensors listed on the far right of Table 3.2 is determined by the use of Equation 3.61. As a simple example, consider a crystal that possesses tetrad symmetry about the z axis. The old (unprimed) and new (primed) axes are as shown in Figure 3.14b. The symmetry operator $\underset{\approx}{A}$ is obtained from the coordinate transformation equation $x'_\alpha = a_{\alpha\beta} x_\beta$. In matrix form the transformation equation inferred from Figure 3.14b is

$$\begin{pmatrix} x'_1 \\ x'_2 \\ x'_3 \end{pmatrix} = \begin{pmatrix} 0 & 1 & 0 \\ -1 & 0 & 0 \\ 0 & 0 & 1 \end{pmatrix} \begin{pmatrix} x_1 \\ x_2 \\ x_3 \end{pmatrix}$$

where $\underset{\approx}{A}$ is the 3×3 matrix in the above equation.

Setting $\left[\underset{\approx}{A}, \underset{\approx}{\chi} \right] = 0$ in accordance with (3.61), we obtain

$$\begin{pmatrix} 0 & 1 & 0 \\ -1 & 0 & 0 \\ 0 & 0 & 1 \end{pmatrix} \begin{pmatrix} \chi_{xx} & \chi_{xy} & \chi_{xz} \\ \chi_{xy} & \chi_{yy} & \chi_{yz} \\ \chi_{xz} & \chi_{yz} & \chi_{zz} \end{pmatrix} = \begin{pmatrix} \chi_{xx} & \chi_{xy} & \chi_{xz} \\ \chi_{xy} & \chi_{yy} & \chi_{yz} \\ \chi_{xz} & \chi_{yz} & \chi_{zz} \end{pmatrix} \begin{pmatrix} 0 & 1 & 0 \\ -1 & 0 & 0 \\ 0 & 0 & 1 \end{pmatrix}$$

The above set of nine equations, only six of which are independent, is solved simultaneously to obtain the $\chi_{\alpha\beta}$. For example, the first equation, beginning at the upper left, is

$$-\chi_{xy} = \chi_{xy}$$

which implies $\chi_{xy} = 0$. A second equation is of the form

$$-\chi_{yy} = -\chi_{xx}$$

which implies that the first two diagonal elements in χ are equal. This procedure is continued until all elements are known, yielding

$$\chi = \begin{pmatrix} \chi_{xx} & 0 & 0 \\ 0 & \chi_{xx} & 0 \\ 0 & 0 & \chi_{zz} \end{pmatrix}$$

in agreement with the entry in the Table 3.2. The foregoing example thus illustrates the relationship between the symmetry properties of the medium and the form of the susceptibility tensor $\chi_{\alpha\beta}$.

Although in the text we restrict ourselves primarily to the isotropic case for simplicity, the information presented in this section provides the basis for generalization to anisotropic media. The susceptibility concept can be extended to nonlinear processes. For example, in third harmonic generation a polarization is generated at a frequency three times that of the applied field. In this case the polarization and field are related by a fourth-rank susceptibility tensor defined by the relation

$$\tilde{P}_\alpha(3\omega) = \epsilon_0 \chi_{\alpha\beta\gamma\delta} \tilde{E}_\beta(\omega) \tilde{E}_\gamma(\omega) \tilde{E}_\delta(\omega)$$

where the relevant frequencies associated with each term are shown in parentheses.

In the most general case the polarization is expanded as a power series in the field with the susceptibility tensors of various ranks playing the role of expansion coefficients. In many treatments of nonlinear processes, expansions of this type are taken as the starting point for analysis.[17] The susceptibility tensors are then treated as phenomenological quantities to be determined by experiment.

3.4 TRANSIENT BEHAVIOR OF THE ELECTRIC DIPOLE TRANSITION—COUPLED AMPLITUDE EQUATIONS AND RATE EQUATIONS

Since the absorption, dispersion, and saturation properties of the electric dipole transition as discussed in Section 3.2 are steady-state phenomena, a simple steady-state approach is adequate for their description. However, under certain conditions, as in the solid state laser, dramatic exchanges of energy take place between the dipole system and radiation field and lead to large-scale fluctuations in the absorption and saturation properties of the medium. This results in the transient buildup and decay of fields, both in time and space. To study such phenomena, it is necessary to include the transient behavior of the coupled set of equations (3.1)–(3.3).

Fortunately a complete transient solution to the coupled equations is not necessary. Instead, approximations can often be introduced that simplify the problem considerably. In the interaction that takes place between the radiation field and the electric dipole system there are generally two distinct time scales, one established by the microwave or optical oscillation period, the other determined by the decay times associated with various loss mechanisms. It is these decay times that control the transfer of energy back and forth between the field and dipole system. In a laser, for example, the oscillation period is on the order of 10^{-15} sec, whereas decay times are typically 10^{-10} sec or longer. It will be seen that it is possible to separate out the slow variations in the amplitudes of quantities from their basic high-frequency behavior by an averaging procedure and thus focus attention on the longer time scale. Such a technique allows us to follow the transfer of energy between the medium and the radiation field on a time scale that is long compared with the oscillation period.

Similarly, there are generally two distinct spatial scales, one associated with the wavelength of the radiation, the other associated with the buildup or decay of fields over a distance that is large compared with the wavelength. These can also be separated out by means of averaging techniques that are often referred to descriptively as stroboscopic averaging, since fine-scale variations are ignored and

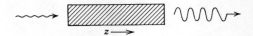

figure 3.15 Amplification of signal resulting from population inversion $N_2 > N_1$.

attention is focused on large-scale behavior only. The approach we shall employ is simply a quasisteady-state or variation-of-parameters approach in which the coefficients of steady-state quantities are allowed to vary slowly in time or space.

3.4.1 Traveling-Wave Amplification

As a first example, we shall consider a problem appropriate to a traveling-wave laser amplifier (see Figure 3.15). Assume that a resonant optical signal is incident on a medium whose population difference has been inverted by some means, that is, the medium has been prepared so that the number of atoms or molecules per unit volume in the upper state exceeds that in the lower ($N_2^e > N_1^e$). Methods for accomplishing such inversion are discussed in Chapter 4. It is intuitively evident that, at least initially, the wave will begin to grow because it sees a negative absorption constant that results from the inverted population difference per unit volume $(N_2 - N_1)^e > 0$ (see Equation 3.13). As the fields become strong, however, it is reasonable to expect that some type of saturation effect will occur.

For an isotropic medium the appropriate equations are, from (3.1)–(3.3),

$$\ddot{P} + \frac{2}{T_2}\dot{P} + \Omega^2 P = -\frac{2\Omega}{\hbar}L\frac{|\mu_{12}|^2}{3}NE$$

$$\frac{\partial N}{\partial t} + \frac{N - N^e}{T_1} = \frac{2}{\hbar\Omega}\dot{P}\cdot E \qquad (3.62)$$

$$\nabla \times (\nabla \times E) + \frac{\eta\mathscr{A}}{c}\frac{\partial E}{\partial t} + \frac{\eta^2}{c^2}\frac{\partial^2 E}{\partial t^2} = -\mu_0\frac{\partial^2 P}{\partial t^2}$$

where $N \equiv N_2 - N_1$.

Now assume that the polarization, electric field, and population difference per unit volume vary as

$$P = \tfrac{1}{2}\tilde{P}(z)e^{i(\Omega t - kz)} + \text{c.c.}$$

$$E = \tfrac{1}{2}\tilde{E}(z)e^{i(\Omega t - kz)} + \text{c.c.} \qquad (3.63)$$

$$N = N(z)$$

We are considering propagating wave solutions that are steady-state oscillations in time of frequency Ω but are allowed to vary slowly in space through the complex amplitudes $\tilde{P}(z)$, $\tilde{E}(z)$. The quantity k is the propagation constant, $k \equiv \Omega\eta/c$. The time varying components in N, which are small, have been neglected.

Substitution of (3.63) into (3.62) yields the following set of coupled-amplitude equations:

$$\tilde{P}(z) = \frac{iT_2}{\hbar} L \frac{|\mu_{12}|^2}{3} N\tilde{E}$$

$$\frac{N - N^e}{T_1} = \frac{i}{2\hbar}(\tilde{P} \cdot \tilde{E}^* - \tilde{P}^* \cdot \tilde{E}) \qquad (3.64)$$

$$2ik\frac{\partial \tilde{E}}{\partial z} + i\mathscr{A}k\tilde{E} = \Omega^2 \mu_0 \tilde{P}$$

In deriving the above equations, we have taken $\partial^2 \tilde{E}/\partial z^2 \ll k\, \partial \tilde{E}/\partial z$ in accordance with the slow-variation assumption. Solution of the coupled-amplitude equations yields the amplitudes and phases of the slowly varying components of the polarization and field and the magnitude of the population difference, all as a function of distance into the medium.

Elimination of the polarization $\tilde{P}(z)$ followed by multiplication of the field equation by $\tilde{E}^*$ and addition to its complex conjugate yield

$$\frac{\partial}{\partial z}|\tilde{E}(z)|^2 + \mathscr{A}|\tilde{E}(z)|^2 = \frac{\Omega T_2}{\hbar \epsilon_0 c \eta} L \frac{|\mu_{12}|^2}{3} N(z)|\tilde{E}(z)|^2 \qquad (3.65)$$

where

$$N(z) = \frac{N^e}{1 + (T_1 T_2/\hbar)L(|\mu_{12}|^2/3)|\tilde{E}(z)|^2} \qquad (3.66)$$

The quantity $|\tilde{E}|^2$ is proportional to the time-averaged energy density in the traveling-wave field. Equation 3.65 is referred to as a rate equation, since it exhibits the rate at which energy is transferred from the medium to the field as the field progresses through the medium. In contrast to the coupled-amplitude equations (3.64), the rate equation no longer contains information about the phase of the field since only the magnitude squared is involved. However, to determine the growth of the energy density it is not necessary to consider phase information.

The combination of (3.65) and (3.66) leads to the nonlinear differential equation

$$\frac{dI}{dz} = \left[\gamma_0 \frac{I(0) + I_{\text{sat}}}{I + I_{\text{sat}}} - \mathscr{A}\right]I = g(I)I \qquad (3.67)$$

where I is the average power per unit area carried by the wave, $I = \eta \epsilon_0 c\, |\tilde{E}|^2/2$, I_{sat} is the saturation parameter defined by (3.37), $I(0)$ is the power at the input plane $z = 0$, γ_0 is the gain at the input plane due to the inverted population difference,

$$\gamma_0 = \frac{\hbar \Omega N^e}{2T_1} \frac{1}{I(0) + I_{\text{sat}}} \qquad (3.68)$$

and $g(I)$ is the gain constant.

At this point it is useful to reconsider the spatial average assumption upon which (3.62) and hence (3.65), (3.66), and (3.67) are based. As discussed in

Section 2.4.3, we have assumed that the spatial average of the product of the matrix elements and the population difference per unit volume is equal to the product of the averages, as given by (2.39):

$$\overline{(\mu_\alpha \mu_\beta^*) N_V (\rho_{11} - \rho_{22})} = \overline{(\mu_\alpha \mu_\beta^*)}(N_1 - N_2) \qquad (2.39)$$

When saturation effects are considered, such as those expressed by (3.66), the population difference is a function of the matrix elements. Therefore a simultaneous solution of (3.65) and (3.66) would, in general, necessitate the substitution of an unaveraged equation of the type of (3.66) into an unaveraged (3.65), and then the composite equation would have to be averaged. Of course if all the molecules are identical, as in a homogeneous crystal, then even though there is saturation, (2.39) applies. These considerations relate not only to (3.65) and (3.66), but also to the rate equations developed later in this chapter and used throughout the text.

Let us consider the error that is introduced in the gain constant $g(I)$ by assuming that (2.39) holds for the case that all orientations of the molecule are equally likely. In Appendix 6 it is shown that (2.39) introduces an error that underestimates the effect of saturation. However, for both very large and very small values of I/I_{sat}, where I_{sat} is defined by (3.37), this error approaches zero. The maximum error is about 30%, and for most purposes this is not significant, and thus we shall continue to use (2.39). If it is deemed necessary, a correction factor may be introduced, by using the ratio of Equation A6.2 to Equation A6.1 as given in Appendix 6, which multiplies the gain constant.

An exact solution to (3.67) can be obtained by the separation-of-variables technique (see Problem 3.14), but let us simply consider the equation from a qualitative standpoint. There are three regions of interest set by the gain parameter γ_0.

1. Region I, high gain, $\gamma_0 > \mathscr{A}$. In this region the total gain $g(I)$ given in (3.67) is positive at the entrance plane, and therefore I begins to grow exponentially with distance. However, as I increases, the first term in brackets in (3.67) decreases because of the influence of I in the denominator, until finally the two terms become equal. The wave then stops growing with distance because the gain per unit length just balances the loss per unit length as shown in Figure 3.16. The final value of I, denoted by $I(\infty)$; is therefore given by the condition $g(I) = 0$ in (3.67), or

$$I(\infty) = \frac{\gamma_0}{\mathscr{A}}\left[I(0) + I_{\text{sat}}\right] - I_{\text{sat}} \qquad (3.69)$$

2. Region II, medium gain, $\mathscr{A} < \gamma_0 < \mathscr{A} I_{\text{sat}}/(I(0) + I_{\text{sat}})$. In this region the total gain $g(I)$ is initially negative at the entrance plane, and therefore I begins to decay exponentially. The decay of I in the denominator of the first term in brackets in (3.67) leads to an increasing gain until finally the gain matches the loss and the

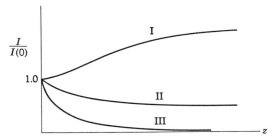

figure 3.16 Growth or decay of intensity with distance in a traveling-wave laser. Curves I, II, and III correspond to high, medium, and low gain regions of operation, respectively. In Region I the gain per unit length caused by an inverted population difference overcomes the loss per unit length sufficiently to provide for amplification of the input beam. In Region II the gain is not sufficient to yield overall amplification, but is sufficient to overcome loss at a certain level of operation and thus permit unattenuated propagation. In Region III the gain is not sufficient to overcome loss at any level, and therefore the input intensity decays to zero.

wave travels unattenuated through the medium at a value determined by (3.69). This situation is also shown in Figure 3.16.

3. Region III, low gain, $\gamma_0 < \mathscr{A} I_{sat}/[I(0) + I_{sat}]$. In this region the wave begins to decay as in Region II, but the gain $g(I)$ remains negative and the wave decays to zero because the first term in $g(I)$ is never large enough to cancel the second.

3.4.2 Cavity Amplification

The traveling-wave laser amplifier served as an example of a solution to the electric dipole equations that is steady state in time but varies in space. Amplification in a cavity maser or laser, however, is described in terms of solutions that have a fixed spatial standing wave distribution but are transient in time. For an isotropic medium the appropriate equations are, from (3.1)–(3.3),

$$\ddot{\mathbf{P}} + \frac{2}{T_2}\dot{\mathbf{P}} + \Omega^2 \mathbf{P} = -\frac{2\Omega}{\hbar} L \frac{|\mu_{12}|^2}{3} N\mathbf{E}$$

$$\dot{N} + \frac{N - N^e}{T_1} = \frac{2}{\hbar\Omega}\dot{\mathbf{P}} \cdot \mathbf{E} \tag{3.70}$$

and

$$\ddot{\mathbf{E}} + \frac{1}{\tau_c}\dot{\mathbf{E}} + \omega_c^2\,\mathbf{E} = -\frac{1}{\epsilon}\ddot{\mathbf{P}}$$

where we assume that the polarization, which is driven by the cavity field, has the same spatial distribution as the cavity normal mode field, so that the integral on the right-hand side of (3.3b) can be dropped, and $N = N_2 - N_1$.

Before examining the transient solution, we note that Equations 3.70 couple together two oscillators, one a polarization oscillator of frequency Ω, the other a cavity field oscillator of frequency ω_c. Therefore, in the event that the resonance transition frequency Ω and cavity frequency ω_c are not identical, the coupled system oscillates at some frequency ω that falls somewhere between the two; that is, frequency pulling occurs. The frequency of oscillation is found by considering the steady-state solutions.

Assuming solutions of the form

$$P = \tfrac{1}{2}\tilde{P}e^{i\omega t} + \text{c.c.}$$

and similarly for E, combination of the polarization and field equation yields

$$\left[(\omega^2 - \Omega^2)(\omega^2 - \omega_c^2) - \frac{2\omega^2}{T_2 \tau_c}\right] + i\left[\frac{\omega}{\tau_c}(\omega^2 - \Omega^2) + \frac{2\omega}{T_2}(\omega^2 - \omega_c^2)\right]$$

$$= \frac{2\Omega}{\hbar} L \frac{|\mu_{12}|^2}{3} \frac{\omega^2}{\epsilon} N$$

where N is the steady-state value for the inverted population difference per unit volume. Since N is real, the imaginary term must vanish with the results shown in Figure 3.17. It is apparent from the figure that the oscillation frequency ω is pulled strongly toward the narrower of the two overlapping lines. For this reason, the case shown in Figure 3.17b, which is applicable to the ammonia cavity maser, for example, is a good candidate for a frequency standard since the frequency is set by the molecular resonance rather than the cavity. This is not true for the case shown in Figure 3.17a.

To obtain the transient solution, assume that the polarization and electric field vary as

$$P = \tfrac{1}{2}\tilde{P}(t)e^{i\Omega t} + \text{c.c.}$$

$$E = \tfrac{1}{2}\tilde{E}(t)e^{i\Omega t} + \text{c.c.}$$
(3.71)

where $\tilde{P}(t)$ and $\tilde{E}(t)$ are slowly varying complex amplitudes, Ω is the microwave or optical frequency of interest, and the optical phase of the polarization or electric field is contained in the complex amplitude. It is assumed that the resonator is tuned to the transition frequency so that $\omega_c = \Omega = \omega$.

Differentiating (3.71) with respect to time, substituting the results into the equations (3.70), and equating coefficients of $e^{i\Omega t}$, we find that

$$\dot{\tilde{P}} + \frac{1}{T_2}\tilde{P} = \frac{i}{\hbar} L \frac{|\mu_{12}|^2}{3} N\tilde{E} \tag{3.72}$$

and

$$\dot{\tilde{E}} + \frac{1}{2\tau_c}\tilde{E} = -\frac{i\Omega}{2\epsilon}\tilde{P} \tag{3.73}$$

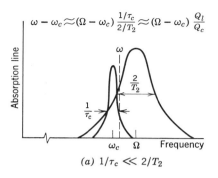
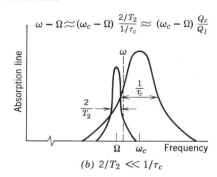

figure 3.17 Frequency pulling effects for cases where molecular and cavity bandwidths differ substantially. The case shown in (a) is in general applicable to the laser, whereas (b) is applicable to the ammonia cavity maser discussed in Chapter 4. The pulled oscillation frequency, ω, lies closest to the narrower of the two lines. Q_l and Q_c are, respectively, the molecular line and cavity Q's defined by $Q_l \equiv \Omega T_2/2$, $Q_c \equiv \omega_c \tau_c$.

The equation for the slowly varying components of the population difference per unit volume N, found by equating zero-frequency terms, is

$$\dot{N} + \frac{N - N^e}{T_1} = \frac{i}{2\hbar}(\tilde{\mathbf{P}} \cdot \tilde{\mathbf{E}}^* - \tilde{\mathbf{P}}^* \cdot \tilde{\mathbf{E}}) \tag{3.74}$$

In deriving the above equations, we have used the condition that the molecular and cavity linewidths obey $2/T_2 \ll \Omega$ and $1/\tau_c \ll \Omega$, respectively. We have also taken $\dot{\tilde{\mathbf{P}}} \ll \Omega \tilde{\mathbf{P}}$ and $\dot{\tilde{\mathbf{E}}} \ll \Omega \tilde{\mathbf{E}}$, since the complex amplitudes are assumed to vary slowly with respect to the oscillation frequency. In addition, high-frequency terms in the population difference are neglected.

Equations 3.72–3.74 constitute the set of coupled-amplitude equations that exhibit the slow variations in the amplitudes of polarization, electric field, and population difference. As in the traveling-wave laser case, further simplification is possible if we are willing to discard phase information about the field.

3.4.3 Laser Cavity Oscillator

We shall first consider parameter values typically characteristic of the laser cavity oscillator. In this case the molecular or atomic linewidth $2/T_2$ is generally much broader than the cavity bandwidth $1/\tau_c$. This implies that $\partial/\partial t \ll 2/T_2$, as can be verified from (3.72) and (3.73) by assuming N essentially constant and evaluating the time constant of the combined equations. Under this condition $\dot{\tilde{\mathbf{P}}}$ can be neglected with respect to $\tilde{\mathbf{P}}/T_2$ in (3.72), and the following equations result:

$$\dot{N} + \frac{N - N^e}{T_1} = -\frac{T_2}{\hbar^2} L \frac{|\mu_{12}|^2}{3} N |\tilde{\mathbf{E}}|^2 \tag{3.75}$$

$$|\dot{\tilde{\mathbf{E}}}|^2 + \frac{1}{\tau_c}|\tilde{\mathbf{E}}| = \frac{\Omega T_2}{\hbar \epsilon} L \frac{|\mu_{12}|^2}{3} N |\tilde{\mathbf{E}}|^2 \tag{3.76}$$

The quantity $|\tilde{\mathbf{E}}|^2$ is proportional to the time-averaged energy density in the cavity field.

If we introduce the substitution

$$\varphi \equiv \frac{\epsilon |\tilde{\mathbf{E}}|^2}{2\hbar\Omega} \qquad (3.77)$$

Equations 3.75 and 3.76 assume the form

$$\dot{N} + \frac{N - N^e}{T_1} = -\frac{2\Omega T_2}{\hbar\epsilon} L \frac{|\mu_{12}|^2}{3} N\varphi \qquad (3.78)$$

and

$$\dot{\varphi} + \frac{\varphi}{\tau_c} = \frac{\Omega T_2}{\hbar\epsilon} L \frac{|\mu_{12}|^2}{3} N\varphi \qquad (3.79)$$

The quantity φ is proportional to the electromagnetic energy density (energy per unit volume) and conforms to the spatial variation characteristic of the cavity mode distribution. The variable φ is sometimes loosely referred to as the photon density, since for a single mode the integral of $\varphi = \epsilon |\tilde{\mathbf{E}}|^2/2\hbar\Omega$ over the cavity volume yields the number of photons in a mode when field quantization is considered. This is misleading, however, because φ is in fact simply a classical variable that has well-defined spatial variations such as nodes and crests with quarter-wavelength separation. It is not rigorously correct to think of the photons as being bunched at the crests, however, since the fundamentals of quantum mechanics prevent the location of individual photons within a region of space smaller than a wavelength because of the uncertainty principle.[18]

Equations 3.78 and 3.79 are rate equations, since they exhibit the rates at which energy is transferred between the cavity fields and medium. As in the traveling-wave case, the rate equations (3.78) and (3.79) differ from the coupled-amplitude equations (3.72)–(3.74) in that they do not contain phase information about the fields.

To facilitate application of the above rate equations, it is convenient to normalize the variables to steady-state values by introducing $\bar{N} = N/N_0$, $\bar{\varphi} = \varphi/\varphi_0$, where the subscript 0 denotes steady-state values obtained by setting time derivatives in (3.78) and (3.79) equal to zero. The steady-state values for a nonzero energy density are found to be

$$N_0 = \frac{\hbar\epsilon}{\Omega \tau_c T_2 L(|\mu_{12}|^2/3)} \qquad (3.80)$$

$$\varphi_0 = \frac{\tau_c}{T_1} \frac{N_0}{2} (\bar{N}^e - 1) \qquad (3.81)$$

In terms of the normalized variables, (3.78) and (3.79) become

$$\dot{\bar{N}} + \frac{\bar{N} - \bar{N}^e}{T_1} = -\frac{\bar{N}^e - 1}{T_1} \bar{N}\bar{\varphi}$$

$$\dot{\bar{\varphi}} + \frac{\bar{\varphi}}{\tau_c} = \frac{1}{\tau_c} \bar{N}\bar{\varphi} \quad (3.82)$$

We can easily verify that the values $\bar{N} = \bar{\varphi} = 1$ correspond to steady-state solutions as required by the normalization condition.

The above normalized rate equations are in the form widely used in the literature in the analysis of lasers.[19] These equations will be taken as the starting point for the analysis of the laser presented in Chapter 4 and will be discussed in detail there.

3.4.4 Maser Cavity Oscillator

We shall now examine the coupled amplitude equations (3.72)–(3.74) in terms of the parameters characteristic of, for example, the ammonia cavity maser. As compared with the laser, the roles of the molecular and cavity linewidths are reversed, since the cavity bandwidth is the broader of the two for the maser; that is, $2/T_2 \ll 1/\tau_c$. This implies that $\partial/\partial t \ll 1/\tau_c$, and therefore the term $\dot{E}$ in (3.73) can be dropped, yielding, for the resonant case,

$$\dot{N} + \frac{N - N^e}{T_1} = -\frac{2}{\tau_c} \varphi$$

$$\dot{\varphi} + \frac{2}{T_2} \varphi = \frac{2\Omega \tau_c}{\hbar \epsilon} L \frac{|\mu_{12}|^2}{3} N\varphi$$

Normalization by the procedure applied in the laser case leads to

$$\dot{\bar{N}} + \frac{\bar{N} - \bar{N}^e}{T_1} = -\frac{\bar{N}^e - 1}{T_1} \bar{\varphi}$$

$$\dot{\bar{\varphi}} + \frac{2}{T_2} \bar{\varphi} = \frac{2}{T_2} \bar{N}\bar{\varphi} \quad (3.83)$$

with the normalization constants (that is, steady-state values) identical with those found in the laser case, (3.80) and (3.81). Once again the values $\bar{N} = \bar{\varphi} = 1$ correspond to the steady-state solution.

The maser rate equations differ from those obtained for the laser in two respects. First, the equation for the energy density in the cavity fields is governed by the molecular time constant T_2 rather than the cavity lifetime τ_c. Second, the driving term on the right-hand side of the population difference equation (3.83) is proportional simply to $\bar{\varphi}$ rather than to the product $\bar{N}\bar{\varphi}$ as in the case of the laser. The rate equations in this form correspond to the reaction-field principle used by Anderson in his treatment of the magnetic resonance amplifier[20] and have also been derived by Tang[21] in a manner similar to the procedure used here. These equations can be taken as the starting point for the analysis of the transient behavior of the ammonia cavity maser.[22]

3.4.5 Transition Probability

Equation 3.75, which is a rate equation for the inverted population difference per unit volume in the laser case, can be written in the form

$$\frac{\partial N_2}{\partial t} + \frac{N_2 - N_2^e}{T_1} = -WN_2 + WN_1 \qquad (3.84a)$$

$$\frac{\partial N_1}{\partial t} + \frac{N_1 - N_1^e}{T_1} = WN_2 - WN_1 \qquad (3.84b)$$

where

$$W = \left(\frac{T_2}{2}\right)\left(\frac{L(|\mathbf{\mu}_{12}|^2/3)|\tilde{\mathbf{E}}|^2}{\hbar^2}\right) \qquad (3.85)$$

To proceed from (3.75) to (3.84), we need only apply the conservation condition $N_1 + N_2 = N_V$, a constant, where N_V is the number of atoms per unit volume. In the derivation of (3.84a) and (3.84b) it had been assumed that the signal was exactly on resonance; that is, $\omega = \Omega$. For the general case of off-resonant signals, T_2 in (3.85) must be replaced by $\pi g(\omega, \Omega)$, where $g(\omega, \Omega)$ is the lineshape function.

The first term on the left-hand side of (3.84a) or (3.84b) keeps track of the net changes in the level population; the second term accounts for nonradiative interactions with the surrounding medium; and the terms on the right-hand side correspond to changes in the level population as a result of interaction with the field.

The form of the equation suggests that W is a rate or probability per unit time that an atom or molecule will make a transition from one level to another as a result of the presence of radiation. The per-atom probability rate W for upward transitions is seen to be identical with the probability rate for downward transitions. An upward transition constitutes an absorption process, and a downward transition corresponds to stimulated emission.

In terms of the transition rate W, we find that the rate equations for the laser case, (3.78) and (3.79), can be written

$$\dot{N} + \frac{N - N^e}{T_1} = -2WN \qquad (3.86a)$$

$$\dot{\varphi} + \frac{\varphi}{\tau_c} = WN \qquad (3.86b)$$

We have here determined the transition rate W for a two-level electric dipole resonant transition. The concept of a transition probability per unit time is a very

3.5 PARAMAGNETIC RESONANCE ABSORPTION

The phenomena discussed in the previous sections in connection with the electric dipole transition all have their counterparts in the magnetic dipole spin ½ case with little modification. Rather than repeat the entire development, we shall simply begin the development by considering the steady-state absorption and dispersion characteristics of the spin ½ transition as was done for the electric dipole case.

The appropriate equations, derived in Chapter 2, are the Bloch equations for the medium, (2.83) and the field equation (2.87):

$$\dot{M}_x + \frac{M_x}{T_2} = \gamma(\mathbf{B} \times \mathbf{M})_x \tag{3.87}$$

$$\dot{M}_y + \frac{M_y}{T_2} = \gamma(\mathbf{B} \times \mathbf{M})_y \tag{3.88}$$

$$\dot{M}_z + \frac{M_z - M_z^e}{T_1} = \gamma(\mathbf{B} \times \mathbf{M})_z \tag{3.89}$$

$$\nabla \times (\nabla \times \mathbf{B}) + \frac{\eta^2}{c^2}\frac{\partial^2 \mathbf{B}}{\partial t^2} = \mu_0 \nabla \times (\nabla \times \mathbf{M}) \tag{3.90}$$

where, for simplicity, we have neglected the loss term in the field equation. If we apply a static magnetic field in the z direction and an alternating field in the plane transverse to the dc field, the x and y components of the magnetization play a role similar to that of the polarization $\mathbf{P}$ in the electric dipole case, and the z component corresponds to the population difference. For a transverse plane wave propagating in the z direction we assume solutions of the form

$$M_\alpha = \tfrac{1}{2}\tilde{M}_\alpha e^{i(\omega t - kz)} + \text{c.c.}, \qquad \alpha = x, y$$

$$M_z = M_z^e$$

Equations 3.87 and 3.88 for the transverse components of the magnetization reduce to

$$\left(i\omega + \frac{1}{T_2}\right)\tilde{M}_x + \Omega\tilde{M}_y = \gamma\mu_0 M_z^e \tilde{H}_y \tag{3.91}$$

$$-\left(i\omega + \frac{1}{T_2}\right)\tilde{M}_y + \Omega\tilde{M}_x = \gamma\mu_0 M_z^e \tilde{H}_x \tag{3.92}$$

where we recall that the static z component of magnetic field defines the radian transition frequency $\Omega = \gamma B_{0z}$. In deriving (3.91) and (3.92), we have also used relationship (2.86):

$$\mathbf{B} = \mu_0(\mathbf{H} + \mathbf{M}) \tag{3.93}$$

It is convenient to consider the magnetization in terms of circularly polarized components of the form

$$\mathbf{M} = \tfrac{1}{2}\tilde{M}_+(\mathbf{1}_x - i\mathbf{1}_y)e^{i(\omega t - k_+ z)}$$
$$+ \tfrac{1}{2}\tilde{M}_-(\mathbf{1}_x + i\mathbf{1}_y)e^{i(\omega t - k_- z)} + \text{c.c.}$$

where

$$\tilde{M}_+ = \tfrac{1}{2}(\tilde{M}_x + i\tilde{M}_y)$$
$$\tilde{M}_- = \tfrac{1}{2}(\tilde{M}_x - i\tilde{M}_y)$$

and similarly for H. Combination of (3.91) and (3.92) then provides an expression for the magnetic susceptibility for a circularly polarized wave,

$$\chi_+(\omega) = \frac{\tilde{M}_+}{\tilde{H}_+} = \pi\gamma\mu_0 M_z^e\, \tilde{g}_L(\omega, \Omega) \tag{3.94}$$

where $\tilde{g}_L(\omega, \Omega)$ is the complex Lorentzian line given by (3.7). As in the electric dipole case, the susceptibility has both real and imaginary parts,

$$\chi_+(\omega) = \chi_+'(\omega) + i\chi_+''(\omega) \tag{3.95}$$

which account for the dispersion and absorption properties of the transition, respectively.

From the wave equation (3.90) a relationship may be obtained between k_+ and $\chi_+(\omega)$ similar to (3.11a) for the electric dipole

$$k_+^2 = \frac{\eta^2\omega^2}{c^2}[1 + \chi_+(\omega)] \tag{3.96}$$

For $\chi_+ \ll 1$, use of the square root expansion $\sqrt{1 + \chi_+} \approx 1 + \chi_+/2$ leads to

$$k_+ = k_+' + ik_+'' \approx \frac{\eta\omega}{c}\left[1 + \frac{\chi_+'(\omega)}{2}\right] + i\frac{\omega\chi_+''(\omega)}{2c}$$

As in the electric dipole case, the absorption constant is given by $\Gamma_+ = -2k_+''$. For $\omega \approx \Omega$, we have from (3.94),

$$\Gamma_+ = \frac{\eta\pi\gamma^2\mu_0\hbar\Omega}{2c}(N_2 - N_1)^e g_L(\omega, \Omega) \tag{3.97}$$

where $g_L(\omega, \Omega)$ is the Lorentzian lineshape factor given by (3.14), $(N_2 - N_1)$ is the population difference per unit volume (for the spin $\frac{1}{2}$ transition level 2 is the lower energy state), and the expression for M_z^e given after (2.82) has been used to eliminate M_z^e. Absorption of the circularly polarized wave $H_+ = H_x + iH_y$ is strongest near the transition frequency Ω.

If the development presented above is carried out for the susceptibility associated with the counter-rotating components, $\chi_-(\omega) = \tilde{M}_-/\tilde{H}_-$, it is found that $(\Omega + \omega)$ must be substituted for $(\Omega - \omega)$ in the denominator of (3.94), indicating the impossibility of a resonant interaction with the H_- field component under the propagation conditions assumed. Therefore, if a linearly polarized wave of frequency $\omega \approx \Omega$ is applied to the medium, one of the two circularly polarized components rotating in opposite directions will be near resonance, the other far from resonance, resulting in selective absorption of one circularly polarized component over another.

The above treatment of the steady-state absorption properties of the spin $\frac{1}{2}$ transition indicates how the development carried out for the electric dipole case can be extended to the spin $\frac{1}{2}$ system. The study of saturation and the development of rate equations proceed as in the electric dipole case.

REFERENCES

1. A. Von Hippel, *Dielectrics and Waves*, M.I.T. Press, Cambridge, Mass., 1954, p. 170.
2. C. H. Townes and A. L. Schawlow, *Microwave Spectroscopy*, McGraw-Hill, New York, 1955, pp. 300ff, 345.
3. C. H. Townes, "The Ammonia Spectrum and Line Shapes near 1.25-cm Wavelength," *Phys. Rev.*, **70**, 665, 1946.
4. See Ref. 2, p. 74.
5. See Ref. 1, p. 137.
6. J. C. Slater, *Quantum Theory of Atomic Structure*, McGraw-Hill, New York, 1960, pp. 156ff.
7. A. M. Portis, "Electronic Structure of F Centers: Saturation of the Electron Spin Resonance," *Phys. Rev.*, **91**, 1071, 1953.
8. B. A. Lengyel, *Introduction to Laser Physics*, Wiley, New York, 1966, pp. 44ff.
9. *Handbook of Mathematical Functions*, ed. by M. Abramowitz and I. Stegun, National Bureau of Standards Applied Mathematics Series 55, 2nd Printing, November 1964, pp. 297ff.
10. A. Messiah, *Quantum Mechanics*, North-Holland Publishing Company, Amsterdam, 1961, Vol. 1, p. 469.
11. A. E. Siegman and J. W. Allen, "Pump Power Dependence of Ruby Laser Starting and Stopping Time," *IEEE Jour. Quant. Elect.*, **QE-1**, 386, 1965.
12. N. Bloembergen, *Nonlinear Optics*, W. A. Benjamin, New York, 1965, pp. 68, 179ff., 192.
13. J. F. Nye, *Physical Properties of Crystals*, University Press, Oxford, 1957, pp. 74, 182.
14. See Ref. 13, p. 10.
15. See Ref. 13, p. 278.
16. C. Kittel, *Introduction to Solid-State Physics*, Wiley, New York, 1966, p. 12.

17. P. A. Franken and J. F. Ward, "Optical Harmonics and Nonlinear Phenomena," *Rev. Mod. Phys.*, **35**, 23, 1963.
18. D. Bohm, *Quantum Theory*, Prentice-Hall, Englewood Cliffs, N.J., pp. 91ff.
19. H. Statz and G. A. deMars, *Quantum Electronics*, ed. by C. H. Townes, Columbia University Press, New York, 1960, p. 530.
20. P. W. Anderson, "The Reaction Field and Its Use in Some Solid-State Amplifiers," *J. Appl. Phys.*, **28**, 1049, 1957.
21. C. L. Tang, "On Maser Rate Equations and Transient Oscillations," *J. Appl. Phys.*, **34**, 2935, 1963.
22. E. T. Jaynes and F. W. Cummings, "Comparison of Quantum and Semiclassical Radiation Theories with Application to the Beam Maser," *Proc. IEEE*, **51**, 89, 1963.

PROBLEMS

3.1 Let a flux of R particles per unit area per unit time fall on a slab of material of area A and thickness Δx.
(*a*) Assuming that there are N_V target centers per unit volume, each with an effective stopping area or cross-section σ_c, derive an expression for the number of particles per unit area per unit time, ΔR, removed by the slab.
(*b*) If the particles are photons of energy $\hbar\omega$, what is the power absorbed per unit volume, $\mathscr{P}$, in terms of the cross-section σ_c and power per unit area I?
(*c*) Derive the exponential decay law and determine the absorption coefficient Γ in terms of the cross-section σ_c.
3.2 Derive (3.17), the scattering cross-section for the classical harmonic oscillator.
3.3 Derive (3.20).
3.4 Obtain a sum rule relationship for the electric quadrupole operator xy. Prove that

$$[xy, [\mathscr{H}_0, xy]] = \frac{\hbar^2}{m}(x^2 + y^2)$$

where $\mathscr{H}_0 = p^2/2m + \mathscr{V}$. From this equality obtain a sum rule relationship.
3.5 Describe an experiment for the determination of T_1, using the phenomenon of saturation. Give the appropriate formulas for obtaining T_1.
3.6 From Figure 3.7 and Equation 3.31, sketch χ' for the Gaussian line.
3.7 Show that $A^t = A^{-1}$ for a rotation operator.
3.8 The piezoelectric tensor relates polarization P_α to stress $\sigma_{\beta\gamma}$:

$$P_\alpha = d_{\alpha\beta\gamma}\sigma_{\beta\gamma}$$

The term $d_{\alpha\beta\gamma}$ is a third-rank tensor where $d_{\alpha\beta\gamma} = d_{\alpha\gamma\beta}$. Determine relationships between the elements of the piezoelectric tensor if the medium has $\bar{4}$ symmetry about the γ axis.

3 RESONANT PROCESSES

3.9 Radiation is incident on a material with a Lorentzian lineshape of width $2/T_2$ and frequency Ω. If the radiation has a Gaussian energy distribution of width $\Delta\omega_G$, determine the absorption characteristic when

$$(a)\ \Delta\omega_G \ll \frac{2}{T_2}$$

$$(b)\ \Delta\omega_G \gg \frac{2}{T_2}$$

3.10 Write an expression for the Helmholtz free energy function for a polarizable medium in an electromagnetic field. (The Helmholtz function is defined in any introductory book on thermodynamics.) From the fact that this function is an exact differential, show that the susceptibility tensor is symmetric.

3.11 Show that it is always possible to diagonalize a real second-rank symmetric tensor.

3.12 Find the principal axes for the susceptibility tensor of the monoclinic crystal given in Table 3.2.

3.13 Rewrite (3.38) for the case where the levels are degenerate. When the levels are degenerate, is more or less incident power required to saturate the transition?

3.14 A static magnetic field of 3000 Wb/m² is applied to a paramagnetic material in the positive z direction. An rf magnetic field propagating in the positive z direction is applied at the spin $\frac{1}{2}$ resonant frequency. This field is linearly polarized. The parameter values are as follows:

$$\text{resonance linewidth} = 1\ \text{MHz}$$
$$\eta = 1 \quad \gamma = e/m \quad T = 300°K$$
$$N_V = 10^{22}\ \text{electrons/cm}^3$$
$$l = \text{material length} = 10\ \text{cm}$$

A linearly polarized wave may be considered as the sum of two circularly polarized waves.

(a) Which direction of circular polarization is absorbed?
(b) What is the absorption constant Γ for this polarization?
(c) What is the polarization of the transmitted field?
(d) What is the ratio of transmitted power to incident power?
(e) If a metallic mirror is placed at the far end, what is the absorption constant for this reflected wave?
(f) If the wave is reflected by a prism, as shown, what is the absorption constant for this reflected wave?

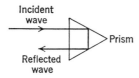

3.15 The growth of the intensity I with distance in a traveling-wave laser is given by the nonlinear differential equation (3.67).

(a) Solve the equation by the separation-of-variables technique and show that it is of the form

$$\frac{z}{1/\mathscr{A}} = \frac{I_{\text{sat}}}{I(\infty)} \ln\left[\frac{I/[I - I(\infty)]}{I(0)/[I(0) - I(\infty)]}\right] - \ln\left(\frac{I - I(\infty)}{I(0) - I(\infty)}\right)$$

where $I(0)$ is the intensity at the entrance plane and $I(\infty)$ is the asymptotic value that I approaches at large distances.

(b) Sketch the solutions in the form of Figure 3.16, showing asymptotic values for $I(\infty)/I_{\text{sat}}$, for the following two cases:

(i) $\gamma_0 = 1.5\mathscr{A}$, $\quad \dfrac{I(0)}{I_{\text{sat}}} = \tfrac{1}{3}$

(ii) $\gamma_0 = (7/8)\mathscr{A}$, $\quad \dfrac{I(0)}{I_{\text{sat}}} = \tfrac{1}{3}$

where γ_0 is defined by (3.68).

LASERS

4

4.1 INTRODUCTION

One of the most important and interesting applications of quantum theory has been the development of the optical oscillator, or "laser." Lasers provide radiation from the far infrared to the ultraviolet portion of the spectrum, with peak pulsed powers as much as a gigawatt and continuous powers exceeding hundreds of watts. With the development of the laser it is now possible to generate signals with very high spectral purity with the frequency spread of the oscillation approximately one part in 10^{14} of the center frequency.

The fundamental idea of the laser is to provide a medium with a negative absorption constant capable of amplifying electromagnetic radiation. This can be accomplished by having a dipole transition with the population of the higher energy state exceeding the population of the lower state, a condition known as inversion.† Various methods may be used to achieve inversion, and some of these methods are discussed in Section 4.2. A minimum inverted population difference is required to provide a gain that is greater than the losses of the system, and this leads to a threshold requirement, as considered in Section 4.3. Steady-state output power and transient behavior of the laser will be considered in Sections 4.4 and 4.5, respectively. Extremely high peak pulsed powers are obtained by a rapid alteration in the circuit losses, and this operation, known as "Q-switching," is discussed in Section 4.6.

In Sections 4.3 to 4.6 examples will be given using the parameters of ruby, because the ruby laser has interesting transient behavior and can be operated Q-switched. The parameters of the neodymium laser and the helium-neon gas laser will be discussed in the two succeeding sections in order to provide a comparison between the behavior of several different types of optical oscillators.

† For the ensuing discussion and the remainder of Chapter 4, the ratio N_i/g_i will be referred to as the population per unit volume of state i, where N_i is the actual population per unit volume and g_i is the degeneracy of the state.

This chapter dealing with lasers is intended to be fundamental in nature rather than exhaustive. Therefore, certain aspects of optical oscillators will not be considered and approximations will be made to simplify the treatment. A more detailed study of lasers can be found in the books and articles listed at the end of the chapter.

4.2 POPULATION INVERSION

In the previous chapter it was noted that the dipole transition absorbs electromagnetic radiation when the population of the lower energy state exceeds the population of the higher energy state. This is the situation that exists in thermal equilibrium. If the population of the upper state exceeds that of the lower, the absorption constant Γ becomes negative, and there is amplification of the electromagnetic wave. A transition for which the occupation of the upper energy level is greater than that for the lower level is said to have an inverted population. In the laser or optical maser, the population inversion exists between two levels with a resonant frequency in the infrared or visible part of the spectrum, while the transition of interest for the maser is at microwave frequencies.

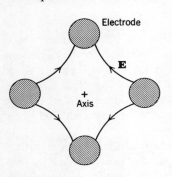

figure 4.1 The quadrupole focuser used to separate molecules in the upper and lower energy states for ammonia gas.

A variety of methods are used to provide population inversions. In 1954 the first maser was developed[6] in which an inverted population was achieved in ammonia gas by a spatial separation of energy states. Figure 4.1 illustrates the four-electrode or quadrupole configuration that was used to focus the molecules in the upper state along the axis and to force the lower-state molecules away from the axis. In Figure 4.1 the electric field **E** is zero along the axis and increases as the radius increases. A static electric field applied to the induced electric dipole transition perturbs the energy of the states so that the energy of the upper state increases in a high field region whereas the energy of the lower state decreases in a high field region. The force on a particle tends to reduce the energy of the molecule so that lower state particles are forced into stronger fields and upper state molecules are forced into weaker fields. This results in a focusing of upper state molecules along the axis where there is zero field, whereas particles in the lower state move outward from the axis.

Figure 4.2 shows how this spatial separation of states is applied to the design of an oscillator based on population inversion. Ammonia gas is passed through an orifice, and the beam of ammonia molecules is directed along the axis of the quadrupole focuser. As a consequence of the previously described forces on the

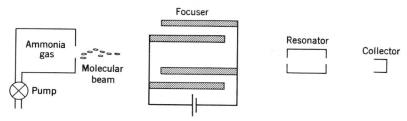

figure 4.2 A schematic diagram of the ammonia maser. The resonator is tuned to the transition frequency, which is $\simeq 23.8\ GHz$.

molecules, essentially all the molecules in the vicinity of the axis are in the higher energy state when the molecules leave the focuser. The decay time of excited molecules is much longer than the transit time from the focuser to the resonator, and thus almost every molecule entering the resonator is in the upper energy state. If the resonator is tuned to the dipole transition frequency, which is $\simeq 23.8$ GHz, electromagnetic radiation in the resonator will be amplified by the inverted population of the molecular beam. When the gain is equal to or greater than the attenuation of the radiation resulting from losses in the cavity, then self-sustaining oscillations occur. There is a threshold condition for oscillation which requires that the gain provided by the molecular system must compensate for circuit losses. A more detailed discussion of the threshold requirements for oscillation will be presented in Section 4.3.

The spatial separation of states that is used in the ammonia maser is not applicable to solids or liquids, but there are features of the ammonia maser that are common to all masers and lasers. These features are as follows.

1. An excitation mechanism. It is the excitation mechanism that provides the population inversion, and in the ammonia maser this role is fulfilled by the quadrupole focuser.
2. An active medium. The active medium is the gas, liquid, or solid that sustains the inverted population and, for the ammonia maser, that is the beam of ammonia molecules.
3. A circuit. Electromagnetic radiation from the active medium is coupled to the external environment by means of a circuit. At infrared or optical frequencies the circuit is usually an interferometer resonator, and for the microwave ammonia maser a cavity resonator is used.

In 1958[7] the optical maser was proposed, and in 1960[8] the first laser was in operation. This was a pulsed ruby laser, where the active ion was chromium in an aluminum oxide host lattice. Population inversion of the chromium ions (Cr^{3+}) is achieved by optical pumping, which for ruby consists of the absorption of green and blue light over a fairly broad range of frequencies and the subsequent decay of the excited ions to a lower energy state. This lower energy state, corresponding to a red wavelength transition to the ground state, has a long relaxation

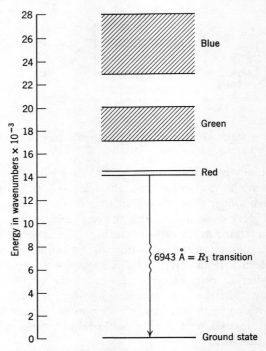

figure 4.3 Energy levels for Cr^{3+} in aluminum oxide (Al_2O_3). Laser action usually occurs at a wavelength of 6943 Å which is known as the R_1 transition. There is also a slightly higher energy transition R_2 with a wavelength of 6930 Å. To the right of the energy level diagram are the colors corresponding to the various energies. Relaxation from the green and blue bands to the R levels occurs by the excitation of lattice vibrations rather than by photon emission.

time (a metastable state), and it is possible to obtain a population inversion between this excited level and the ground state. Figure 4.3 illustrates the energy diagram appropriate for the ruby laser. Energy is given in wavenumbers (cm^{-1}), where the wavenumber is the reciprocal of the wavelength in cm of the transition. Since energy $= h\nu = hc \times$ (wavenumber), an energy of one wavenumber corresponds to approximately 2×10^{-16} ergs, 2×10^{-23} J, or 1.2×10^{-4} eV.

In a typical configuration, a xenon flash lamp is discharged near a cylindrical ruby rod. The light from this lamp is absorbed in the green and blue energy bands shown in Figure 4.3, causing excitation of the Cr^{3+} ion to the corresponding energy levels. Relaxation times from higher energy bands to two levels about 15×10^3 cm^{-1} above the ground state is short compared to the relaxation time from the latter levels to the ground state. This results in an accumulation of excited

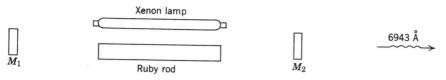

figure 4.4 Ruby laser configuration. Optical pumping is provided by a xenon flash lamp, and an interferometer resonator is formed by the mirrors M_1 and M_2.

ions in the 15×10^3 cm^{-1} levels and a population inversion between these levels and the ground state. Laser action usually occurs at the R_1 transition to the ground state at a wavelength of 6943 Å.

Figure 4.4 illustrates an arrangement for obtaining population inversion in ruby. In the original laser an interferometer resonator was formed by having optically polished ends for the ruby rod that were planar and parallel, whereas in Figure 4.4 the resonator is formed by external mirrors M_1 and M_2. At least one of the mirrors is partially transmitting so as to allow some of the laser radiation to be emitted. As in the case of the ammonia maser, oscillations occur when the gain resulting from the population inversion is greater than the attenuation resulting from losses in the cavity. The excitation mechanism is the optical pumping provided by the xenon lamp, ruby is the active medium, and the circuit is the interferometer resonator.

In 1961[9] the first continuous laser action was obtained in a mixture of helium and neon gases with an electrical discharge maintained in the gas mixture to produce free electrons and ions. Collisions between the electrons and helium atoms cause an excitation of some atoms to a metastable energy state approximately 165×10^3 cm^{-1} above the ground state. The excited helium atoms transfer their energy to neon atoms by collision, the neon atoms having an excited state that is only slightly less than 165×10^3 cm^{-1} above the ground state. This process is known as a resonant collision, and the slight difference in energy between the helium and neon levels is absorbed in kinetic energy of the atoms. Population inversion then exists between excited neon levels and lower energy states of neon, and thus laser action is provided in both the infrared and the red part of the spectrum.

Figure 4.5 is an energy diagram illustrating the resonant collision process in the helium-neon gas mixture. The levels in neon designated as $3s$ and $2s$ in Figure 4.5 are populated by the resonant collision process, and at room temperature the $2p$ and $3p$ levels are essentially unoccupied. Therefore, population inversion may exist between the $2s$ and $2p$ states, between the $3s$ and $2p$ states and between the $3s$ and $3p$ states. The wavelengths most commonly obtained from these transitions are 3.391 μ, 1.152 μ, and 0.6328 μ. Relaxation from the $1s$ state to the ground state occurs primarily by collisions between the excited atoms and

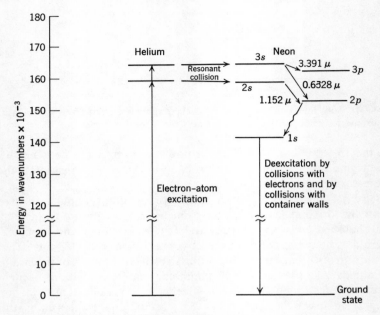

figure 4.5 Energy diagram for the helium neon resonant collision process. Although the excited neon levels are shown as single states, each state actually consists of a number of closely spaced energy levels.[4] Paschen notation has been used to designate the neon states. Three of the more common laser outputs at 0.6328 μ, 1.152 μ, and 3.391 μ are illustrated in the diagram.

electrons and by collisions with the walls of the container in which the discharge occurs. The 3.391 μ emission competes with the 0.6328 μ emission, since both of these laser wavelengths tend to deplete the population of the 3s level.

The essential features of the helium-neon gas laser are the excitation mechanism, which is the resonant collision; the active medium, which is neon; and a circuit obtained by using an interferometer resonator.

Three methods have been described thus far for providing an inverted population, but other techniques exist and have been applied to different types of lasers. Optical oscillation in noble gases such as argon occur in a discharge by energizing the argon ion to metastable level through collisions with free electrons. Laser action in semiconductors, such as gallium arsenide, may be obtained by injecting electrons into the conduction band. This provides a population inversion for electrons between the bottom of the conduction band and the top of the valence band or possibly between the conduction band and impurity levels just above the valence band. At infrared wavelengths it is possible, by means of the resonant collision process, to obtain inversion between molecular vibrational and rotational levels. This has been accomplished in CO_2, a laser that is efficient and provides high average powers. Further discussion of these excitation mechanisms and additional ways to obtain inversion are presented in Chapter 10 of Birnbaum's book.[1]

A condition of population inversion is not sufficient to guarantee optical oscillations in a resonator, for it is necessary to obtain a large enough inversion to overcome the circuit losses. This leads to a threshold condition in terms of the minimum inversion and pump power required for oscillation, which we shall now take up in the following section.

4.3 THRESHOLD REQUIREMENTS

In order to discuss threshold requirements for lasers quantitatively, we shall consider an electric dipole transition that has been prepared by some excitation mechanism so that the population of the upper state exceeds the population of the lower state in the absence of radiation at the transition frequency. The normalized rate equations that will be used for the analysis are the same as those given in the previous chapter, with the condition that the normalized equilibrium population difference per unit volume $\bar{N}^e$ will be taken positive. A positive value for $\bar{N}^e$ may be accomplished by any of the methods for obtaining population inversion described in Section 4.2. The normalized rate equations for $2/T_2 \gg 1/\tau_c$ as given by (3.82) are

$$\dot{\bar{N}} + \frac{\bar{N} - \bar{N}^e}{T_1} = \frac{1 - \bar{N}^e}{T_1} \bar{N}\bar{\varphi} \tag{4.1}$$

$$\dot{\bar{\varphi}} + \frac{\bar{\varphi}}{\tau_c} = \frac{1}{\tau_c} \bar{N}\bar{\varphi} \tag{4.2}$$

where $\bar{N}$ is the normalized population difference per unit volume, $\hbar\Omega\bar{\varphi}$ is the normalized energy density, and τ_c is the time constant for electromagnetic energy decay. T_1 is the lifetime of excited states in the presence of the excitation source and is not necessarily the same as the lifetime without the excitation. The dependence of T_1 on the pump energy is considered in Problem 4.3 at the end of the chapter.

Equations 4.1 and 4.2 are the rate equations appropriate when the molecular resonance linewidth is much broader than the circuit linewidth, and this is a good approximation for almost all laser configurations since the molecular resonance is typically at least one or two orders of magnitude larger than the interferometer bandwidth. For example, the linewidth at room temperature for ruby is approximately 3.3×10^{11} Hz. The interferometer linewidth is determined by the fraction of power lost for each transit of the light in the interferometer. If P_l/P is the fractional power lost in a one-way transit of the light, then

$$\Delta\omega_c \simeq \frac{2}{T_r} \frac{P_l}{P}$$

where $\Delta\omega_c$ is the resonator linewidth and T_r is the one-way transit time of the light. For $P_l/P = 0.25$ and $T_r = 10^{-9}$ sec (corresponding to an interferometer dimension of approximately 30 cm), we see that the linewidth is 8×10^7 Hz. Thus, for this example the molecular resonance linewidth is more than three orders of magnitude greater than the circuit linewidth.

The equations apply to a homogeneously broadened resonance, such as the ruby laser operating at room temperature. In Section 4.8 the helium-neon gas laser will be considered as an example of an inhomogeneously broadened resonance.

Spontaneous emission terms involving the decay of an excited state molecule to the lower state with the emission of a photon in the absence of applied radiation do not appear. In Chapter 6, where field quantization is considered, spontaneous emission terms will be included in the rate equations. Only during the initial buildup of laser oscillations will the energy density arising from spontaneous emission be comparable to the energy density produced by the laser action, for under steady-state conditions the laser energy density will be larger by many orders of magnitude.

In the analysis that follows we shall assume that oscillations occur in a single mode of the interferometer resonator, and the spatial variation of the population difference and energy density will be neglected. Between the mirrors of an interferometer there can be modes with different configurations in the plane transverse to the axis of the resonator, as well as modes with greater or fewer phase variations along the resonator axis. By placing additional surfaces between the interferometer mirrors, single-mode operation can be obtained, and we shall assume that these additional surfaces are present. Population difference and energy density are functions of the coordinates because of the spatial dependence of the field in the resonator, but in solving the rate equations we shall consider only the spatial average of these variables. This assumption simplifies the analysis, and a more detailed analysis including spatial variation gives results that do not differ significantly from the results obtained using spatial averages.[11]

We can arrive at the threshold requirement from Equation 4.2. The term $\bar{\varphi}/\tau_c$ represents the rate of loss of energy resulting from cavity losses, and for oscillations to occur the generation rate of energy resulting from induced emission must be greater than the loss rate. Hence the right-hand side of (4.2) must be larger than $\bar{\varphi}/\tau_c$ for $\bar{\varphi}$ to grow, and it is thus required that $\bar{N} > 1$. At the threshold condition for oscillation, the electromagnetic energy density is vanishingly small, so that from (4.1) the steady-state value for $\bar{N}$ is $\bar{N}^e$. The oscillation condition is therefore that $\bar{N}^e > 1$. This method for deriving the threshold condition emphasizes the requirement that the energy generation rate must exceed the loss rate, but it is also necessary to guarantee that there is a stable steady-state solution.

An alternative derivation of the threshold condition may be obtained by considering the stability of the steady-state solutions to the rate equations. If the time derivatives in (4.1) and (4.2) are set equal to zero, two independent

solutions are obtained:
$$\bar{N} = \bar{\varphi} = 1 \tag{4.3a}$$
or
$$\bar{N} = \bar{N}^e \quad \bar{\varphi} = 0 \tag{4.3b}$$

The first solution corresponds to a steady state in which there is a nonzero energy density, whereas the energy density is zero for the second solution. An arbitrary transient will settle at one of these steady-state solutions, depending on which solution is stable in a given case. The stability of the steady-state solutions may be checked by perturbing the variables from these values and determining whether the system returns to the original steady-state condition. Considering (4.3a) first, let

$$\bar{\varphi}(t) \equiv 1 + \delta p(t)$$
$$\bar{N}(t) \equiv 1 + \delta \nu(t)$$

where $\delta \ll 1$. To first order in δ, from (4.1) and (4.2),

$$\dot{\nu} + \frac{\bar{N}^e}{T_1} \nu = \frac{1 - \bar{N}^e}{T_1} p$$
$$\dot{p} = \frac{1}{\tau_c} \nu \tag{4.4}$$

If an exponential time dependence of the form $e^{\alpha t}$ is assumed for p and ν, an algebraic expression in α is obtained. If all the roots of this equation have negative, real parts, then the perturbation from steady state will decay to zero and the solution is stable. On the other hand, positive, real parts of the solution for α indicate instability. Substitution of the exponential time dependence for ν and p into Equation 4.4 yields an expression for α:

$$\alpha^2 + \frac{\bar{N}^e}{T_1} \alpha - \frac{1 - \bar{N}^e}{T_1 \tau_c} = 0$$

The parameter α has only negative, real parts to its solutions when

$$\bar{N}^e > 1 \tag{4.5}$$

and therefore (4.3a) is a stable solution only when (4.5) is satisfied.

A procedure similar to the one given above yields the result that for $\bar{N}^e < 1$ solution (4.3b) is stable. To obtain steady-state laser action it is therefore necessary that (4.5) be satisfied, since this leads to a stable solution with a nonzero value for energy density. A value of unity for $\bar{N}^e$ is the threshold condition for obtaining oscillation, since a larger value produces oscillation and a smaller value does not.

The unnormalized value for the equilibrium population difference per unit volume, N^e, is given by $N^e = N_0 \bar{N}^e$, where N_0 is the normalization constant. From

Equation 3.80, in the previous chapter, we have

$$N_0 = \frac{3\hbar\epsilon}{L|\mu_{12}|^2 \Omega \tau_c T_2 F} \tag{4.6}$$

where we have introduced a filling factor F to account for the case when the active medium does not fill the entire resonator. (Refer to the paragraph on filling factor following (2.55); see also Problem 4.12.) Since $\bar{N}^e = 1$ at threshold, the unnormalized equilibrium population difference per unit volume for oscillation to occur is $N^e > N_0$. In terms of the actual equilibrium populations per unit volume of the two levels N_1^e and N_2^e, for the case of degeneracy it is necessary to replace N_1^e by N_1^e/g_1. Therefore, since

$$N^e = \frac{N_2^e}{g_2} - \frac{N_1^e}{g_1}$$

the oscillation requirement is

$$\frac{N_2^e}{g_2} - \frac{N_1^e}{g_1} > \frac{3\hbar\epsilon}{L|\mu_{12}|^2 \Omega \tau_c T_2 F} \tag{4.7}$$

where g_1 and g_2 are the degeneracy factors and F is the filling factor. If the levels have the same degeneracy, then $g_2 = g_1$ and N_2^e must exceed N_1^e for oscillation to occur. If the degeneracy of the lower state is greater than the degeneracy of the upper state, the inequality of (4.6) may be met without having $N_2^e > N_1^e$. Note that as the circuit loss diminishes, corresponding to increasing τ_c, the population difference per unit volume $(N_2^e/g_2) - (N_1^e/g_1)$ required to reach threshold decreases; this result is expected because reduced cavity loss requires less gain from the active medium.

It is possible to calculate the power required from the excitation source to maintain a specified population difference. In the absence of the pump, the population difference per unit volume relaxes toward the thermal equilibrium distribution N_T^e, with a time constant T_1. It is the role of the excitation source to maintain a steady-state population difference per unit volume N^e, where $N^e \neq N_T^e$. There is a constant loss of power from the active medium resulting from the T_1 relaxation, and this power is supplied by the pump. Since N is a population difference per unit volume, a two unit increment in N is an energy change of $\hbar\Omega$ per unit volume, and thus

$$\frac{\hbar\Omega}{2} \frac{N^e - N_T^e}{T_1}$$

is the power per unit volume required from the pump. Therefore

$$\mathscr{P}_p = \frac{\hbar\Omega}{2} \left(\frac{N^e - N_T^e}{T_1} \right) \tag{4.8}$$

where $\mathscr{P}_p$ is the pump power per unit volume required to maintain an equilibrium population difference per unit volume equal to N^e. From (4.7) the threshold pump power per unit volume is

$$\mathscr{P}_{th} = \frac{\hbar\Omega}{2T_1}\left[\frac{3\hbar\epsilon}{L\,|\mu_{12}|^2\,\Omega\tau_c T_2 F} - N_T^e\right] \qquad (4.9)$$

If the lower energy state is the ground state for the ion, as is the case for the ruby laser, then $N_T^e \simeq -N_V$, where N_V is the number of active ions per unit volume. That is, in thermal equilibrium almost all the ions are in the lower energy state.

The expression for power given by (4.8) is the minimum pump power required. Additional power is lost in the transitions from the absorption bands, which are populated by the excitation source, to the upper laser level, and the power resulting from these transitions is converted to heat in the active medium.

Typical parameter values for the ruby laser are

$$N_0 = \frac{3\hbar\epsilon}{L\,|\mu_{12}|^2\,\Omega\tau_c T_2 F} \simeq 10^{18}/\text{cm}^3$$

$$N_V \simeq 2 \times 10^{19}/\text{cm}^3$$

$$T_1 \simeq 4 \times 10^{-3} \text{ sec}$$

$$\hbar\Omega \simeq 2.7 \times 10^{-19}$$

where the cavity lifetime τ_c has been taken to be 2×10^{-9} sec. These numbers give a threshold pump power density of approximately 600 W/cm³. The power required to supply the resonator losses is 5% of the total power supplied, and the remainder is used to overcome the thermal equilibrium distribution. If the lower level is several κT in energy above the ground state then $N_T^e \simeq 0$, and the pump power would have to supply only the cavity losses and therefore could be greatly reduced. This is the basis of operation of the four-level neodymium laser described in Section 4.7.

4.4 STEADY-STATE POWER OUTPUT

The power emitted by the laser is equal to the rate of loss of stored energy resulting from transmission through the resonator mirrors. Considering only mirror losses, we have

$$U = U_0 e^{-t/\tau_m} \qquad (4.10)$$

where τ_m is the electromagnetic energy lifetime resulting from mirror transmission, U is the stored energy per unit volume, and U_0 is the energy at $t = 0$. We can evaluate the constant τ_m by noting that for a time interval given by the round trip transit time of the energy in the resonator, the stored energy is reduced by the factor $\rho_1\rho_2$, where ρ_1 and ρ_2 are the reflectivities of the mirrors. Therefore, from (4.10) we have

$$\rho_1\rho_2 = e^{-2\eta l/\tau_m c}$$

where c/η is the velocity of light, η is the refractive index, and l is the distance between resonator mirrors. Solving for τ_m,

$$\tau_m = -\frac{2\eta l}{c \ln(\rho_1\rho_2)} \tag{4.11}$$

The output power unit volume of the active medium, $\mathscr{P}_0$, is equal to $-\dot{U}$, which according to (4.10) is equal to

$$\mathscr{P}_0 = \frac{U}{\tau_m} \tag{4.12}$$

where

$$U = \varphi\hbar\Omega$$

If the threshold condition is exceeded, it has been shown that $\bar{\varphi} = 1$, and thus the steady-state electromagnetic energy density equals $\varphi_0 \hbar\Omega$. From (3.81) we have

$$\varphi = \varphi_0 = \frac{\tau_c}{T_1}\frac{N_0}{2}(\bar{N}^e - 1) \tag{4.13}$$

where N_0 is given by (4.6). If we combine (4.12) and (4.13), the expression for output power per unit volume is

$$\mathscr{P}_0 = \frac{\hbar\Omega}{\tau_m}\frac{\tau_c}{T_1}\frac{N_0}{2}(\bar{N}^e - 1) \tag{4.14}$$

where τ_m is evaluated from (4.11).

As an example, let us consider a ruby laser with a volume of 1 cm³, and the pump power is of sufficient magnitude to raise $\bar{N}^e$ to 20% above the threshold value (that is, $\bar{N}^e = 1.2$). If the scattering and diffraction losses within the resonator are taken to be equal to the loss resulting from mirror transmission, then $\tau_c/\tau_m = \frac{1}{2}$. According to the ruby laser parameters given in Section 4.3, the output power is approximately 3 watts.

4.5 TRANSIENT BEHAVIOR

The transient behavior is important when the laser is used in pulsed operation, since the pulse duration may be comparable to the time constants of the oscillator. A steady-state situation is never reached during the pulse duration for most pulsed ruby lasers. We shall restrict our analysis to a single-mode laser, which exhibits regular oscillatory behavior. The multimode case results in rather complicated transient behavior, as has been discussed in the literature.[12]

From Equations 4.1 and 4.2, the rate equations, it may be seen that the transient behavior can be divided into four regions on the time axis. The boundary conditions at $t = 0$ are taken to be $\bar{\varphi} \ll 1$ and $\bar{N} = 1$, and the laser is assumed to be above threshold so that $\bar{N}^e > 1$. The transient behavior is as follows.

1. From Equation 4.1, neglecting the term on the right-hand side because $\bar{\varphi} \ll 1$, we see that $\dot{\bar{N}}$ is positive and $\bar{N}$ grows exponentially in time. From (4.2), since $\bar{N} > 1$, $\dot{\bar{\varphi}} > 0$ and so $\bar{\varphi}$ increases.
2. $\bar{N}$ is approximately unity, so that when $\bar{\varphi}$ becomes >1 from (4.1), $\dot{\bar{N}}$ is negative and $\bar{N}$ starts to decrease.
3. When $\bar{N}$ becomes <1, then from (4.2) $\dot{\bar{\varphi}}$ is negative and $\bar{\varphi}$ decreases.
4. As $\bar{\varphi}$ becomes <1, from (4.1) we see that $\dot{\bar{N}}$ is positive so that $\bar{N}$ increases. When $\bar{N}$ becomes >1 the conditions are the same as in step 1, and so the cycle repeats itself.

Figure 4.6 illustrates the time evolution of $\bar{N}$ and $\bar{\varphi}$ with the time scale divided into the four steps of the cycle. Figure 4.7 is a trace of an oscilloscope photograph of the light emitted by a ruby laser, with a time scale of one microsecond per division. Each light pulse on the oscilloscope trace resulted from a cycle described by steps 1 through 4.

We can make a quantitative analysis of the transient behavior by using the approximation that $\bar{N}$ is close to unity during the cycle. The validity of this approximation may be tested when the solution is determined. Taking the ratio of (4.2) to (4.1) and only keeping terms to first power in the difference between $\bar{N}$ and unity, we have

$$\frac{d\bar{\varphi}}{d\bar{N}} = \frac{T_1(\bar{N} - 1)\bar{\varphi}}{\tau_c(1 - \bar{N}^e)(\bar{\varphi} - 1)} \tag{4.15}$$

Equation 4.15 is separable in the variables and may be integrated to give

$$\frac{T_1}{2\tau_c(\bar{N}^e - 1)}(\bar{N} - 1)^2 = \ln\frac{\bar{\varphi}}{\bar{\varphi}_i} - (\bar{\varphi} - \bar{\varphi}_i) \tag{4.16}$$

where $\bar{\varphi}_i$ is the normalized energy density when $\bar{N} = 1$.

The $\bar{\varphi}\bar{N}$ plane may be divided into four sections so that $d\bar{\varphi}/d\bar{N}$ has a definite sign in each section. This is illustrated in Figure 4.8, where the straight lines indicate a positive slope in Sections 1 and 3 and a negative slope in Sections 2 and 4. The arrows on the straight line segments show the direction of increasing time as determined from the rate equations. At the boundaries between the sections the slope is either zero or infinite. The segments of the $\bar{\varphi}\bar{N}$ plane designated by numerals 1 through 4 correspond to the respective intervals along the time axis in Figure 4.6.

Steady-state is given by the point $\bar{\varphi} = \bar{N} = 1$. At the beginning of a laser spike, $\bar{\varphi} \simeq 0$ and $\bar{N} \simeq 1$, and thus we see from Figure 4.8 that to reach steady-state a counterclockwise, decreasing spiral must be followed, centered around the point

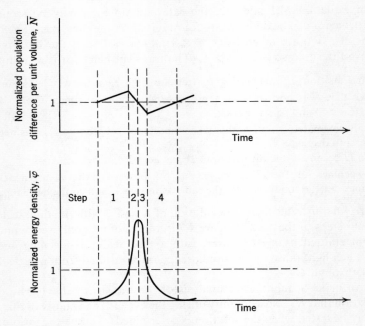

figure 4.6 Normalized population difference per unit volume and energy density as functions of time, showing the four steps of the cycle for the generation of a light pulse.

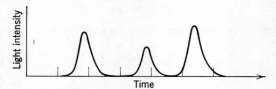

figure 4.7 An oscilloscope presentation of the spiking of a ruby laser with a time scale of 1 μsec per division.

$\bar{\varphi} = \bar{N} = 1$ and converging on this point. Steady-state cannot be reached by remaining solely within Section 1 of the $\bar{\varphi}\bar{N}$ plane, because this would violate the slope requirement in Section 1. All four sections of the plane must be traversed in the contour of a decreasing spiral. As a function of time, therefore, the emitted light will overshoot the steady-state value in spikes of diminishing height as steady-state is approached.

Figure 4.9 shows a curve in the $\bar{\varphi}\bar{N}$ plane for a ruby laser with the parameter values given in Section 4.3 with $\bar{N}^e = 2$. This value for $\bar{N}^e$ corresponds to a 10%

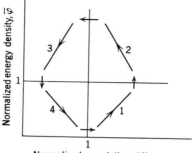

figure 4.8 The straight lines indicate the slope of the $\bar{\varphi}$-$\bar{N}$ curve in different segments of the $\bar{\varphi}$-$\bar{N}$ plane. Arrows indicate the direction of increasing time.

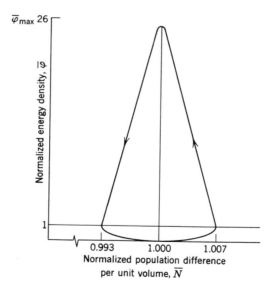

figure 4.9 A plot of $\bar{\varphi}$ versus $\bar{N}$ for the parameter values

$$\bar{\varphi}_i = 10^{-10}$$

$$\frac{2\tau_c(\bar{N}^e - 1)}{T_1} = 2 \times 10^{-6}$$

population inversion, that is, $N^e/N_V = 0.1$, where N_V is the number of active atoms per unit volume. (For the steady-state example of Section 4.4 we chose $\bar{N}^e = 1.2$, whereas for the pulsed problem we are taking $\bar{N}^e = 2$. We do so because experimentally it is easier to obtain a higher population inversion on a pulsed basis than for continuous operation.) The initial value for the normalized energy density, $\bar{\varphi}_i$, is taken to be the spontaneous emission energy density in a single cavity mode, giving a value $\bar{\varphi}_i \simeq 10^{-10}$. The curve in Figure 4.9 is quite insensitive to the value for $\bar{\varphi}_i$, so that a change in $\bar{\varphi}_i$ by a factor of 10^5 changes the peak energy density only by a factor of 2. A more comprehensive treatment of spontaneous emission will be presented in Chapter 6.

The maximum value for photon density, $\bar{\varphi}_{max}$, occurs when $\bar{N} = 1$. Noting that $\bar{\varphi}_{max} \gg \bar{\varphi}_i$, we have, from (4.16),

$$\bar{\varphi}_{max} = \ln \frac{\bar{\varphi}_{max}}{\bar{\varphi}_i}$$

With $\bar{\varphi}_i = 10^{-10}$, this gives $\bar{\varphi}_{max} = 26$, which means that at the peak of the laser spike the power will be 26 times the steady-state value. For the ruby laser parameters of Section 4.3, $\bar{N}^e = 2$, $\tau_c/\tau_m = \frac{1}{2}$ and a ruby volume of 3 cm³, from (4.14) the peak power is approximately 1.3 kW.

Maximum and minimum values for $\bar{N}$ are obtained from (4.16) by letting $\bar{\varphi} = 1$ and using $\bar{\varphi}_i \ll 1$:

$$\bar{N}_\pm = 1 \pm \left[\frac{2\tau_c(\bar{N}^e - 1)}{T_1}\right]^{1/2} [-1 - \ln \bar{\varphi}_i]^{1/2}$$

where $\bar{N}_+$ and $\bar{N}_-$ are the maximum and minimum values of $\bar{N}$, respectively. With the given parameter values, $\bar{N}_\pm = 1 \pm 6.6 \times 10^{-3}$. Population difference varies less than 1% about the steady-state value; this fact supports the initial assumption that $\bar{N} \simeq 1$ during the laser spike.

If the rate equations are solved for the time dependence of $\bar{N}$ and $\bar{\varphi}$, curves of the form shown in Figure 4.6 are obtained. The duration of the pulse is on the order of one microsecond. This time interval is much less than T_1, the relaxation time for excited ions, and thus essentially all of the change in energy in the active medium during a laser spike is delivered to the circuit. Therefore, from Figure 4.9, the energy $\mathscr{W}$ to the circuit is

$$\mathscr{W} = \frac{\hbar\Omega}{2}(\bar{N}_+ - \bar{N}_-)N_0 V \text{ Joules} \qquad (4.17)$$

where V is the volume of the active medium. For the ruby laser example, $\bar{N}_+ - \bar{N}_- = 13.2 \times 10^{-3}$, whereupon the emitted energy in a spike is

$$\mathscr{W} \simeq 2.7 \times 10^{-3} \text{ Joules}$$

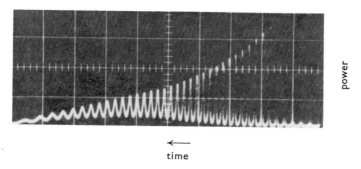

figure 4.10 Power emitted by a ruby laser as a function of time, as obtained by Roess.[13]

with $V = 3$ cm³, and assuming that half the energy is lost within the interferometer and half is emitted.

With the assumption that $\bar{N} \simeq 1$, the trajectory of $\bar{\varphi}$ versus $\bar{N}$ is a closed path as shown in Figure 4.9. This means that the energy density returns to the noise level at the end of each spike and the next spike will not be correlated in time to the previous one. Such a situation is illustrated by the oscilloscope trace of Figure 4.7. An exact solution of the rate equations, however, shows that $\bar{\varphi}$ does not return to $\bar{\varphi}_i$ at the end of a spike, but is greater than $\bar{\varphi}_i$. On the $\bar{\varphi}\bar{N}$ plane the contour is not closed but rather spirals inward toward the $\bar{\varphi} = \bar{N} = 1$ point, and as a function of time the energy density appears as a sequence of spikes of diminishing amplitude. Figure 4.10 illustrates this time dependence of power in a ruby laser.

The type of behavior shown in Figure 4.10 results when $\bar{\varphi}$ remains well above the noise level at the end of the laser spike, so that the following spike is correlated to the preceding one. From the rate equations it is possible to determine that spikes of diminishing amplitude, as illustrated in the above figure, will occur if the cavity lifetime is long and the pump power is well above the threshold value.

4.6 Q-SWITCHING

It is possible to obtain peak output powers many orders of magnitude greater than the steady-state output power from a laser over time intervals ranging from 10 to 100 nanoseconds.[14,15] Peak output power from a ruby laser, for example, can be made as high as a gigawatt. The method for producing these very intense light pulses, known as Q-switching or Q-spoiling, consists of varying the resonator losses in a period of time that is short compared to the relaxation time of excited states. Some of the laser active media for which Q-switched operation has been achieved are ruby, neodymium, and carbon dioxide.

In Section 4.5 it was noted that the maximum population difference obtained during a laser spike barely exceeded the threshold value. The reason for this is that oscillations commence shortly after threshold is reached, thereby depleting the excited states faster than the excitation source can replenish the supply. The equilibrium population difference per unit volume N^e that would exist in the absence of oscillation, may, however, be much larger than the maximum population difference that actually exists before the onset of oscillation. If oscillation could be delayed until N^e is reached, more energy would be stored in the active medium and a significant increase in emitted power could result. Oscillations will not occur if the losses of the system are made so high that the threshold condition cannot be met. A sudden reduction in cavity losses will then allow oscillations, with much more energy initially stored in the active medium. If a laser medium has a relaxation time on the order of milliseconds, it is possible to reduce the resonator losses in a short time compared to the relaxation time so that most of the stored energy is delivered to the cavity.

Q-switching, which refers to the sudden increase in resonator Q, can be accomplished in several ways. For example, an element known as a Kerr cell may be placed within the resonator that introduces optical loss into the cavity as a function of a dc electric field that is applied to the element. Initially, the population is inverted with the electric field adjusted so that the resonator loss is high enough to prevent oscillation. When N^e is reached, the electric field is then suddenly changed so as to minimize the cavity loss, resulting in the emission of a high intensity pulse of output power.

Another method for Q-switching consists of using a rotating prism as one mirror of the interferometer. Figure 4.11 illustrates the rotating-prism Q-switch. In the high loss configuration shown in Figure 4.11a the prism does not form a resonator with the fixed mirror. In Figure 4.11b the light from the fixed mirror

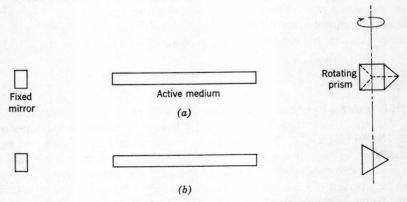

figure 4.11 The configuration for a rotating prism, Q-switched laser: (a) high loss; (b) low loss.

is reflected back onto the mirror by the prism and so forms a high-Q cavity. Population inversion is accomplished when the Q is low, as in 4.11a, and oscillations occur when the resonator is as shown in Figure 4.11b.

Still another technique for Q-switching is the use of a saturable dye inside the interferometer. The saturable dye is a material for which the absorption constant decreases with increasing light intensity caused by saturation. Dye attenuation is chosen so that with no incident light the loss is slightly less than that required to prevent laser oscillation when $N = N^e$. Therefore the threshold condition is met when $N \simeq N^e$ and as the laser power increases the dye loss decreases, thereby producing the desired Q-switching effect. In this manner the population difference per unit volume is approximately N^e prior to the onset of oscillation.

It is convenient to use the $\bar{\varphi}\bar{N}$ plane to analyze the Q-switched laser. Once the pulse is initiated the energy density of the optical field will be much greater than the steady-state density, so that $\bar{\varphi} \gg 1$. Using $\bar{\varphi} \gg 1$ in the normalized rate equations, (4.1) and (4.2), we obtain the relationship

$$\frac{d\bar{\varphi}}{d\bar{N}} = -\frac{T_1}{\tau_c(\bar{N}^e - 1)} \frac{\bar{N} - 1}{\bar{N}} \tag{4.18}$$

This equation was obtained by taking the ratio of the rate equations, thereby eliminating the dependence on time. Equation 4.18 is separable in the variables, and upon integration we have

$$\bar{\varphi} = \frac{T_1}{\tau_c(\bar{N}^e - 1)} \left[\ln \frac{\bar{N}}{\bar{N}^e} + \bar{N}^e - \bar{N} \right] \tag{4.19}$$

where the boundary condition is $\bar{\varphi} = 0$ at $\bar{N} = \bar{N}^e$. Figure 4.12 shows curves of $\bar{\varphi}$ versus $\bar{N}$ for different values of $\bar{N}^e$, and the arrows indicate the direction of increasing time. Maximum energy density occurs when the normalized population difference is unity, and so $\bar{\varphi}_{max}$ is obtained from (4.18) by letting $\bar{N} = 1$. With

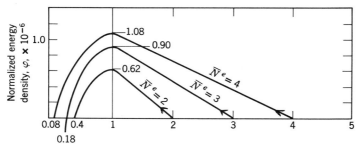

figure 4.12 Normalized photon density as a function of normalized population difference in Q-switched operation for the parameter value: $T_1/\tau_c = 2 \times 10^6$.

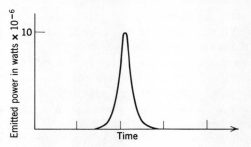

figure 4.13 An oscilloscope trace of a Q-switched pulse, with a time scale of 100 nanoseconds per division.

$\bar{N}^e = 2$, the maximum energy density is 0.62×10^6 times greater than the steady-state value. For the parameter values of the example considered in Sections 4.2–4.5, the peak emitted power is about 10 MW with $\bar{N}^e = 2$ and a volume of 1 cm³.

The energy delivered to the cavity fields is equal to the loss of energy in the active medium during a pulse of light, as given by Equation 4.17. For $\bar{N}^e = 2$, the emitted energy is approximately 0.1 J/cm³, based on the condition that the energy lost in the interferometer and the emitted energy are equal. An approximate calculation of pulse duration may be obtained by assuming a triangular pulse shape. The time between half-power points is then the ratio of the total emitted energy to the peak emitted power, which, for the previous example, is 1×10^{-8} sec.

Energy density as a function of time for the Q-switched laser may be obtained from the rate equations by using the boundary conditions that at $t = 0$, $N = N^e$, and φ is given by the spontaneous emission photon density. Figure 4.13 is a trace of a typical oscilloscope photograph for a Q-switched pulse showing emitted power versus time, with a time scale of 100 nanosec per division.

4.7 THE FOUR-LEVEL LASER

In Section 4.3 we noted that when the lower level of the laser transition is the ground state for the ion, most of the pump power at threshold is used to overcome the thermal equilibrium population difference. The four-level laser requires less pump power to maintain the threshold population difference because the lower laser level is several κT above the ground state and therefore the thermal equilibrium population difference per unit volume N_T^e is approximately zero. Figure 4.14 illustrates the four pertinent energy levels for the neodymium ion, Nd^{3+}, in calcium tungstate, $CaWO_4$.

At room temperature the two laser levels are essentially unoccupied and so from (4.9) with $N_T^e = 0$, the pump power per unit volume required to maintain

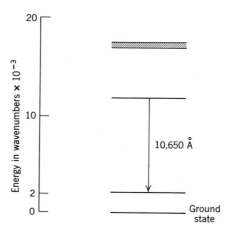

figure 4.14 Energy diagram for Nd^{3+} in $CaWO_4$. Light is absorbed by the striped band, and laser emission occurs to a lower level that is approximately $10\kappa T$ above the ground state at room temperature.

the threshold population difference is

$$\mathscr{P}_{th} = \frac{3\hbar^2 \epsilon}{2L |\boldsymbol{\mu}_{12}|^2 T_1 \tau_c T_2 F} \tag{4.20}$$

For the parameter values applicable to Nd^{3+},

$$N_0 = \frac{3\hbar\epsilon}{L |\boldsymbol{\mu}_{12}|^2 \Omega \tau_c T_2 F} \simeq 2.2 \times 10^{16}/cm^3$$

$$T_1 = 10^{-4} \text{ sec}$$

$$\hbar\Omega \simeq 1.7 \times 10^{-19} \text{ J}$$

the pump power per unit volume at threshold is approximately 19 W/cm³, which is about 3% of the value required for the ruby laser. The cavity lifetime has been taken to be 2×10^{-9} sec.

4.8 THE HELIUM-NEON GAS LASER

In Section 4.2 there was a discussion of the manner in which population inversion is obtained for the He-Ne laser; now the threshold requirements and steady-state output power will be determined. An important distinction between the gas laser and the solid-state ruby and neodymium lasers is that each active molecule in the gas laser does not have the same transition frequency because of the motion of the gas molecules. If a given molecule is moving with velocity v_i

in the direction of a propagating electromagnetic wave, the interaction frequency is

$$\Omega_i = \omega_0\left(1 + \frac{v_i}{c}\right)$$

where ω_0 is the transition frequency of the stationary particle. As discussed in Section 3.2.5, this frequency shift effect, known as the Doppler shift, gives rise to an absorption line for the transition that consists of an ensemble of Lorentzian lines centered at different frequencies, Ω_i. The number of molecules per unit volume, $d\mathcal{M}$, in the transition frequency range $d\Omega_i$ is determined by the thermal equilibrium velocity distribution in a gas and is given by (3.23) and (3.24), which when combined yield

$$d\mathcal{M} = N_V \frac{\sqrt{(4/\pi)\ln 2}}{\Delta\omega_G} \exp\left[-4\ln 2\frac{(\Omega_i - \omega_0)^2}{(\Delta\omega_G)^2}\right] d\Omega_i \qquad (4.21)$$

where N_V is the total number of molecules per unit volume and $\Delta\omega_G$ is the linewidth in rad/sec that corresponds to the frequency difference between the points on either side of the peak at which $d\mathcal{M}/d\Omega_i$ drops to one-half its line-center value. The constant factor multiplying the exponential on the right-hand side of Equation 4.21 is chosen so that the integral over Ω_i from $-\infty$ to ∞ yields the number of molecules per unit volume. Figure 4.15 is a plot of $d\mathcal{M}/d\Omega_i$ as a function of frequency.

In Chapter 3 we stated that an absorption line for which different molecules have different transition frequencies is known as an inhomogeneously broadened line, in contrast to a homogeneously broadened line, for which all molecules have the same transition frequency. In a gas the overall absorption line is an inhomogeneously broadened Gaussian line of bandwidth $\Delta\omega_G$ centered at ω_0, composed of an ensemble of homogeneously broadened Lorentzian lines, each of width $\Delta\omega_L = 2/T_2$.

In Section 2.4 of Chapter 2, equations were obtained for an electric dipole transition relating the dipole moment of a molecule $\langle\mu\rangle$ to the difference in

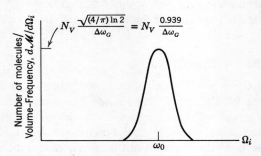

figure 4.15 The Gaussian lineshape produced by Doppler broadening.

occupation probabilities for the two levels:

$$\langle\ddot{\mu}\rangle + \frac{2}{T_2}\langle\dot{\mu}\rangle + \Omega_i^2\langle\mu\rangle = -\frac{2|\mu_{12}|^2\Omega_i}{3\hbar}E(\Delta\rho) \tag{4.22}$$

$$(\dot{\Delta\rho}) + \frac{\Delta\rho - (\Delta\rho)^e}{T_1} = \frac{2}{\hbar\Omega_i}E\langle\dot{\mu}\rangle \tag{4.23}$$

where $\Delta\rho = \rho_{22} - \rho_{11}$ and, for simplicity, a one-dimensional problem is being considered. For a homogeneous line the polarization P is obtained by multiplying $\langle\mu\rangle$ by the number of molecules per unit volume, since every molecule has the same transition frequency. For an inhomogeneously broadened line, however, the polarization is obtained by integrating $\langle\mu\rangle \, d\mathcal{M}$:

$$P = \int \langle\mu\rangle \, d\mathcal{M} = \int_0^\infty d\Omega_i \langle\mu\rangle N_V \frac{0.939}{\Delta\omega_G}\exp\left[-\frac{4\ln 2}{(\Delta\omega_G)^2}(\Omega_i - \omega_0)^2\right] \tag{4.24}$$

We see that P is given by a convolution integral of the Lorentzian lineshape of $\langle\mu\rangle$ with the Gaussian function. The circuit equation is the same as for the homogeneously broadened line, namely,

$$\ddot{E} + \frac{1}{\tau_c}\dot{E} + \omega_c^2 E = -\frac{1}{\epsilon_0}\ddot{P} \tag{4.25}$$

where τ_c is the electromagnetic energy decay time and ω_c is the resonant frequency of the resonator. It has been assumed that the active medium fills the entire resonator volume. Equations 4.22–4.25 are the appropriate differential equations for a laser with a Gaussian broadened resonance.

Single-mode operation will be assumed, and exponential notation is used so that, for example, $E = \frac{1}{2}\tilde{E}e^{i\omega t} + \text{c.c.}$, where ω is the oscillation frequency and the $\sim$ over a variable designates the complex amplitude of the variable, which is a constant in the steady-state. From (4.23) the time average of $(\Delta\rho)$ is given by

$$\frac{(\Delta\rho) - (\Delta\rho)^e}{T_1} = \frac{i\omega}{2\hbar\Omega_i}(\langle\tilde{\mu}\rangle\tilde{E}^* - \langle\tilde{\mu}\rangle^*\tilde{E}) \tag{4.26}$$

It is possible to eliminate $\langle\tilde{\mu}\rangle$ from (4.22) and (4.26), giving an expression for $(\Delta\rho)$ in terms of the electric field:

$$(\Delta\rho) = \frac{(\Delta\rho)^e}{1 + \frac{|\tilde{E}|^2}{|\tilde{E}|^2_{\text{sat}}}\frac{g_L(\omega,\Omega_i)}{T_2/\pi}} \tag{4.27}$$

where $|\tilde{E}|^2_{\text{sat}}$ is given by

$$|\tilde{E}|^2_{\text{sat}} = \frac{3\hbar^2}{|\mu_{12}|^2 T_1 T_2}$$

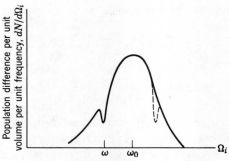

figure 4.16 The depression in population difference near the oscillation frequency in an inhomogeneously broadened resonance.

In the derivation of (4.27) the approximation has been made that $\Omega_i + \omega \simeq 2\omega$.

If typical laser parameters are substituted into (4.27), it is seen that the denominator of the right-hand side is approximately unity except where Ω_i is nearly equal to ω. This means that $(\Delta\rho)^e$ is reduced below its equilibrium value near the oscillation frequency over a frequency interval that is equal to the homogeneous linewidth $2/T_2$. The population difference per unit volume per unit frequency interval $dN/d\Omega_i$ exhibits a similar behavior, since

$$\frac{dN}{d\Omega_i} = (\Delta\rho)\frac{d\mathcal{M}}{d\Omega_i}$$

Therefore, $dN/d\Omega_i$ is the product of $(\Delta\rho)$ as given by Equation 4.27 and the Gaussian exponential function. The solid line in Figure 4.16 illustrates the reduction in $dN/d\Omega_i$ in the vicinity of an oscillation at frequency ω.

For a homogeneously broadened resonance, the population difference is depressed uniformly since every molecule has the same transition frequency, but for inhomogeneous broadening the population difference is reduced only over the extent of the homogeneous linewidth. This effect, as illustrated in Figure 4.16, has been termed hole burning.[16]

Figure 4.16 applies to a single propagating wave, but in a resonator two waves propagate in opposite directions. A wave propagating to the right will interact with molecules of a given velocity, and a wave propagating to the left will interact with different molecules, having the same velocity magnitude but oppositely directed. Therefore, in a resonator a wave propagating in one direction will see another hole burned an equal frequency interval on the other side of ω_0. The hole burned by the wave propagating in the opposite direction is shown by the dashed line in Figure 4.16. If the oscillation frequency is close to ω_0, one hole will have a strong influence on the other, but if $\omega_0 - \omega$ is greater than several homogeneous linewidths, the holes do not interact with regard to determining threshold and power output. With $\omega \simeq \omega_0$, the waves propagating in opposite directions interact with the same molecules, and there is a reduction of output power near the line center. If $\omega = \omega_0$, then only a single hole is burned at the center of the Gaussian line, and the analysis may be performed in terms of the total resonator field rather than in terms of two propagating waves.

To obtain threshold conditions and emitted power, it is necessary to evaluate the polarization. This is accomplished by eliminating $(\Delta \rho)$ between (4.22) and (4.27), giving an expression for the dipole moment in terms of the electric field:

$$\langle \tilde{\mu} \rangle = \frac{T_2^2 (iK/\omega)[(1/T_2) + i(\Omega_i - \omega)](\Delta \rho)^e \tilde{E}}{T_2^2 (\Omega_i - \omega)^2 + 1 + |\tilde{E}|^2/|\tilde{E}|^2_{\text{sat}}} \quad (4.28)$$

$$K = \frac{|\mu_{12}|^2 \Omega_i}{3\hbar} \quad (4.29)$$

Polarization is obtained from $\int \langle \tilde{\mu} \rangle \, d\mathcal{M}$, which involves integrating the product of the Gaussian function and $\langle \tilde{\mu} \rangle$, obtained from (4.28). This integration is simplified by noting that $\langle \tilde{\mu} \rangle$ is large only where $\Omega_i \simeq \omega$, and the Gaussian function is slowly varying so that it may be taken outside the integral with Ω_i set equal to ω. If the time axis is chosen so that $\tilde{E}$ is real, it will be seen that only Im $\{P\}$ is needed to determine threshold conditions and power output, where Im refers to the "imaginary part of." Performing Im $\int \langle \tilde{\mu} \rangle \, d\mathcal{M}$, Im $\{P\}$ has the following value:

$$\text{Im}\{P\} = -\frac{0.939 N_V}{\omega \Delta \omega_G} K(\Delta \rho)^e \frac{\pi \tilde{E}}{\sqrt{1 + |\tilde{E}|^2/|\tilde{E}|^2_{\text{sat}}}} \exp\left(-\frac{2.8}{(\Delta \omega_G)^2}(\omega - \omega_0)^2\right) \quad (4.30)$$

where K is evaluated at $\Omega_i = \omega$. For typical laser parameters, $\omega^2 T_2^2 \gg 1 + |\tilde{E}|^2/|\tilde{E}|^2_{\text{sat}}$ and this inequality has been used to simplify the expression for Im $\{P\}$ given by (4.30).

Equation 4.25, the circuit equation, is yet to be considered. The term $\dot{E}/\tau_c$ on the left-hand side of this equation represents the decay of energy associated with the circuit losses, and if steady-state oscillations occur, this energy loss rate must equal the rate at which energy is introduced into the resonator by the active medium. With $\tilde{E}$ chosen to be real, this balance of energy flow is obtained from (4.25) by equating imaginary coefficients of $e^{i\omega t}$. If the oscillation is considered to occur at the center of the Gaussian line, that is, $\omega = \omega_0$, then from the circuit equation and (4.30) we have

$$\frac{1}{\tau_c} = \frac{0.939 \pi N_V}{\epsilon_0 \Delta \omega_G} \frac{K(\Delta \rho)^e}{\sqrt{1 + |\tilde{E}|^2/|\tilde{E}|^2_{\text{sat}}}} \quad (4.31)$$

At threshold the laser field is vanishingly small, so that $\tilde{E} = 0$, and therefore $(\Delta \rho)^e_{\text{th}}$ required for threshold is obtained from (4.31) in terms of the laser parameters and the photon lifetime by setting $\tilde{E}$ equal to zero.

The pump power per unit volume required to maintain the inverted population at threshold is given by (4.8), where $N_T^e = 0$ since the laser levels are essentially unoccupied at room temperature (the ground level is not involved in the transition), and $N^e = \int (\Delta\rho_{\text{th}})^e \, d\mathcal{M} = N_V(\Delta\rho)^e_{\text{th}}$. Using the value for $(\Delta\rho)^e_{\text{th}}$ obtained from (4.31), and substituting for K as given by (4.29), the threshold pump power per unit volume is

$$\mathscr{P}_{\text{th}} = \frac{\hbar^2 \epsilon_0 \Delta\omega_G}{0.626 \pi \tau_c T_1 |\mathbf{\mu}_{12}|^2} \tag{4.32}$$

Typical laser parameters for the He-Ne laser are

$$\omega \simeq 3 \times 10^{15} \text{ rad/sec}$$

$$\Delta\omega_G \simeq 5 \times 10^9 \text{ rad/sec}$$

$$T_1 \simeq 10^{-7} \text{ sec}, \qquad \tau_c \simeq 10^{-7} \text{ sec}$$

$$|\mathbf{\mu}_{12}| \simeq 1.4 \text{ debye} = 1.4 \times \frac{10^{-29}}{3} \text{ coul m}$$

With these parameter values the pump power per unit volume to maintain the threshold population inversion for an oscillation at the center of the Gaussian line is 1.0 mW/cm^3. A comparison between (4.32) and (4.20) indicates that the pump power has approximately the same dependence upon the laser parameters for the gas laser as for the solid state laser; however, the pump power is approximately four orders of magnitude less for the gas oscillator. For all lasers $T_1 |\mathbf{\mu}_{12}|^2$ is approximately constant, as will be shown in Chapter 6.

The difference in threshold power results from a difference in the ratio of the absorption linewidth to the energy decay time, $\Delta\omega_G/\tau_c$. In the solid state laser the energy decay time is generally two orders of magnitude shorter than for the gas, because of much higher scattering losses. Ten percent scattering loss per pass is typical for a 10-cm length solid, whereas a 1-meter gas laser has negligible scattering loss and will have only 1% loss per pass with a 99% reflecting mirror. This results in a decay lifetime that is 100 times greater in the gas laser. In addition, the absorption linewidth of the He-Ne Gaussian line is two orders of magnitude less than the linewidth for the solid-state oscillator. Therefore the ratio $\Delta\omega_G/\tau_c$ is four orders of magnitude less for He-Ne, and the pump power is reduced by this factor.

The power emitted by the laser may be calculated from (4.31). If this equation is solved for $|\tilde{E}|^2$, we have

$$|\tilde{E}|^2 = |\tilde{E}|^2_{\text{sat}} \left[\left(\frac{N^e}{N^e_{\text{th}}} \right)^2 - 1 \right] \tag{4.33}$$

where $N^e = N_V(\Delta\rho)^e$ is the equilibrium population difference per unit volume established by the pump, and $N^e_{\text{th}} = N_V(\Delta\rho)^e_{\text{th}}$ is the threshold value for the population difference per unit volume. As previously given by (4.12), the emitted power $\mathscr{P}_0$ per unit volume is

$$\mathscr{P}_0 = \frac{\varphi \hbar \omega}{\tau_m} = \frac{\epsilon_0 |\tilde{E}|^2}{2\tau_m} \tag{4.34}$$

where τ_m is the photon lifetime resulting from mirror transmission. From (4.33) and (4.34), an expression is obtained relating the output power per unit volume to the laser parameters:

$$\mathscr{P}_0 = \frac{3\epsilon_0 \hbar^2}{\tau_m |\mathbf{\mu}_{12}|^2 T_1 T_2} \left[\left(\frac{N^e}{N_{\text{th}}^e} \right)^2 - 1 \right] \tag{4.35}$$

With the parameter values given for (4.32), and with $\tau_m = \tau_c = 10^{-7}$ sec, $T_2 = 4 \times 10^{-8}$ sec, and a volume of 100 cm^3, if the laser is pumped so that $N^e = 1.4 N_{\text{th}}^e$, the emitted power is 19 mW.

REFERENCES

Some general references on lasers are:

1. G. Birnbaum, *Optical Masers*, Academic Press, New York, 1964.
2. W. E. Lamb, Jr., "Theory of an Optical Maser," *Phys. Rev.*, **134**, No. 6A, A1429–A1450, June 15, 1964.
3. B. A. Lengyel, *Introduction to Laser Physics*, Wiley, New York, 1966.
4. A. Yariv, *Quantum Electronics*, Wiley, New York, 1967, Chapters 15–17.
5. W. V. Smith and P. P. Sorokin, *The Laser*, McGraw-Hill Book Co., 1966.

Additional articles referred to in the text:

6. J. P. Gordon, H. J. Zeiger, and C. H. Townes, "The Maser-A Type of Microwave Amplifier, Frequency Standard, and Spectrometer," *Phys. Rev.*, **99**, 1264–1274, August 1955.
7. A. L. Schawlow and C. H. Townes, "Infrared and Optical Masers," *Phys. Rev.*, **112**, 1940–1949, December 1958.
8. T. H. Maiman, "Stimulated Optical Radiation in Ruby Masers," *Nature*, **187**, 493–494, August 1960.
9. A. Javan, W. B. Bennett, Jr., and D. R. Herriott, "Population Inversion and Continuous Optical Maser Oscillation in a Gas Discharge Containing a He-Ne Mixture," *Phys. Rev. Letters*, **6**, 106–110, February 1961.
10. See, for example, *American Institute of Physics Handbook*, McGraw-Hill, New York, 1957, Section 7-58.
11. A. L. Mikaeliane, M. L. Ter-Mikaeliane, Y. G. Turkov, and V. V. Djatchenko, "The Utilization of Semiclassical and Rate Equations for the Design of Steady-State Optical Masers," paper presented at the 1966 International Quantum Electronics Conference, Phoenix, Arizona, April 12–15, 1966.
12. C. L. Tang, H. Statz, and G. deMars, Jr., "Spectral Output and Spiking Behavior of Solid-State Lasers," *J. Appl. Phys.*, **34**, No. 8, 2289–2295, August 1963.
13. D. Roess, "Single-mode Operation of a Room-Temperature CW Ruby Laser," *Appl. Phys. Letters*, **8**, No. 5, 109–111, March 1966.
14. R. W. Hellwarth, "Control of Fluorescent Pulsations," in *Advances in Quantum Electronics*, ed. by J. Singer, Columbia University Press, New York, 1961, pp. 334–341.
15. W. G. Wagner and B. A. Lengyel, "Evolution of the Giant Pulse in a Laser," *J. Appl. Phys.*, **34**, 7, 2040–2046, July 1963.
16. W. R. Bennett, Jr., "Hole Burning Effects on a He-Ne Optical Maser," *Phys. Rev.*, **126**, No. 2, 580–593, April 15, 1962.

PROBLEMS

4.1 Consider a quadrupole focuser for the ammonia maser where the radial component of the electrostatic field is of the form $3 \times 10^4 r^4$ V/cm. The force on the particle is given by $\boldsymbol{F} = -\nabla <\mathscr{H}>$.

(a) Using the parameters for the 23.87 GHz transition in ammonia, determine the radial force component for upper and lower state molecules as a function of radius.

(b) Make phase plane plots, $\dot{r}$ versus r, for upper and lower state molecules. From these plots sketch r as a function of position for upper and lower state molecules. Assume that the initial molecular velocity is in the axial direction.

4.2 For the three-level laser, power is absorbed from a pump source to maintain an equilibrium population difference per unit volume N^e that is other than the thermal equilibrium value N_T^e. The absorbed power is proportional to N_1, the number of molecules per unit volume in state $|1\rangle$ and $|E_p|^2$, where E_p is the pump field. The corresponding rate equation is

$$\frac{\hbar\Omega}{2}\left[\dot{N} + \frac{N - N_T^e}{T_1}\right] = KN_1 |E_p|^2$$

where $N = N_2 - N_1$ and K is a constant. Given that $N_1 + N_2 \cong N_V =$ number of active ions per unit volume, and $N_T^e \cong -N_V$:

(a) Obtain an expression for N^e.

(b) Determine the constant K for the following conditions: The pump radiation is absorbed in the transition $|1\rangle \to |3\rangle$ with a matrix element μ_{13} and Lorentzian linewidth $\Delta\omega_{13}$. Relaxation from $|3\rangle \to |2\rangle$ is much faster than the $|2\rangle \to |1\rangle$ relaxation. The medium is isotropic and the levels are nondegenerate. Figure P4.2 illustrates the pumping process.

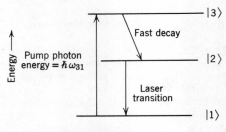

figure P4.2 The three-level laser.

4.3 In the presence of the pump the decay time from $|2\rangle$ to $|1\rangle$, T_1', differs from the decay time T_1 in the absence of the pump. For $N^e = 0.2N_V$, $N_T^e = -N_V$, and $T_1 = 4 \times 10^{-3}$ sec, determine T_1', where N_V is the number of active ions per unit volume.

4.4 In this problem we wish to consider saturation effects when there is inhomogeneous broadening, such as the hole burning effect exhibited by the He-Ne laser. From (3.28) in Chapter 3, show that the absorption constant Γ for the Gaussian broadened line is given by

$$\Gamma(\omega, \omega_0) = \int_{-\infty}^{\infty} g_G(\Omega_i, \omega_0) \Gamma_L(\omega, \Omega_i) \, d\Omega_i,$$

where g_G is the Gaussian function given by (3.24) and Γ_L is the absorption constant for the Lorentzian line given by (3.13) and (3.14).

(a) At the center of the Gaussian line where $\omega = \omega_0$, show that

$$\Gamma(\omega_0, \omega_0) = \frac{\pi L}{\hbar \epsilon_0 c \eta} \frac{|\mu_{12}|^2}{3} (N_1 - N_2)^e \frac{\omega_0 [(4/\pi) \ln 2]^{1/2}}{b \Delta \omega_G} \exp\{z^2(1 - \operatorname{erf} z)\}$$

where

$$z \equiv \frac{2(\ln 2)^{1/2}}{T_2 \Delta \omega_G} \left(1 + \frac{I}{I_{\text{sat}}}\right)^{1/2} \quad b \equiv \left(1 + \frac{I}{I_{\text{sat}}}\right)^{1/2} \quad \Delta \omega_G = \text{linewidth}$$

(b) In the inhomogeneous limit where $\Delta \omega_G \to \infty$, show that

$$\Gamma = \Gamma_{\text{unsat}} \frac{1}{(1 + I/I_{\text{sat}})^{1/2}}$$

and in the homogeneous limit where $\Delta \omega_G \to 0$, show that

$$\Gamma = \Gamma_{\text{unsat}} \frac{1}{(1 + I/I_{\text{sat}})}$$

where Γ_{unsat} is the absorption constant for $I \to 0$.

4.5 For each example given below, state whether the broadening is homogeneous or inhomogeneous and why. Broadening due to:
(a) Radiative decay by spontaneous emission (natural broadening).
(b) A velocity distribution (Doppler broadening).
(c) Collisions with like molecules in a gas (Holtzmark broadening).
(d) Collisions with a foreign gas (Lorentz broadening).
(e) An electric field produced by other ions and electrons (Stark effect broadening).
What is the essential difference between homogeneous and inhomogeneous broadening?

4.6 From the discussion presented in Section 3.4.2 we may determine some frequency pulling effects on a laser. For

$$\frac{\Omega}{2\pi} = \text{resonant frequency of the active medium} = 2.6 \times 10^{15} \text{ Hz}$$

$$\frac{\omega_c}{2\pi} = \text{circuit resonant frequency} = \frac{\Omega}{2\pi} + 100 \text{MHz}$$

$$\tau_c = 2 \times 10^{-9} \text{ sec}$$

$$2/T_2 = 2\pi \times 3.3 \times 10^{11} \text{ Hz}$$

(a) Determine the frequency of the laser.
(b) Show that the steady-state value for the population difference per unit volume when $\omega_c \neq \Omega$ is given by

$$\frac{N}{N_0} \simeq 1 + \left(\frac{\omega_c - \Omega}{2/T_2}\right)^2$$

where N_0 is the on-resonance population difference per unit volume given by (3.80).

4.7 If the two laser levels have degeneracies g_1 and g_2, modify the following expressions accordingly:
(a) Equation 4.9 for the threshold pump power density.
(b) Equation 4.14 for the steady-state output power.
(c) Equation 4.17 for the energy in a spike.
(d) Equation 4.19 for the Q-switched laser.

4.8 For the three-level laser obtain an expression for steady-state output power, P_{out}, in terms of P_p/P_{th}, where

$$P_p = \text{pump power}$$

$$P_{\text{th}} = \text{threshold pump power}$$

Show that the dependence of P_{out} on P_p is as illustrated in the figure. Obtain expressions for θ and P_{max}. (Neglect the change of T_1 with pump power, as discussed in Problem 4.3.)

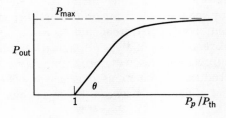

4.9 For the example of the ruby laser, calculate the error caused in determining the oscillation frequency if it is taken that $E_{\text{macroscopic}} = E_{\text{local}}$. Neglect anisotropy.

4.10 Using the parameters given for the Nd^{3+} laser, calculate peak power per unit volume, energy output per unit volume, and pulse duration for the Q-switched laser.

4.11 A helium-neon laser is designed with one mirror that is totally reflecting, and the output mirror is made to have a variable reflectivity. There is 1% loss per one-way pass in the interferometer when the output mirror is totally reflecting. The equilibrium population difference per unit volume N^e is held constant at 1.2 times the threshold population difference when the output mirror is totally reflecting. For the parameters of the helium-neon laser, obtain an expression for output power as a function of the reflectivity of the output mirror. This expression may be derived from a simultaneous solution of (4.31), (4.33), and (4.34). Plot output power as reflectivity varies from 1.0 to 0.9.

4.12 Modify the right hand side of (3.3b) when the polarization **P** does not fill the entire cavity volume. This, in turn, leads to a modification of (3.79) from which N_0 was obtained. Evaluate the factor that multiplies (4.6) when the medium does not fill the entire resonator. This is the filling factor.

NONLINEAR EFFECTS IN QUANTIZED MEDIA

5

5.1 INTRODUCTION

The subjects that have been discussed in Chapters 2, 3, and 4 involved interactions between radiation and matter in which the frequency of the radiation is approximately equal to a transition frequency in the medium. If the field were quantized, so that the electromagnetic energy were expressed in terms of photon energy, we would categorize the previously presented interactions as single-photon effects. That is, a single photon could be absorbed and the medium would undergo a transition to a higher energy state, or, alternatively, a single photon could be emitted and there would be a corresponding energy loss in the medium. Figure 5.1 illustrates the energies involved in the single-photon interaction.

If we examine the equations of motion for the density operator, or if we consider the semiclassical equations given in Chapter 2, we see that these equations are nonlinear. Products of variables are involved in the equations, and so in addition to the single-photon effects considered in Chapters 2, 3, and 4, various nonlinear processes will occur. For example, it has been observed[1,2] that for certain materials there is a resonant absorption of electromagnetic radiation at one-half the transition frequency. This interaction is termed two-photon absorption, because in terms of the photon energy of the radiation it is necessary for two photons to be absorbed for the transition to occur in the medium. This energy relationship is shown in Figure 5.2. It should be noted that energy is conserved for two-photon absorption just as it is for single-photon absorption, because the absorbed energy equals the energy change of the medium.

Other two-photon processes include the absorption of radiation at two different frequencies so that the sum of the frequencies equals the transition frequency, and Raman oscillation, which is the absorption of radiation at a frequency greater than the transition frequency and the emission of radiation at the difference

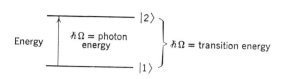

figure 5.1 The single-photon interaction process, where the photon energy is approximately equal to the transition energy.

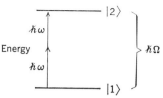

figure 5.2 Two-photon absorption, in which a transition occurs by interaction with radiation at one-half the transition frequency.

frequency. Figure 5.3 illustrates these effects. On the basis of the photon energies associated with each of these processes, it is seen that energy is conserved.

Interactions similar to those that have been illustrated for two-photon processes also occur for larger photon numbers. Three-photon absorption is shown in Figure 5.4a; three-photon absorption of different frequencies is shown in 5.4b; and a possible three-photon Raman process is illustrated in 5.4c. The three-photon Raman effect shown in Figure 5.4c involves the absorption of two photons at frequency ω_1, the emission of a photon at ω_2, and a transition of the medium from state $|1\rangle$ to state $|2\rangle$.

There are several reasons for studying nonlinear processes in quantized media. Multiple-photon effects give information regarding material properties such as the parity of states or the oscillator strengths of the transitions.[3] For example, if the eigenstates have a definite parity, then two-photon absorption occurs between states of the same parity. This nonlinear effect may therefore be used to study the coupling between the states, whereas the electric dipole transition cannot be used for this purpose because it is between states of opposite parity. Another purpose for this investigation is that although we shall specifically consider nonlinearities in quantized media, nonlinear effects have basic features in common regardless of the particular nonlinear element. We shall learn that the likelihood of

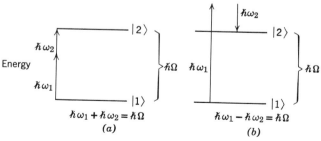

figure 5.3 Other two-photon processes: (a) the energy relationships for absorption at two different frequencies; (b) the energy diagram for Raman oscillation.

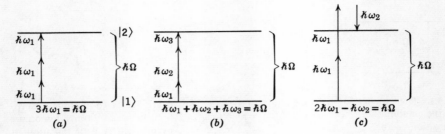

figure 5.4 Some possible three-photon processes: (a) three-photon absorption; (b) absorption at different frequencies; (c) Raman oscillation.

three-photon absorption followed by emission at the transition frequency varies as the cube of the incident radiation. This is true for third harmonic generation in any nonlinear material, until, of course, saturation effects enter to disturb the cube-law relationship. Still another motivation for studying these nonlinear processes is that much of the macroscopic nonlinear behavior of materials results from the multiple-photon effects between energy levels or energy bands. Throughout the frequency spectrum, nonlinear elements have extensive applications as harmonic generators, modulators, detectors, and parametric oscillators. Finally, the study of multiple-photon phenomena provides an interesting example for the illustration of perturbation procedures as applied to quantized media. In Section 8.4.2 there is a discussion of the indirect transition in semiconductors, which is a phonon-photon process that has features in common with the multiple-photon effect. The procedure that is used to analyze indirect transitions is essentially the same as that used for two photon transitions.

The advent of the laser with its intense optical output has generated increasing interest in multiple-photon processes because an n-photon effect has a transition probability that varies as the nth power of the incident radiation. Therefore these processes are more pronounced when the intensity of the light source is increased. Some of the experiments involving multiple-photon processes using a laser source are listed in the references at the end of the chapter.[4,5]

In this chapter we shall discover the conditions under which these multiple-photon processes occur, and we shall evaluate the likelihood or strength of these events as measured by the transition probability or scattering cross-section. Section 5.2 is devoted to parity considerations and transition probabilities. Scattering cross-section is evaluated for two-photon absorption between states of the same parity and for three-photon absorption between states of opposite parity in Section 5.3. Harmonic generation resulting from multiple-photon absorption is considered in Section 5.4, and Section 5.5 deals with the dependence of the transition frequency on the intensity of an electromagnetic field. Raman oscillation is considered in Section 5.6.

5.2 PARITY CONSIDERATIONS AND TRANSITION PROBABILITIES

The transition probability for multiple-photon absorption will be evaluated by a perturbation procedure applied to the equations of motion for the density operator. We assume that an interaction or perturbation term $\mathcal{H}'$ is added to the Hamiltonian and that this term is much smaller than $\mathcal{H}_0$, which is the unperturbed Hamiltonian of the medium. The Hamiltonian may be written in the form

$$\mathcal{H} = \mathcal{H}_0 + \lambda \mathcal{H}' \tag{5.1}$$

where the parameter λ has been introduced to indicate that the coefficient of λ is a perturbation. Each matrix element of the density operator may be expanded in terms of powers of λ, and the power of λ indicates the order of the perturbation:

$$\rho_{ij} = \sum_p \lambda^p \rho_{ij}^{(p)} \tag{5.2}$$

A first-order perturbation on ρ_{ij} is given by $\rho_{ij}^{(1)}$, which is proportional to $\mathcal{H}'$; a second-order perturbation ρ_{ij} is given by $\rho_{ij}^{(2)}$, which is proportional to $(\mathcal{H}')^2$; and so on.

The transition probability W has been defined as the probability per unit time of a transition from one eigenstate to another resulting from induced transitions. Therefore, if we consider that a system is in state $|1\rangle$ at $t = 0$, so that

$$\rho_{ij}(t=0) = \delta_{i1}\delta_{j1} \tag{5.3}$$

then the transition probability from state $|1\rangle$ to state $|k\rangle$ is

$$W = \overline{\frac{\partial \rho_{kk}}{\partial t}} \tag{5.4}$$

where the overbar indicates a time average over the period of the frequency of the perturbation. The time average is introduced because W is intended to describe changes in population over time intervals that are long compared to the period of a harmonic perturbation.

Since we are interested only in induced transitions between eigenstates and not in transitions resulting from relaxation processes, the appropriate equation of motion for the diagonal elements of the density operator is given by (1.31) with W_{kj} set equal to zero:

$$i\hbar \frac{\partial \rho_{jj}}{\partial t} = [\lambda \mathcal{H}', \rho]_{jj} \tag{5.5}$$

If the expansion for ρ_{jj} in terms of powers λ as given by (5.2) is introduced into (5.5), and like powers of λ are equated, we obtain

$$i\hbar \frac{\partial \rho_{jj}^{(p)}}{\partial t} = [\mathcal{H}', \rho^{(p-1)}]_{jj} \tag{5.6}$$

as the expression for evaluating the pth-order perturbation on ρ_{jj}. In a similar manner, from (1.30) we find the pth-order perturbation on the off-diagonal matrix elements of the density operator to be

$$i\hbar\left(\frac{\partial}{\partial t} + i\omega_{ij} + \frac{1}{\tau_{ij}}\right)\rho_{ij}^{(p)} = [\mathscr{H}', \rho^{(p-1)}]_{ij} \qquad \text{for } i \neq j \qquad (5.7)$$

Let us now proceed to evaluate transition probabilities by means of (5.6) and (5.7). From (5.6) we have

$$i\hbar \frac{\partial \rho_{jj}^{(0)}}{\partial t} = 0 \qquad (5.8)$$

and from (5.7)

$$i\hbar\left(\frac{\partial}{\partial t} + i\omega_{ij} + \frac{1}{\tau_{ij}}\right)\rho_{ij}^{(0)} = 0 \qquad \text{for } i \neq j \qquad (5.9)$$

With the condition, as given by (5.3), that the system is initially in state $|1\rangle$, Equations 5.8 and 5.9 yield

$$\rho_{ij}^{(0)} = \delta_{i1}\delta_{j1} \qquad (5.10)$$

If the commutator in (5.6) is written as a summation, then the first-order perturbation $\rho_{jj}^{(1)}$ may be expressed in the form

$$i\hbar \frac{\partial \rho_{jj}^{(1)}}{\partial t} = \sum_k (\mathscr{H}'_{jk}\rho_{kj}^{(0)} - \rho_{jk}^{(0)}\mathscr{H}'_{kj}) \qquad (5.11)$$

The matrix elements $\rho_{jk}^{(0)}$ are given by (5.10), whereupon we have

$$i\hbar \frac{\partial \rho_{jj}^{(1)}}{\partial t} = 0 \qquad (5.12)$$

The initial condition given by (5.3) in conjunction with (5.12) yields

$$\rho_{jj}^{(1)} = 0 \qquad (5.13)$$

From (5.7), the first-order perturbations for the off-diagonal matrix elements are given by

$$i\hbar\left(\frac{\partial}{\partial t} + i\omega_{ij} + \frac{1}{\tau_{ij}}\right)\rho_{ij}^{(1)} = \sum_k \left(\mathscr{H}'_{ik}\rho_{kj}^{(0)} - \rho_{ik}^{(0)}\mathscr{H}'_{kj}\right) \qquad i \neq j$$

and from (5.10) we have

$$i\hbar\left(\frac{\partial}{\partial t} + i\omega_{i1} + \frac{1}{\tau_{i1}}\right)\rho_{i1}^{(1)} = \mathscr{H}'_{i1} \qquad (5.14)$$

Since ρ is Hermitian, $\rho_{1i}^{(1)} = (\rho_{i1}^{(1)})^*$. All other first-order off-diagonal matrix elements are zero. To determine transition probability for a sinusoidal excitation, $\mathscr{H}'$ is taken to be

$$\mathscr{H}' = \frac{\tilde{\mathscr{H}}'}{2}(e^{i\omega t} + e^{-i\omega t}) \qquad (5.15)$$

and the corresponding steady-state solution for $\rho_{i1}^{(1)}$ is found from (5.14) to be

$$\rho_{i1}^{(1)} = \frac{\tilde{\mathcal{H}}'_{i1} e^{i\omega t}}{2i\hbar(i\omega + i\omega_{i1} + 1/\tau_{i1})} + \frac{\tilde{\mathcal{H}}'_{i1} e^{-i\omega t}}{2i\hbar(-i\omega + i\omega_{i1} + 1/\tau_{i1})} \tag{5.16}$$

To determine the transition probability W it is necessary to evaluate $\overline{\partial \rho_{jj}/\partial t}$, which has been shown to be zero up to and including first-order terms. Therefore the lowest-order value for W is given by $\overline{\partial \rho_{jj}^{(2)}/\partial t}$, which may be found from (5.6). From (5.6) we have

$$i\hbar \frac{\partial \rho_{jj}^{(2)}}{\partial t} = \sum_k (\mathcal{H}'_{jk}\rho_{kj}^{(1)} - \rho_{jk}^{(1)}\mathcal{H}'_{kj}) = \mathcal{H}'_{j1}\rho_{1j}^{(1)} - \rho_{j1}^{(1)}\mathcal{H}'_{1j} \tag{5.17}$$

where the last step in (5.17) resulted from the fact that $\rho_{1j}^{(1)}$ and $\rho_{j1}^{(1)}$ are the only nonzero first-order terms. Substituting the expression for $\rho_{j1}^{(1)}$ as given by (5.16) into (5.17), and using

$$\mathcal{H}'_{1j} = \frac{\tilde{\mathcal{H}}'_{1j}}{2}(e^{i\omega t} + e^{-i\omega t})$$

we obtain

$$i\hbar \frac{\partial \rho_{jj}^{(2)}}{\partial t} = \left(-\frac{|\tilde{\mathcal{H}}'_{j1}|^2}{4i\hbar(-i\omega - i\omega_{j1} + 1/\tau_{j1})} - \frac{|\tilde{\mathcal{H}}'_{j1}|^2}{4i\hbar(i\omega - i\omega_{j1} + 1/\tau_{j1})} - \text{c.c.} \right) + B \tag{5.18}$$

where B is a sum of terms with a time dependence $e^{\pm 2i\omega t}$. A time average of both sides of (5.18) eliminates B, and upon combining the remaining terms on the right-hand side, we have

$$i\hbar \overline{\frac{\partial \rho_{jj}^{(2)}}{\partial t}} = \frac{i|\tilde{\mathcal{H}}'_{j1}|^2}{2\hbar \tau_{j1}} \left(\frac{1}{(\omega - \omega_{j1})^2 + 1/\tau_{j1}^2} + \frac{1}{(\omega + \omega_{j1})^2 + 1/\tau_{j1}^2} \right) \tag{5.19}$$

When $\omega \simeq \omega_{j1}$, the first term on the right-hand side of (5.19) becomes large, and this is the resonance effect associated with single-photon absorption. With $\omega \simeq \omega_{j1}$, the last term on the right-hand side of (5.19) may be neglected so that the transition probability for single-photon absorption, $W_{(1)}$, is given by

$$W_{(1)} = \overline{\frac{\partial \rho_{jj}^{(2)}}{\partial t}} = \frac{|\tilde{\mathcal{H}}'_{j1}|^2}{2\hbar^2} \pi g_L(\omega) \tag{5.20}$$

where $g_L(\omega)$ is the Lorentzian lineshape factor given by (3.22). The value for $W_{(1)}$ for a two-level electric dipole transition in an isotropic medium where

$$|\mathcal{H}'_{21}|^2 = \frac{|\mu_{12}|^2}{3}|\tilde{E}|^2$$

is therefore equal to

$$W_{(1)} = \frac{|\mathbf{\mu}_{12}|^2 |\tilde{E}|^2}{6\hbar^2} \pi g_L(\omega) \tag{5.21}$$

On resonance (that is, $\omega = \omega_{21}$), $\pi g_L(\omega_{21}) = \tau_{21} = T_2$, and $W_{(1)}$ is then identical to (3.85) with $L = 1$, which is the value that L has for an isolated molecule. Equation 5.21 was derived for a single molecule, and so $g(\omega)$ is the Lorentzian lineshape function. For the many-molecule case where the line is Doppler broadened, (5.21) is still valid with $g_L(\omega)$ replaced by (3.24), the Gaussian lineshape function. On resonance, for a Gaussian function

$$g_G(\omega_{21}) = \frac{2}{\Delta \omega_G} \left(\frac{\ln 2}{\pi} \right)^{1/2}$$

where $\Delta \omega_G$ is the linewidth.

The conditions that have been imposed in the derivation of the transition probability are the following:

1. A single-frequency source has been assumed. This assumption is valid if the spectral spread of the incident radiation is narrow compared to the transition linewidth, as would almost certainly be the case with a laser source. If this condition is not satisfied, the incident signal may be expressed as a Fourier transform and the transition probability is obtained from an integral in the frequency domain.

2. We have assumed that the ratio of the transition frequency to the linewidth is much greater than unity. In going from (5.19) to (5.20) by saying that $[(\omega - \omega_{j1})^2 + 1/\tau_{j1}^2]^{-1}$ becomes large near resonance, we require that $\omega_{j1} \tau_{j1} \gg 1$, which is equivalent to the statement that the ratio of frequency to linewidth must be much greater than unity.

3. Transition probability was obtained from an average in time over the period of the harmonic perturbation. For this averaging to be valid, the time for a transition, as given by the reciprocal of the transition probability, must be long compared to the period of the harmonic perturbation, that is, $\omega/2\pi W \gg 1$, where $2\pi/\omega$ is the period of oscillation and W is the transition probability. According to (5.20), this means that

$$\frac{|\tilde{\mathcal{H}}'_{j1}|^2}{\hbar^2} \frac{\pi^2 g_L(\omega)}{\omega} \ll 1 \tag{5.22}$$

Equation 5.22 places an upper limit on the intensity of the incident radiation for a specified transition frequency and dipole matrix element. For optical transitions, this condition is satisfied even with focused, Q-switched laser beams as the source of radiation. At microwave frequencies this condition is almost always fulfilled, but it is possible with kilowatt oscillators and resonant cavity circuitry to violate this requirement.

Any perturbation procedure requires that the power series in λ converges, and it is found that this condition is satisfied if the inequality expressed by (5.22) is satisfied. In obtaining transition probabilities for multiple-photon absorption, all the preceding conditions will be assumed to hold.

In a manner similar to that used for calculating the transition probability for single-photon absorption, from higher-order perturbation terms the transition probability for two-photon absorption between $|1\rangle$ and $|j\rangle$ may be evaluated. If the frequency of the incident radiation is one-half a transition frequency ω_{j1}, then the resonance effect for two-photon absorption is given by terms in the density matrix that have the factor

$$\frac{1}{i(2\omega - \omega_{j1}) + 1/\tau_{j1}} \quad (5.23)$$

The expression given by (5.23) is large when $\omega \simeq \omega_{j1}/2$. According to (5.7), a factor of this form appears when ρ_{1j} is driven by a term with a time dependence $e^{2i\omega t}$ or ρ_{j1} is driven by a term with a time dependence $e^{-2i\omega t}$. If we write out $\rho_{1j}^{(2)}$ as given by (5.7), we see that there are driving terms with an $e^{2i\omega t}$ time dependence:

$$i\hbar\left(\frac{\partial}{\partial t} + i\omega_{1j} + \frac{1}{\tau_{1j}}\right)\rho_{1j}^{(2)} = \sum_l (\mathcal{H}'_{1l}\rho_{lj}^{(1)} - \rho_{1l}^{(1)}\mathcal{H}'_{lj}) \quad (5.24)$$

Since

$$\mathcal{H}'_{ij} = \frac{\tilde{\mathcal{H}}'_{ij}}{2}(e^{i\omega t} + e^{-i\omega t})$$

and since $\rho_{1l}^{(1)}$ has a time dependence $e^{\pm i\omega t}$, as given by (5.16), then the product $\rho_{1l}^{(1)}\mathcal{H}'_{lj}$ includes terms of the form $e^{2i\omega t}$. If (5.16) is substituted into (5.24), the expression for the coefficient of the $e^{2i\omega t}$ term in ρ is found to be

$$i\hbar\left(2i\omega - i\omega_{j1} + \frac{1}{\tau_{j1}}\right)\rho_{1j}^{(2)} = \frac{1}{4i\hbar}\sum_l \frac{\tilde{\mathcal{H}}'_{1j}\tilde{\mathcal{H}}'_{1l}}{(i\omega - i\omega_{l1} + 1/\tau_{l1})} \quad (5.25)$$

where we have used the fact that $\omega_{1l} = -\omega_{l1}$ and $\tau_{1l} = \tau_{l1}$. The denominator in the summation on the right-hand side of (5.25) may be simplified by noting that $|\omega - \omega_{l1}|\tau_{l1} \gg 1$, provided that there is no direct absorption resulting from a transition frequency at the radiation frequency. Then, solving for $\rho_{1j}^{(2)}$ from (5.25),

$$\rho_{1j}^{(2)} = \frac{1}{4\hbar^2(2\omega - \omega_{j1} - i/\tau_{j1})}\sum_l \frac{\tilde{\mathcal{H}}'_{1j}\tilde{\mathcal{H}}'_{1l}}{(\omega - \omega_{l1})} \quad (5.26)$$

Since ρ is Hermitian, $\rho_{j1}^{(2)} = [\rho_{1j}^{(2)}]^*$.

We now have terms with the resonance denominator for two-photon absorption, and the transition probability from the ground state to state $|j\rangle$ is evaluated from $\overline{\partial\rho_{jj}/\partial t}$. The expression for $\partial\rho_{jj}^{(3)}/\partial t$ as given by (5.6) involves the product of $\rho_{1j}^{(2)}$ and $\mathcal{H}'_{j1}$, where the former has an $e^{2i\omega t}$ time dependence, and the latter varies as $e^{\pm i\omega t}$, so that $\overline{\partial \rho_{jj}^{(3)}/\partial t}$ has no dc term and therefore the transition probability resulting from $\overline{\partial \rho_{jj}^{(3)}/\partial t}$ is zero. Hence it is necessary to evaluate $\overline{\partial \rho_{jj}^{(4)}/\partial t}$. From (5.6),

$$i\hbar \frac{\partial \rho_{jj}^{(4)}}{\partial t} = \sum_q (\mathcal{H}'_{jq}\rho_{qj}^{(3)} - \rho_{jq}^{(3)}\mathcal{H}'_{qj}) \tag{5.27}$$

where $\rho_{qj}^{(3)}$ is given by (5.7):

$$i\hbar\left(\frac{\partial}{\partial t} + i\omega_{qj} + \frac{1}{\tau_{qj}}\right)\rho_{qj}^{(3)} = \sum_f (\mathcal{H}'_{qf}\rho_{fj}^{(2)} - \rho_{qf}^{(2)}\mathcal{H}'_{fj}) \tag{5.28}$$

Since only $\rho_{1j}^{(2)}$ and $\rho_{j1}^{(2)}$ have the desired resonance denominator for two-photon absorption to state $|j\rangle$, only these terms need be included in (5.28), and thus

$$i\hbar\left(\frac{\partial}{\partial t} + i\omega_{qj} + \frac{1}{\tau_{qj}}\right)\rho_{qj}^{(3)} = \rho_{1j}^{(2)}(\mathcal{H}'_{q1} - \mathcal{H}'_{jj}\delta_{q1}) \tag{5.29}$$

A further simplification results by noting that the dc terms in (5.27) result from terms in $\rho_{qj}^{(3)}$ varying at $e^{\pm i\omega t}$. Since we are interested in terms that vary as $e^{2i\omega t}$ in $\rho_{ij}^{(2)}$, the only terms of interest on the right-hand side of (5.29) vary as $e^{i\omega t}$ because $\mathcal{H}'$ varies as $e^{\pm i\omega t}$. Therefore $\partial/\partial t \to i\omega$ and with $|\omega + \omega_{qj}|\tau_{qj} \gg 1$, we have

$$\rho_{qj}^{(3)} = -\frac{\rho_{1j}^{(2)}(\tilde{\mathcal{H}}'_{q1} - \tilde{\mathcal{H}}'_{jj}\delta_{q1})}{2\hbar(\omega + \omega_{qj})} \tag{5.30}$$

Combining (5.27) and (5.30), and writing out the time dependence of $\mathcal{H}'_{q1}$ as $(\tilde{\mathcal{H}}'_{q1}/2)(e^{i\omega t} + e^{-i\omega t})$, the dc terms in $i\hbar(\partial \rho_{jj}^{(4)}/\partial t)$ are

$$i\hbar \frac{\partial \rho_{jj}^{(4)}}{\partial t} = \sum_q \frac{1}{4\hbar(\omega - \omega_{q1})}(\tilde{\mathcal{H}}'_{jq}\tilde{\mathcal{H}}'_{q1}\rho_{1j}^{(2)} - \tilde{\mathcal{H}}'_{qj}\tilde{\mathcal{H}}'_{1q}\rho_{j1}^{(2)}) \tag{5.31}$$

Equation 5.31 was obtained from the combination of (5.27) and (5.30), and by letting

$$\frac{1}{\omega + \omega_{qj}} = \frac{1}{\omega + (\omega_{q1} + \omega_{1j})} = \frac{1}{\omega + \omega_{q1} - \omega_{j1}}$$

$$= \frac{1}{\omega + \omega_{q1} - 2\omega} = \frac{1}{\omega - \omega_{q1}}$$

The transition probability for two-photon absorption $W_{(2)}$ is given by

$$W_{(2)} = \overline{\frac{\partial \rho_{jj}^{(4)}}{\partial t}}$$

and is obtained by substituting (5.26) into (5.31):

$$W_{(2)} = \frac{1}{16i\hbar^4} \left(\frac{1}{2\omega - \omega_{j1} - i/\tau_{j1}} \sum_q \frac{\tilde{\mathcal{H}}'_{jq}\tilde{\mathcal{H}}'_{q1}}{\omega - \omega_{q1}} \sum_l \frac{\tilde{\mathcal{H}}'_{lj}\tilde{\mathcal{H}}'_{1l}}{\omega - \omega_{l1}} \right.$$

$$\left. - \frac{1}{2\omega - \omega_{j1} + i/\tau_{j1}} \sum_q \frac{\tilde{\mathcal{H}}'_{qj}\tilde{\mathcal{H}}'_{1q}}{\omega - \omega_{q1}} \sum_l \frac{\tilde{\mathcal{H}}'_{jl}\tilde{\mathcal{H}}'_{l1}}{\omega - \omega_{l1}} \right)$$

$$= \frac{\pi g_L(2\omega)}{8\hbar^4} \left| \sum_q \frac{\tilde{\mathcal{H}}'_{jq}\tilde{\mathcal{H}}'_{q1}}{\omega - \omega_{q1}} \right|^2 \qquad (5.32)$$

where

$$g_L(2\omega) = \frac{1}{\pi} \frac{1/\tau_{j1}}{(2\omega - \omega_{j1})^2 + 1/\tau_{j1}^2}$$

is the normalized Lorentzian curve. As for the case of single-photon absorption, $g_G(2\omega)$ is given by (3.24) for a Gaussian lineshape.

We see that $W_{(2)}$ varies as the fourth power of the field or the square of the radiation intensity, if $\mathcal{H}'$ varies linearly with the electric or magnetic field as for the electric dipole, electric quadrupole, and magnetic dipole interactions. On the other hand, $W_{(1)}$ varies linearly with the radiation intensity.

In Chapter 2 there was a discussion of the parity requirements for single-photon absorption. For states of definite parity, if $\mathcal{H}'$ is an electric dipole interaction, $|1\rangle$ and $|j\rangle$ must be of opposite parity for $\tilde{\mathcal{H}}'_{1j}$ to be nonzero, since the dipole operator has negative parity. Electric quadrupole and magnetic dipole transitions occur between states of the same parity. Let us examine the parity conditions for two-photon absorption as given by (5.32). If $\mathcal{H}'$ is an electric dipole interaction, then when $|1\rangle$ has positive parity $|q\rangle$ must have negative parity for $\tilde{\mathcal{H}}'_{q1}$ to be nonzero, and similarly $|j\rangle$ must have positive parity for $\tilde{\mathcal{H}}'_{jq}$ to be nonzero. Therefore, two-photon absorption involving the electric dipole interaction may only occur between states of the same parity. If $\mathcal{H}'$ represents the electric quadrupole or magnetic dipole, two-photon absorption will again occur between states of the same parity, but for optical frequency transitions these transition probabilities are approximately twelve orders of magnitude weaker than for the electric dipole. Between states of opposite parity the most intense two-photon absorption results from a combination of electric dipole and electric quadrupole or magnetic dipole transitions. That is, $\mathcal{H}'_{q1}$ may be an electric dipole transition with $|q\rangle$ having a parity different from the parity for $|1\rangle$ and $\mathcal{H}'_{jq}$ may be an electric quadrupole with $|j\rangle$ having the same parity as $|q\rangle$. Then $|j\rangle$ and $|1\rangle$ have opposite parity. For optical-frequency transitions, two-photon absorption between states of opposite parity is approximately six orders of magnitude weaker than two-photon absorption between states of the same parity because of the necessity to include the electric quadrupole for the latter.

If only two levels are involved in the two-photon absorption, as would be the case when $q = 1$ or $q = j$ is the only nonzero term in (5.32), then for electric dipole interactions it is necessary that $|1\rangle$ or $|j\rangle$ be a state of mixed parity. For example, if $q = 1$ is the only nonzero term, $\tilde{\mathcal{H}}'_{11}$ is a factor in the expression for $W_{(2)}$. This means that $|1\rangle$ must be a state of mixed parity, since the electric dipole operator has negative parity. In an isolated atom the eigenstates have a definite parity so that two-photon absorption will not occur for the electric dipole interaction with only two levels involved. However, in a crystal the inversion symmetry of the Hamiltonian may be lost by the presence of a local field, and thus the states can be of mixed parity. Alternatively, the same results might be accomplished with the application of a dc field to the medium.

Transition probability for three-photon absorption between $|1\rangle$ and $|j\rangle$ is evaluated from $\overline{\partial \rho_{jj}^{(6)}/\partial t}$, in which we seek a resonant term of the form

$$\frac{1}{i(3\omega - \omega_{j1}) + 1/\tau_{j1}}$$

The procedure for evaluating $W_{(3)}$ is analogous to the procedure used to obtain $W_{(2)}$, with the result

$$W_{(3)} = \frac{\pi g_L(3\omega)}{32\hbar^6} \left| \sum_{q,k} \frac{\tilde{\mathcal{H}}'_{jq}\tilde{\mathcal{H}}'_{qk}\tilde{\mathcal{H}}'_{k1}}{(2\omega - \omega_{q1})(\omega - \omega_{k1})} \right|^2 \quad (5.33)$$

The derivation of $W_{(3)}$ is not presented because the techniques involved here have been presented in deriving (5.32). We see that $W_{(3)}$ varies as the cube of the incident power for electric dipole, electric quadrupole, and magnetic dipole interactions.

We may conclude from (5.33) that for electric dipole transitions between states of definite parity, three-photon absorption will occur only if $|1\rangle$ and $|j\rangle$ are of opposite parity. Three-photon absorption may occur in a two-level system between states of opposite parity. That this is so can be seen by letting $k = j$ and $q = 1$ be the only nonzero terms in (5.33). Under these conditions,

$$W_{(3)} = \frac{\pi g_L(3\omega)}{32\hbar^6} \left| \frac{\tilde{\mathcal{H}}'_{j1}\tilde{\mathcal{H}}'_{1j}\tilde{\mathcal{H}}'_{j1}}{2\omega(\omega - \omega_{j1})} \right|^2 \quad (5.34)$$

and $\tilde{\mathcal{H}}'_{1j}$ is not zero if $|1\rangle$ and $|j\rangle$ are of opposite parity. Therefore the lowest multiple-photon electric dipole absorption effect in a two-level system with states of definite parity is three-photon absorption between states of opposite parity.

A general expression for m-photon absorption between $|1\rangle$ and $|j\rangle$ is given by (5.35):

$$W_{(m)} = \frac{\pi g_L(m\omega)}{2^{2m-1}\hbar^{2m}}$$

$$\times \left| \sum_{q,k,\ldots s,p} \frac{\tilde{\mathcal{H}}'_{jq}\tilde{\mathcal{H}}'_{qk} \cdots \tilde{\mathcal{H}}'_{sp}\tilde{\mathcal{H}}'_{p1}}{[(m-1)\omega - \omega_{q1}][(m-2)\omega - \omega_{k1}] \cdots [2\omega - \omega_{s1}][\omega - \omega_{p1}]} \right|^2$$

$$(5.35)$$

If $\mathscr{H}'$ represents only electric dipole interactions, then $W_{(m)}$ varies as the mth power of the radiation, and for states of definite parity $|1\rangle$ and $|j\rangle$ must be of the same parity for m even, and of opposite parity for m odd.

5.3 SCATTERING CROSS-SECTIONS FOR MULTIPLE-PHOTON ABSORPTIONS

The scattering cross-section σ_c has been defined in Section 3.2 as having the value

$$\sigma_c = \frac{\mathscr{P}}{N_V I} \tag{5.36}$$

where $\mathscr{P}$ is the average power absorbed per unit volume, N_V is the number of molecules per unit volume, and I is the average power per unit area of the incident radiation. Scattering cross-section is the effective area over which a molecule absorbs the incident radiation, and thus σ_c times the incident power density is the power absorbed by a molecule.

For single-photon absorption the scattering cross-section for a transition with an oscillator strength close to unity is of the same order as the physical area of the molecule. For multiple-photon absorption, σ_c is dependent upon the intensity of the incident radiation, and even for relatively intense fields the cross-section is much smaller than the value for single-photon absorption. In this section we shall evaluate σ_c for two- and three-photon interactions, using typical parameter values for electric dipole transitions.

It is possible to express σ_c in terms of the transition probability, W. Since W is the probability of a transition per unit time for a molecule, the rate of energy absorbed by N_V molecules per unit volume is equal to the product of the transition energy $\hbar\omega_{21}$ and $N_V W$. Therefore

$$\mathscr{P} = N_V \hbar \omega_{21} W \tag{5.37}$$

For a propagating wave the incident power density is given by

$$I = \frac{\eta c \epsilon_0 |\tilde{E}|^2}{2} \tag{5.38}$$

when η is the refractive index and c is the free-space velocity of light. From the combination of (5.36), (5.37), and (5.38) we have

$$\sigma_c = \frac{2\hbar\omega_{21}}{\eta c \epsilon_0 |\tilde{E}|^2} W \tag{5.39}$$

Single-photon absorption for the electric dipole transition is evaluated by substituting $W_{(1)}$ from (5.21) into (5.39), yielding

$$\sigma_{c_{(1)}} = \frac{\pi \omega_{21} |\mathbf{\mu}_{12}|^2 g_L(\omega)}{3\eta c \epsilon_0 \hbar} \tag{5.40}$$

which is the same as (3.16) with $L = 1$. Equation 5.40 applies to an isolated molecule, and for a dense medium where η differs significantly from unity, it is necessary to multiply by L, the Lorentz correction factor.

Scattering cross-section for two-photon absorption is determined by substituting $W_{(2)}$ from (5.32) into (5.39). To simplify the evaluation we shall assume that only three levels are involved, so that the transition is $|1\rangle \to |2\rangle$ with $|3\rangle$ as the intermediate level. In (5.32) this means that $q = 3$ and $j = 2$, and for the electric dipole transition in an isotropic medium

$$|\mathscr{H}_{ij}|^2 = \frac{|\mathbf{\mu}_{ij}|^2 |\tilde{E}|^2}{3}$$

so that we have

$$W_{(2)} = \frac{\pi g_L(2\omega)}{8\hbar^4} \frac{|\mathbf{\mu}_{32}|^2}{3} \frac{|\mathbf{\mu}_{31}|^2}{3} \frac{|\tilde{E}|^4}{(\omega - \omega_{31})^2} \tag{5.41}$$

The scattering cross-section for two-photon absorption in a three-level system in which $\omega_{21} = 2\omega$ is obtained from the substitution of (5.41) into (5.39):

$$\sigma_{c_{(2)}} = \frac{\pi \omega g_L(2\omega) |\mathbf{\mu}_{32}|^2 |\mathbf{\mu}_{31}|^2 |\tilde{E}|^2}{18 \eta c \epsilon_0 \hbar^3 (\omega - \omega_{31})^2} \tag{5.42}$$

Three-photon absorption scattering cross-section is calculated from the substitution of $W_{(3)}$ as given by (5.33) into (5.39). For three-photon absorption in a two-level system between states of opposite parity, (5.34) is substituted into (5.39) with $j = 2$ and $\omega_{21} = 3\omega$, yielding

$$\sigma_{c_{(3)}} = \frac{3\pi \omega g_L(3\omega)}{16 \hbar^5 \eta c \epsilon_0} \left(\frac{|\mathbf{\mu}_{21}|^2}{3}\right)^3 \frac{|\tilde{E}|^4}{[2\omega(\omega - \omega_{21})]^2} \tag{5.43}$$

Scattering cross-sections for one-, two-, and three-photon absorption may be evaluated by assigning typical values to the parameters appearing in (5.40), (5.42), and (5.43). Table 5.1 lists these assumed values.

With these parameter values, Table 5.2 lists σ_c for several different processes. From Table 5.2 we see that $\sigma_{c_{(m+1)}} \simeq 10^{-19} |\tilde{E}|^2 \sigma_{c_{(m)}}$, for m-photon absorption. If we consider an incident power density of 10 MW/cm², which is well within the capability of Q-switched lasers, this gives $|\tilde{E}| \simeq 9 \times 10^4$ V/cm with the following corresponding values for σ_c.

$$\sigma_{c_{(2)}} \simeq 1.5 \times 10^{-25} \text{ cm}^2$$

$$\sigma_{c_{(3)}} \simeq 8 \times 10^{-34} \text{ cm}^2$$

A comparison between theoretical and experimental values for W and σ_c for two-, three-, and four-photon absorption in anthracene and naphthalene is presented in one of the references.[3]

5 NONLINEAR EFFECTS IN QUANTIZED MEDIA   145

TABLE 5.1 ASSUMED PARAMETER VALUES FOR THE DETERMINATION OF SCATTERING CROSS-SECTION

Parameter	Symbol	Assumed value
Matrix element	$\|\mu_{ij}\|$	1 debye
Refractive index	η	1.0 (considering an isolated molecule)
Lineshape function	$g_L(m\omega)$ [a]	$10^{-12}/3$ sec
Frequency of incident radiation	ω	2.7×10^{15} rad/sec (ruby laser frequency)
Transition frequency	ω_{21}	$(2.7 \times 10^{15} \times m)$ rad/sec (for an m-photon process)
Transition frequency for the third level in two-photon absorption	ω_{31}	$2\omega_{21}$ (it is assumed that the three levels are equally spaced)

[a] The lineshape function $g(m\omega)$ has been evaluated by considering the on-resonance condition in which $m\omega = \omega_{21}$, so that $g_L(m\omega) = T_2/\pi = 2/\pi(\Delta\omega)$ for a Lorentzian line. With the linewidth $\Delta\omega$ taken to be 10 wavenumbers, $g_L(m\omega)$ has the value given in Table 5.1.

TABLE 5.2 SCATTERING CROSS-SECTIONS FOR MULTIPLE-PHOTON ABSORPTION

Process	σ_c in cm^2	Equation from which σ_c is determined
Single-photon absorption	2.6×10^{-16}	(5.40)
Two-photon absorption in a three-level system	$1.1 \times 10^{-35} \|\tilde{E}\|^2$ ($\tilde{E}$ is measured in V/cm)	(5.42)
Three-photon absorption in a two-level system	$1.2 \times 10^{-54} \|\tilde{E}\|^4$ ($\tilde{E}$ is measured in V/cm)	(5.43)

5.4 HARMONIC GENERATION

In Sections 5.2 and 5.3 we considered multiple-photon absorption, in which m photons were absorbed, each with energy $\hbar\omega$, and a transition occurred in the medium from $|1\rangle$ to $|2\rangle$ so that $m\hbar\omega = \hbar\omega_{21}$. Alternatively, energy may be conserved by the emission of a photon with energy $\hbar\omega_{21}$, with no change of state of the medium. This is harmonic generation, with the radiation of energy at frequency ω_{21}, which is m times the input frequency. The procedure for calculating the intensity of the harmonic power is to evaluate the component of polarization that varies at the transition frequency, and then to determine the electric field at this harmonic frequency.

The dipole moment $\langle \mu \rangle$ is given as

$$\langle \mu \rangle = \text{Tr}\,(\rho\mu)$$

The matrix elements ρ_{12} and ρ_{21} are the only terms in the density operator that are large at the transition frequency, since these are the terms that have a resonance denominator at $\omega = \omega_{21}$. Therefore the component of $\langle \mu \rangle$ that varies at frequency ω_{21} is

$$\langle \mu \rangle_{21} = \rho_{12}\mu_{21} + \rho_{21}\mu_{12} \tag{5.44}$$

where the notation $\langle \mu \rangle_{21}$ is used to designate the component of $\langle \mu \rangle$ at frequency ω_{21}. If we consider states of definite parity, second harmonic generation will not occur by means of electric dipole transitions. The reason for this is that two-photon absorption takes place between states of the same parity so that $\mu_{12} = 0$, and therefore there is no component of polarizaton at ω_{21}. To obtain second harmonic generation with electric dipole interactions, there must be a local field, either from the crystal or externally applied, to upset the inversion symmetry of the Hamiltonian.

Third harmonic generation will occur, however, if $|1\rangle$ and $|2\rangle$ are of opposite parity, since the matrix element μ_{12} is not zero. In this case we are interested in a component of ρ_{12} that varies at three times the frequency of the incident radiation. From (5.28) we see that $\rho_{12}^{(3)}$ contains the desired frequency variation, since $\mathcal{H}'_{af}$ has a time dependence $e^{\pm i\omega t}$ and $\rho_{fj}^{(2)}$ has an $e^{\pm 2i\omega t}$ component, so that $\rho_{12}^{(3)}$ will have a term of the form $e^{\pm 3i\omega t}$. The dipole moment term that oscillates at three times the input frequency is obtained by using $\rho_{12}^{(3)}$ in (5.44). Polarization—given by $N_V \langle \mu \rangle_{12}$, where N_V is the number of atoms (assumed to be identical) per unit volume—is then used to determine the third harmonic electric field from (2.50).

If we are interested in third harmonic generation in a two-level system between states of opposite parity, it is somewhat simpler to perform the calculation from (3.1)–(3.3), where we have equations directly in terms of the polarization. Equations 3.1 and 3.2 are rewritten below, considering a linearly polarized electric field in an isotropic medium

$$\ddot{P} + \frac{2}{T_2}\dot{P} + \Omega^2 P = -\frac{2\Omega}{\hbar} L \frac{|\mu_{12}|^2}{3} NE \tag{5.45}$$

$$\dot{N} + \frac{N - N^e}{T_1} = \frac{2}{\hbar\Omega} \dot{P} E \tag{5.46}$$

where $N = N_2 - N_1$ is the population difference per unit volume. It is assumed that the amplitude of the electric field produced at the third harmonic is much less than the fundamental field amplitude, so that the reaction of the harmonic on the system need not be included. Exponential notation is used:

$$E = \frac{\tilde{E}_\omega}{2} e^{i\omega t} + \text{c.c.} \tag{5.47}$$

$$P = \frac{\tilde{P}_\omega}{2} e^{i\omega t} + \frac{\tilde{P}_{3\omega}}{2} e^{i3\omega t} + \text{c.c.} \tag{5.48}$$

$$N = N_0 + \left(\frac{\tilde{N}_{2\omega}}{2} e^{i2\omega t} + \text{c.c.}\right) \tag{5.49}$$

where the subscript $m\omega$ indicates a component with $e^{im\omega t}$ time dependence.

Polarization at the third harmonic frequency is generated by the following process.

1. To first order in the applied field, we see from (5.45) that a polarization $\tilde{P}_\omega$ results from the presence of $\tilde{E}_\omega$.
2. From (5.46), there is a component of N at 2ω, resulting from the product of P and E.
3. Polarization $\tilde{P}_{3\omega}$ is generated by the product of $\tilde{N}_{2\omega}$, and $\tilde{E}_\omega$ as given by (5.45).

The process is summarized as follows:

$$\tilde{E}_\omega \xrightarrow[(5.45)]{} \tilde{P}_\omega \xrightarrow[(5.46)]{} \tilde{N}_{2\omega} \xrightarrow[(5.45)]{} \tilde{P}_{3\omega}$$

where the equation number giving the relationship between the variables is indicated below the arrow.

Each step in the generation of $\tilde{P}_{3\omega}$ may be evaluated separately. From (5.45), noting that $\Omega \simeq 3\omega$ and $\Omega T_2 \gg 1$, we have

$$\tilde{P}_\omega = -\frac{|\mu_{12}|^2 L N_0}{4\hbar\omega} \tilde{E}_\omega \tag{5.50}$$

From (5.46), noting that $\Omega T_1 \gg 1$, we have

$$\tilde{N}_{2\omega} = \frac{\tilde{P}_\omega \tilde{E}_\omega}{6\hbar\omega} \tag{5.51}$$

Finally, returning to (5.45) and using $\Omega = 3\omega$, we find that

$$\tilde{P}_{3\omega} = \frac{iT_2 |\mu_{12}|^2 L}{6\hbar} \tilde{N}_{2\omega} \tilde{E}_\omega \tag{5.52}$$

The value for $\tilde{N}_{2\omega}$ in (5.52) is given by (5.51), and the value for $\tilde{P}_\omega$ in (5.51) is given by (5.50). Thus, combining (5.50)–(5.52), the expression for $\tilde{P}_{3\omega}$ in terms of $\tilde{E}_\omega$ is

$$\tilde{P}_{3\omega} \simeq -\frac{i |\mu_{12}|^4 L^2 T_2 N_0}{144 \hbar^3 \omega^2} (\tilde{E}_\omega)^3 \tag{5.53}$$

If the incident radiation is considered to be a traveling wave, the polarization at the harmonic frequency is

$$P_{3\omega} = \frac{\tilde{P}_{3\omega}}{2} e^{i3\omega t} e^{-i3kz} + \text{c.c.}$$

where $k = \eta_1 \omega/c$ is the propagation constant and η_1 is the refractive index at the input frequency.

We consider a field with an amplitude coefficient that is a slowly varying function of z:

$$E_{3\omega} = \frac{\tilde{E}_{3\omega}(z)}{2} e^{i3\omega t} e^{-i3kz} + \text{c.c.} \tag{5.54}$$

If (5.54) is substituted into (3.3a) and the loss is considered negligible (that is, $\mathscr{A} = 0$), we obtain

$$6ik \frac{d\tilde{E}_{3\omega}}{dz} + \left[(3k)^2 - \left(\frac{3\omega\eta_3}{c}\right)^2\right] \tilde{E}_{3\omega} = \mu_0 (3\omega)^2 \tilde{P}_{3\omega} \tag{5.55}$$

where η_3 is the refractive index at the third harmonic. In the derivation of (5.55) we have assumed that

$$\frac{d^2 \tilde{E}_{3\omega}}{dz^2} \ll 3k \frac{d\tilde{E}_{3\omega}}{dz}$$

on the basis that $\tilde{E}_{3\omega}$ is slowly varying, so that the second derivative of $\tilde{E}_{3\omega}$ with respect to z has been neglected. The integration of (5.55) yields

$$\tilde{E}_{3\omega} = \frac{\mu_0 (3\omega)^2 \tilde{P}_{3\omega}}{(3k)^2 - (3\omega\eta_3/c)^2} \left[1 - \exp\left(\frac{i[(3k)^2 - (3\omega\eta_3/c)^2]}{6k} z\right)\right] \tag{5.56}$$

where we have used the boundary condition that $\tilde{E}_{3\omega} = 0$ at $z = 0$.

From (5.38), the power density at the third harmonic, $I_{3\omega}$, is

$$I_{3\omega} = \frac{\eta_3 c \epsilon_0 |\tilde{E}_{3\omega}|^2}{2} \tag{5.57}$$

The substitution of (5.53) and (5.56) into (5.57) yields an expression for harmonic power in terms of the input field strength,

$$I_{3\omega} = I_0 \frac{\sin^2 \alpha z}{\alpha^2} \tag{5.58}$$

where

$$\alpha \equiv \frac{(3k)^2 - (3\omega\eta_3/c)^2}{12k} \qquad I_0 \equiv \frac{\eta_3 \mu_0 |\mu_{12}|^8 L^4 T_2^2 N_0^2}{9 \times 2^{11} c k^2 \hbar^6} |\tilde{E}_\omega|^6$$

5 NONLINEAR EFFECTS IN QUANTIZED MEDIA   149

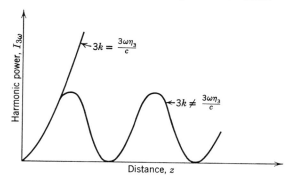

figure 5.5 The dependence of harmonic output power on distance.

The three essential features of (5.58) that are common to all harmonic generation experiments independent of the particular nonlinear medium are as follows.

1. The maximum harmonic density is generated when

$$3k = \frac{3\omega\eta_3}{c} \tag{5.59}$$

where from (5.58) we see that $I_{3\omega}$ increases as the square of distance. If (5.59) is not satisfied, the harmonic power has a sinusoidal variation as shown in Figure 5.5, with a period that increases as $3k$ becomes more nearly equal to $3\omega\eta_3/c$.

The radiation at the harmonic frequency has a propagation constant that is determined by the nonlinear process and is equal to $3k$. On the other hand, the natural propagation constant for radiation incident on the medium at frequency 3ω is $3\omega\eta_3/c$, and maximum power conversion occurs when these two constants are equal. If (5.59) is satisfied, the system is said to be momentum matched or phase matched.

Since $k = \omega\eta_1/c$, if the medium is nondispersive so that the refractive index is independent of frequency, (5.59) is satisfied, but this condition is not generally met because of the dispersive character of most media. It is possible to use birefringent materials and satisfy (5.59) by having the refractive index of the ordinary wave at one frequency equal to the refractive index of the extraordinary wave at the other frequency. This aspect of harmonic generation with traveling waves has been discussed extensively in the literature.[6] The index matching condition as stated in (5.59) applies to the generation of any harmonic, so that for maximum power at the mth harmonic the requirement is that

$$\eta_1 = \eta_m$$

where η_m is the refractive index at m times the input frequency.

2. The power output at the third harmonic varies as the cube of the input power. This relationship will continue to hold until the field strength at the harmonic frequency becomes sufficiently large so that the reaction of the harmonic on the system is not negligible. In general, regardless of the type of nonlinear element that is used, the power at the mth harmonic varies as $(I_\omega)^m$, where I_ω is the input power density.

3. Harmonic generation is not a threshold process. We see from (5.58) that even for vanishingly small fields at the fundamental frequency there is some power generated at the harmonic. This is in contrast with the laser or stimulated Raman oscillator, in which it is necessary that the incident radiation exceed a minimum value for oscillation to occur. The distinction between threshold and nonthreshold processes is that in the former the first term in the power series expansion for the polarization at the output frequency in terms of the electric field at this frequency is linearly proportional to the electric field, whereas for the nonthreshold effect the first term in the expansion is independent of the electric field at the output frequency.

If, as in (5.52), $\tilde{P}_{3\omega}$ does not depend upon $\tilde{E}_{3\omega}$, then from Maxwell's equations, (3.3a) or (3.3b), the electric field $\tilde{E}_{3\omega}$ is evaluated in terms of other variables such as $\tilde{E}_\omega$. This does not involve a threshold. If, as with the laser, we calculate from (3.1) that to first order in the field the polarization varies linearly with the output electric field, this does involve a threshold. When the expression for $\tilde{P}$ is substituted into Maxwell's equations, $\tilde{E}$ cancels on both sides of the equation and an expression is obtained for the propagation constant k for the traveling wave, or for frequency ω for the resonator. To obtain oscillation, it is necessary that k or ω be complex with an appropriate sign to give increasing signal amplitude. With no polarization the signal decays exponentially with distance or time because of the system losses, and for oscillation to occur this requires that the polarization contribute gain that exceeds the magnitude of the system losses. Section 5.6, dealing with Raman oscillation, presents a more detailed discussion of the threshold condition.

With a given set of parameters for the dipole transition the harmonic power may be determined from (5.58). At microwave frequencies this power has been calculated and measured by Fontana, Pantell, and Smith,[7] and at optical frequencies a calculation has been performed by Pao and Rentzepis.[8]

5.5 STARK SHIFT

In this section we shall consider the nonlinear effect that results in the shift of the transition frequency of an electric dipole interaction in the presence of an electromagnetic field. For a static field this is known as the dc Stark shift, and an analogous effect occurs for oscillating fields. If electromagnetic radiation at frequency ω_1 is incident upon a medium, the response to a probing signal at frequency ω is an absorption curve that has been shifted by an amount proportional to the field intensity at frequency ω_1.

5 NONLINEAR EFFECTS IN QUANTIZED MEDIA

To evaluate the magnitude of this shift, let us define the field, polarization, and population difference as

$$E = \frac{\tilde{E}_1}{2} e^{i\omega_1 t} + \frac{\tilde{E}}{2} e^{i\omega t} + \text{c.c.} \tag{5.60}$$

$$P = \frac{\tilde{P}}{2} e^{i\omega t} + \text{c.c.} \tag{5.61}$$

$$N = N_0 + \frac{\tilde{N}_-}{2} e^{i(\omega - \omega_1)t} + \frac{\tilde{N}_+}{2} e^{i(\omega + \omega_1)t} + \cdots \tag{5.62}$$

A component of polarization at frequency ω_1 has not been included in (5.61), because the inclusion of this term does not produce a Stark shift. The only frequency components in N that are relevant to the problem are the terms $e^{i(\omega \pm \omega_1)t}$, and so the other components have not been written out explicitly.

From (5.45), equating coefficients of $e^{i\omega t}$, we have

$$\left(\Omega^2 - \omega^2 + \frac{i2\omega}{T_2}\right)\tilde{P} = -\frac{2\Omega |\mu_{12}|^2}{\hbar} \frac{L}{3}\left(N_0 \tilde{E} + \frac{\tilde{N}_- \tilde{E}_1}{2} + \frac{\tilde{N}_+ \tilde{E}_1^*}{2}\right) \tag{5.63}$$

From (5.46), equating coefficients of $e^{i(\omega + \omega_1)t}$ and $e^{i(\omega - \omega_1)t}$ yields

$$\left(i(\omega - \omega_1) + \frac{1}{T_1}\right)\tilde{N}_- = \frac{i\omega}{\hbar \Omega} \tilde{P} \tilde{E}_1^*$$

$$\left(i(\omega + \omega_1) + \frac{1}{T_1}\right)\tilde{N}_+ = \frac{i\omega}{\hbar \Omega} \tilde{P} \tilde{E}_1 \tag{5.64}$$

A simplification of the mathematics occurs if ω_1 is sufficiently far from ω so that $|\omega - \omega_1| T_1 \gg 1$. The response to the probing signal is of primary interest for $\omega \simeq \Omega$, which means that $\omega_1 \simeq \Omega$ is excluded from the analysis by the condition that $|\omega - \omega_1| T_1 \gg 1$.

The substitution of (5.64) into (5.63) gives the following equation:

$$\left(\Omega^2 - \omega^2 + \frac{2\omega^2 |\mu_{12}|^2 L}{3\hbar^2(\omega^2 - \omega_1^2)} |\tilde{E}_1|^2 + \frac{i2\omega}{T_2}\right)\tilde{P} = -\frac{2\Omega |\mu_{12}|^2 L}{3\hbar} N_0 \tilde{E} \tag{5.65}$$

and an expression for the resonant frequency is obtained by setting the real part of the coefficient of $\tilde{P}$ in (5.65) equal to zero:

$$\omega = \left[\Omega^2 + \frac{2\omega^2 |\mu_{12}|^2 L}{3\hbar^2(\omega^2 - \omega_1^2)} |\tilde{E}_1|^2\right]^{1/2} \tag{5.66}$$

In general, the percent frequency shift is small, so that the square root in (5.66) may be expanded in a power series to give

$$\omega \simeq \Omega + \frac{\Omega |\mu_{12}|^2 L}{3\hbar^2(\Omega^2 - \omega_1^2)} |\tilde{E}_1|^2 \tag{5.67}$$

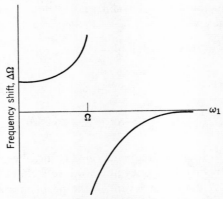

figure 5.6 The shift in the transition frequency, $\Delta\Omega$, for constant incident power density as a function of the frequency of the incident electromagnetic field, ω_1.

In terms of the time-average power per unit area I_1 at ω_1, from (5.38) we have

$$I_1 = \tfrac{1}{2}\eta_1 c\epsilon_0 |\tilde{E}_1|^2 \tag{5.68}$$

where η_1 is the refractive index of the medium at ω_1. Therefore, from (5.67) and (5.68) the frequency shift $\Delta\Omega$ is

$$\Delta\Omega = \frac{2\Omega |\mathbf{\mu}_{12}|^2 L I_1}{3\hbar^2 \eta_1 c\epsilon_0 (\Omega^2 - \omega_1^2)} \tag{5.69}$$

The frequency shift as a function of ω_1 for fixed incident power density is illustrated in Figure 5.6.

Table 5.3 gives examples of frequency shifts for different transition frequencies,

TABLE 5.3 SHIFT IN THE TRANSITION FREQUENCY RESULTING FROM INCIDENT RADIATION AT FREQUENCY ω_1 FOR VARIOUS TRANSITION FREQUENCIES AND POWER DENSITIES

Transition frequency Ω (rad/sec)	Radiation frequency ω_1 (rad/sec)	Power density I_1 (W/cm^2)	Frequency shift $\Delta\Omega$ (rad/sec)
1.4×10^{11} (microwave)	$\omega_1 \ll \Omega$	10^4	0.2×10^8
4×10^{15} (optical)	$\omega_1 \ll \Omega$	10^4	0.7×10^4
4×10^{15}	$\omega_1 \ll \Omega$	10^6	0.7×10^6
1.4×10^{11}	4.1×10^{15}	10^4	-0.23
4×10^{15}	4.1×10^{15} [a]	10^4	-1.4×10^5

[a] Since T_1 is invariably $\gg 10^{-14}$, the condition that $|\omega - \omega_1| T_1 \gg 1$ is satisfied.

several values of ω_1, and various power densities. It is assumed that the refractive index is unity and $|\mu_{12}| = 1$ debye. With the use of a Q-switched laser, measurements of frequency shift have been performed and compared to the predicted value.[9]

5.6 STIMULATED RAMAN OSCILLATIONS

As stated in Section 5.4, stimulated Raman oscillation is a threshold process. The threshold condition for a three-level, two-photon effect as illustrated in Figure 5.7 has been calculated by Javan,[10] and the oscillation condition for a two-level, three-photon process as shown in Figure 5.4c has been determined by Fontana, Pantell, and Smith.[11]

Let us consider, for example, the Raman oscillator illustrated by Figure 5.7.

1. An electric field is assumed of the form

$$E = \frac{\tilde{E}_1}{2} e^{i\omega_1 t} + \frac{\tilde{E}_2}{2} e^{i\omega_2 t} + \text{c.c.}$$

Note that it is necessary to include the driving frequency term at ω_1 and the output frequency at ω_2. The matrix elements of the interaction term in the Hamiltonian $\mathcal{H}'_{ij}$ therefore have frequency components at both ω_1 and ω_2.

2. An equilibrium probability distribution is chosen consistent with the environment. For optical transitions in thermal equilibrium, $\rho_{jj} \simeq \delta_{1j}$, where $|1\rangle$ is the ground state.

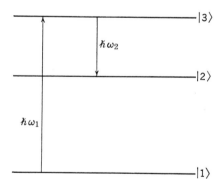

figure 5.7 The energy diagram for the three-level, two-photon Raman oscillator analyzed by Javan.[10] Electric dipole coupling is assumed between states of definite parity, with $|1\rangle$ and $|2\rangle$ having the same parity and $|3\rangle$ having the opposite parity. Energy is introduced into the medium at frequency ω_1, and Raman oscillation occurs at frequency ω_2.

3. The matrix elements of ρ that vary at ω_2 are calculated, and since ρ_{23} has a resonant denominator at ω_2, this is the element of interest. If the perturbation terms of ρ_{23}, as given by (5.7), are written out explicitly, it is found that the component at ω_2 is obtained from $\rho_{23}^{(3)}$ by the following sequence:

$$\mathcal{H}'_{13} \xrightarrow[(5.14)]{} \rho_{13}^{(1)} \xrightarrow[(5.24)]{} \rho_{21}^{(2)} \xrightarrow[(5.28)]{} \rho_{23}^{(3)}$$

where the equation to be used in each step of the evaluation is indicated below the arrow. The proof that the above sequence is the only way to go from the driving field to ρ_{23} by resonant interactions is left to Problem 5.8. If ω_1 equals the transition frequency from $|1\rangle$ to $|3\rangle$, then from (5.14), $\rho_{13}^{(1)}$ is of the form

$$\rho_{13}^{(1)} \sim \mu_{13}\tau_{13}\tilde{E}_1 e^{i\omega_1 t} \tag{5.70}$$

From (5.24), $\rho_{21}^{(2)}$ has a term of the form

$$\rho_{21}^{(2)} \sim \mu_{23}\tau_{12}\tilde{E}_2 e^{i\omega_2 t}(\rho_{13}^{(1)})^* \tag{5.71}$$

From (5.28), $\rho_{23}^{(3)}$ has a term

$$\rho_{23}^{(3)} \sim \mu_{13}\tau_{23}\tilde{E}_1 e^{i\omega_1 t}\rho_{21}^{(2)} \tag{5.72}$$

With the substitution of (5.70) and (5.71) into (5.72) we have

$$\rho_{23}^{(3)} \sim |\mu_{13}|^2 \mu_{23}\tau_{12}\tau_{13}\tau_{23}\tilde{E}_2 |\tilde{E}_1|^2 e^{i\omega_2 t} \tag{5.73}$$

The time dependence of the right-hand side of (5.73) shows that the sequence we followed does lead to a term in ρ_{23} that varies at ω_2.

4. The component of polarization at $e^{i\omega_2 t}$, P_2, is evaluated from

$$P_2 = N_V(\rho_{23}^{(3)}\mu_{32} + \text{c.c.}) \tag{5.74}$$

for N_V identical molecules/vol. From (5.73) and (5.74), P_2 is of the form

$$P_2 \sim N_V |\mu_{13}|^2 |\mu_{23}|^2 \tau_{12}\tau_{13}\tau_{23} |\tilde{E}_1|^2 \tilde{E}_2 e^{i\omega_2 t} \tag{5.75}$$

Defining

$$\tilde{P}_2 = \epsilon_0 \chi \tilde{E}_2$$

where χ is the susceptibility, an exact calculation gives

$$\chi = \frac{iN_V |\mathbf{\mu}_{13}|^2 |\mathbf{\mu}_{23}|^2 \tau_{12}\tau_{13}\tau_{23} |\tilde{E}_1|^2}{36\hbar^3 \epsilon_0} \tag{5.76}$$

5. The term $\tilde{P}_2$ is substituted into Maxwell's equations, (3.3a) and (3.3b), to evaluate $\tilde{E}_2$ in terms of $\tilde{P}_2$. Since $\tilde{P}_2$ is proportional to $\tilde{E}_2$ as shown by (5.75), $\tilde{E}_2$ is canceled on both sides of the equation and the resulting expression involves only $|\tilde{E}_1|^2$, the parameters of the transition, and the loss of the medium and circuit.

As stated in Section 5.4, the threshold for oscillation occurs when the power introduced by the polarization term on the right-hand side of (3.3a) or (3.3b) equals the rate of energy loss by the system. If we consider the example of a resonator at frequency ω_2 with the quantized medium filling the resonator, then P_2 has the same spatial dependence as $\tilde{E}_2$, which, in turn, has the same spatial dependence as the normal mode field. Therefore the equation for E_2 may be written as in (2.55):

$$\ddot{E}_2 + \frac{1}{\tau_c} \dot{E}_2 + \omega_c^2 E_2 = -\frac{1}{\epsilon} \ddot{P}_2 \tag{5.77}$$

where

$$E_2 = \frac{\tilde{E}_2}{2} e^{i\omega_2 t} + \text{c.c.}$$

The loss term is $(1/\tau_c)\dot{E}_2$, so that the threshold condition is

$$\frac{i\omega_2}{\tau_c} \tilde{E}_2 = \frac{\omega_2^2 \epsilon_0}{\epsilon} \chi \tilde{E}_2 \tag{5.78}$$

From (5.76) and (5.78) we have

$$\frac{1}{\tau_c} = \frac{\omega_2 \epsilon_0}{\epsilon} \frac{N_V |\mu_{13}|^2 |\mu_{23}|^2 \tau_{12} \tau_{13} \tau_{23}}{36 \hbar^3} |\tilde{E}_1|^2 \tag{5.79}$$

Equation (5.79) gives the minimum value of $|\tilde{E}_1|^2$ for stimulated Raman oscillation to occur. If $|\tilde{E}_1|^2$ is smaller than the value specified by (5.79), the losses predominate and there is no oscillation, whereas for larger values of $|\tilde{E}_1|^2$ the field at frequency ω_2 grows to some equilibrium value determined by saturation effects. If it is assumed that the radiation at frequency ω_3 makes a single pass through the resonator, then for optical-frequency transitions an incident power density of several W/cm² should be adequate to obtain stimulated Raman oscillation. Rokni and Yatsiv[12] have reported experimental observation of the resonant Raman effect in potassium.

The derivation of (5.79) is based upon the condition that all lineshapes are Lorentzian and that all signals are exactly on resonance. In general, for arbitrary lineshapes and off-resonance signals it is necessary to replace τ_{ij} by $\pi g_{ij}(\omega)$ in the right-hand side of (5.79), where $g_{ij}(\omega)$ is the normalized lineshape function for the $|u_i\rangle \to |u_j\rangle$ transition.

In this section we have considered the *resonant* Raman effect, so called because the incident radiation at frequency ω_1 produces resonant transitions from state $|1\rangle$ to state $|3\rangle$. A portion of Chapter 7 will deal with the nonresonant Raman process, in which state $|3\rangle$ is not present. Some authors use the term "Raman scattering" to refer solely to the scattering of light by optic phonons, the process considered in Chapter 7, but in Section 5.6 we have used a more general definition, to include scattering from electronic states.

REFERENCES

1. V. W. Hughes and L. Grabner, "The Radiofrequency Spectrum of $Rb^{85}F$ and $Rb^{87}F$ by the Electric Resonance Method," *Phys. Rev.*, **79**, 314, 826, 1950.
2. J. Brossel, B. Cagnac, and A. Kastler, "Observations de Résonances Magnétiques à Plusieurs Quanta Sur un Jet d'Atomes de Sodium Orientés Optiquement," *Comptes Rendus*, **237**, 984–986, October 1953.
3. See, for example, R. Pantell, F. Pradere, J. Hanus, M. Schott, and H. Puthoff, "Theoretical and Experimental Values for Two, Three, and Four Photon Absorptions," *J. Chem. Phys.*, **46**, 3507–3511, May 1967.
4. W. Kaiser and C. Garrett, "Two-Photon Excitation in $CaF_2:Eu^{2+}$," *Phys. Rev. Letters*, **7**, 229–231, September 1961.
5. S. Singh and L. Bradley, "Three-Photon Absorption in Naphthalene Crystals by Laser Excitation," **12**, 612–164, June, 1964.
6. See, for example, Chapter 4 of N. Bloembergen, *Nonlinear Optics*, W. A. Benjamin, New York, 1965.
7. J. Fontana, R. Pantell, and R. Smith, "Harmonic Generation Using the Ammonia Inversion Transition," *Proc. IRE*, **50**, 469–470, April 1962.
8. Y. Pao and P. Rentzepis, "Multiphoton Absorption and Optical-Harmonic Generation in Highly Absorbing Molecular Crystals," *J. Chem. Phys.*, **43**, 1281–1286, August 1965.
9. E. Aleksandrov, A. Bonch-Bruevich, N. Kostin, and V. Khodovi, "Frequency Shift of Optical Transition in the Field of a Light Wave," *JETP Lett.* (English translation), **3**, 53–55, January 1966.
10. A. Javan, "Stimulated Raman Effect," *Proc. of the International School of Physics Enrico Fermi Course XXXI*, Academic Press, New York, 1964, pp. 284–305.
11. J. Fontana, R. Pantell, and R. Smith, "Parametric Effects in a Two-Level Electric Dipole System," *J. Appl. Phys.*, **33**, 2085–2086, June 1962.
12. M. Rokni and S. Yatsiv, "Resonance Raman Effect in Free Atoms of Potassium," *Phys. Lett.*, **24A**, 277, February 1967.

PROBLEMS

5.1 Obtain an expression analogous to (5.32) for two-photon absorption where the photon energies differ. (Refer to Figure 5.3a.)

5.2 For a medium with a center of symmetry the electric dipole interaction yields Equations 5.45 and 5.46. Show that only odd harmonics of a driving field are generated from these equations. If a dc field is applied the symmetry is upset. Show that with a dc field it is possible to produce even harmonics.

5.3 From the Bloch equations, (2.83), show that it is possible to generate the second harmonic of a driving field. If the dc magnetic field B_0 is along the z direction, what should be the polarization of the driving field and what will be the polarization of the second harmonic?

5.4 Derive an expression analogous to (5.58) for second harmonic generation as a function of distance for the spin $\frac{1}{2}$ interaction.

5.5 In Section 5.4 on harmonic generation we neglected the depletion of the driving field resulting from the buildup of the third harmonic. For the case of perfect phase matching (that is, $\eta_1 = \eta_3$), from (5.52), (5.55), and an energy conservation condition,

$$|\tilde{E}_\omega|^2 + |\tilde{E}_{3\omega}|^2 = |\tilde{E}_0|^2 = \text{constant}$$

show that

$$|\tilde{E}_{3\omega}|^2 = |\tilde{E}_0|^2 \frac{(Dz)^2}{(Dz)^2 + 1} \tag{P5.1}$$

where

$$D \equiv \frac{\mu_0 |\mu_{12}|^4 L^2 T_2 N_0 |\tilde{E}_0|^2}{24\hbar^3 k}; \qquad k = \frac{\eta_1 \omega}{c}$$

Sketch (P5.1) and compare with the square law relationship obtained from (5.58). Show that (P5.1) reduces to (5.58) for $z \to 0$.

5.6 From first-order perturbation theory, show that the change of an eigenvalue E_i in the presence of a radiation field is given by

$$\Delta E_i = \frac{1}{2\hbar} \sum_k \frac{\omega_{ik}}{\omega_{ik}^2 - \omega^2} |\mathcal{H}'_{ik}|^2 \tag{P5.2}$$

where ω is the frequency of the radiation and $\hbar\omega_{ik} = E_i - E_k$. Show that (P5.2) reduces to the Stark shift equation (5.69) for a two-level system.

5.7 Show that the frequency of maximum absorption for a two-level electric dipole system in the presence of a single, intense signal is

$$\omega \simeq \Omega + \frac{|\mu_{12}|^2 L |\tilde{E}|^2}{12\hbar^2 \Omega} \tag{P5.3}$$

where Ω is the unperturbed transition frequency, and L is the Lorentz correction factor. Compare (P5.3) to (5.67), where the latter equation gives the resonant frequency for a probing signal in the presence of another intense signal.

5.8 Show that the sequence given directly before (5.70),

$$\mathcal{H}'_{31} \to \rho_{13}^{(1)} \to \rho_{21}^{(2)} \to \rho_{23}^{(3)}$$

is the only way to go from the driving field to ρ_{23} by resonant interactions. This may be proved by writing out the terms in the perturbation procedure and assuming that the system is initially in state $|1\rangle$.

FIELD QUANTIZATION
6

6.1 INTRODUCTION

Einstein first introduced the hypothesis that the energy in a radiation field exists in the form of discrete quanta.[1,2] According to this hypothesis the energy at frequency ω is restricted to integral multiples of a basic photon energy unit $\hbar\omega$. This is in contrast to the classical picture of an electromagnetic field in which the energy can assume a continuum of values.

In this chapter we shall consider quantization of the electromagnetic field, in which changes in radiation energy occur in discrete steps in accordance with Einstein's hypothesis. The mathematical formalism for handling discrete energy changes in the radiation field is based on the use of creation and annihilation operators that account for the creation and annihilation of photons. This type of approach is widely used in the description of systems that act as harmonic oscillators. Examples include those systems whose quanta of energy are referred to as photons, phonons, magnons, and plasmons.[3]

The equations describing the interaction between radiation and matter will be derived for the case in which both fields and matter are quantized. A comparison with the results obtained in Chapter 3 on a semiclassical basis reveals that the rate equations describing the interaction in terms of the expectation values of variables of interest are nearly identical for the two approaches. The major difference is the appearance of additional terms in the quantized field approach, which accounts for the process of spontaneous emission. The inclusion of the effects of spontaneous emission is important in the discussion of certain quantum electronics applications. Spontaneous emission determines, for example, the starting conditions for lasers, the spectral properties of radiation under thermal equilibrium conditions, and the noise properties of lasers and parametric amplifiers.

The electromagnetic field is quantized by a procedure identical with that followed in quantizing a mechanical system. First, a Hamiltonian is written in terms of coordinates q_i and conjugate momenta p_i so that Hamilton's equations, as given by

$$\dot{q}_i = \frac{\partial \mathcal{H}}{\partial p_i}, \qquad \dot{p}_i = -\frac{\partial \mathcal{H}}{\partial q_i} \tag{6.1}$$

yield the equations of motion for the system.

Second, to pass from the classical to the quantum mechanical formalism, we consider the variables q_i, p_j to be operators that satisfy the commutation relation

$$[q_i, p_j] = i\hbar \delta_{ij} \tag{6.2}$$

The noncommutativity of a coordinate and its conjugate momentum is a requirement that ranks as a fundamental postulate of quantum mechanics, and it is precisely at this point that Planck's constant enters into quantum mechanics. Equation 6.2 for the commutation relationship between coordinate and momentum applies to both mechanical systems and field variables.

Finally, the solution to Schrödinger's equation must be obtained. When $\mathcal{H}$ is not an explicit function of time, the solutions are given by (1.5):

$$|\varphi_n\rangle = \exp\left(-i\frac{E_n}{\hbar}t\right)|n\rangle$$

where $|n\rangle$ and E_n are the eigenvectors and eigenvalues of the Hamiltonian operator $\mathcal{H}$ that satisfy the time-independent eigenvalue equation

$$\mathcal{H}|n\rangle = E_n|n\rangle \tag{6.3}$$

In Section 6.2 we shall solve (6.3) for the eigenvalues by a method that employs the creation and annihilation operators of the electromagnetic field.

6.2 QUANTIZATION OF CAVITY FIELDS

As an introduction to the quantization procedure, we shall begin with the simple case of an enclosed cavity resonator containing no free charge or current. The walls are assumed to be perfectly conducting and to surround a medium of permittivity ϵ and permeability μ. The field distribution within such a cavity can be expressed as a sum of normal mode solutions

$$\mathbf{E} = -\frac{1}{\sqrt{\epsilon}} \sum_a p_a(t) \mathbf{E}_a(\mathbf{r}), \qquad \mathbf{H} = \frac{1}{\sqrt{\mu}} \sum_a \omega_a q_a(t) \mathbf{H}_a(\mathbf{r}) \tag{6.4}$$

where ϵ is the permittivity of the medium, μ is the permeability, the $\mathbf{E}_a$ and $\mathbf{H}_a$ are the orthonormal functions specified by (2.51) and (2.52).

If we take as the Hamiltonian of the system the energy stored in the resonator, we can show that by identifying $q_a(t)$ and $p_a(t)$ as the coordinates and conjugate momenta, respectively, this choice of Hamiltonian yields the appropriate equations of motion for the system, which are Maxwell's equations. Let

$$\mathscr{H} = \tfrac{1}{2}\int (\epsilon \mathbf{E} \cdot \mathbf{E} + \mu \mathbf{H} \cdot \mathbf{H})\, dV = \tfrac{1}{2}\sum_a (p_a^2 + \omega_a^2 q_a^2) \tag{6.5}$$

where the second expression is obtained by substitution of the normal mode expansion (6.4), followed by application of the orthogonality condition (2.52).

From Hamilton's equations,

$$\dot{q}_a = \frac{\partial \mathscr{H}}{\partial p_a} = p_a$$

Therefore

$$\frac{\partial \mathbf{H}}{\partial t} = \frac{1}{\sqrt{\mu}}\sum_a \omega_a \dot{q}_a \mathbf{H}_a = \frac{1}{\sqrt{\mu}}\sum_a \omega_a p_a \mathbf{H}_a$$

Also, from (2.51),

$$\nabla \times \mathbf{E} = -\frac{1}{\sqrt{\epsilon}}\sum_a p_a \nabla \times \mathbf{E}_a = -\frac{1}{\sqrt{\epsilon}}\sum_a p_a \frac{\omega_a \eta}{c}\mathbf{H}_a$$

From the two preceding equations, we have $\nabla \times \mathbf{E} = -\mu\, \partial \mathbf{H}/\partial t$, which is one of Maxwell's equations, and a similar procedure yields $\nabla \times \mathbf{H} = \epsilon\, \partial \mathbf{E}/\partial t$. The Hamiltonian assumed in (6.5) is therefore an appropriate choice.

The Hamiltonian for the electromagnetic field, (6.5), is identical to the Hamiltonian for a summation of unit mass harmonic oscillators. Therefore, quantization of the electromagnetic field in a cavity reduces to the problem of quantization of a harmonic oscillator. To simplify the notation we shall consider a single normal mode field. With this assumption the summation appearing in the field Hamiltonian (6.5) reduces to a single term, and therefore the Hamiltonian is identical with the unit mass harmonic oscillator Hamiltonian and we shall drop the subscript a on the radian frequency ω,

$$\mathscr{H} = \tfrac{1}{2}(p^2 + \omega^2 q^2) \tag{6.6}$$

We shall now determine the eigenvectors and eigenvalues that satisfy the equation

$$\mathscr{H}\,|n\rangle = E_n\,|n\rangle \tag{6.7}$$

where $\mathscr{H}$ is given by (6.6).

The solution to the eigenvalue equation (6.7) can proceed in more than one way. One standard approach consists of the operator substitution $q_i \equiv q_i$, $p_i \equiv -i\hbar\, \partial/\partial q_i$, followed by solution of the resulting differential equation for the eigenfunctions and eigenvalues.

An alternative approach involves the construction of a set of eigenvectors and the determination of the corresponding eigenvalues by a technique based strictly on the manipulation of the abstract bra and ket formalism. This general approach is called second quantization and is particularly straightforward in its application to the solution of the harmonic oscillator problem.

6 FIELD QUANTIZATION

To begin, it is useful to define the following operators:

$$a = \frac{1}{\sqrt{2\hbar\omega}}(\omega q + ip) \tag{6.8a}$$

$$a^\dagger = \frac{1}{\sqrt{2\hbar\omega}}(\omega q - ip) \tag{6.8b}$$

Since p and q correspond to observables and are therefore Hermitian, it follows that $a^\dagger$ is the adjoint of a. We note that the operators a and $a^\dagger$ are not themselves Hermitian, since $a \neq a^\dagger$, and therefore these operators do not correspond to observables. The operators a and $a^\dagger$, like q and p, do not commute. If we form the commutator of a and $a^\dagger$ by means of (6.8) and the relation $[q, p] = i\hbar$, we find that

$$[a, a^\dagger] = 1 \tag{6.9}$$

In terms of the operators a and $a^\dagger$, q and p can be written

$$q = \left(\frac{\hbar}{2\omega}\right)^{1/2}(a^\dagger + a) \tag{6.10a}$$

$$p = i\left(\frac{\hbar\omega}{2}\right)^{1/2}(a^\dagger - a) \tag{6.10b}$$

Substitution of the above into the Hamiltonian (6.6) followed by application of the commutator relationship (6.9) yields an expression for $\mathcal{H}$ in terms of a and $a^\dagger$,

$$\mathcal{H} = \hbar\omega a^\dagger a + \frac{\hbar\omega}{2} \tag{6.11}$$

Substitution of (6.10) into (6.4) yields expressions for the electric and magnetic fields, also in terms of a and $a^\dagger$, of the form (for a single mode)

$$\mathbf{E} = -i\left(\frac{\hbar\omega}{2\epsilon}\right)^{1/2} \mathbf{E}_a(\mathbf{r})(a^\dagger - a)$$

$$\mathbf{H} = \left(\frac{\hbar\omega}{2\mu}\right)^{1/2} \mathbf{H}_a(\mathbf{r})(a^\dagger + a) \tag{6.12}$$

We are now in a position to determine the eigenvalues of $\mathcal{H}$. As mentioned previously, the procedure that we shall follow differs from the one most often used, for example, in determining the eigenfunctions and eigenvalues for a hydrogenic atom. In that approach one generally solves the eigenvalue equation in the form of a partial differential equation to obtain a set of spatially dependent wave functions. Such wave functions are not directly observable, however, and from the standpoint of matrix mechanics this approach is considered to be simply a mathematical technique for determining the eigenvalues. The alternative approach of second quantization, due originally to Dirac, permits us to solve the eigenvalue problem in terms of the general bra and ket formalism, and we shall follow this approach here. If we operate on the eigenvalue equation (6.7) with a, we obtain

$$a\mathcal{H}\,|n\rangle = aE_n\,|n\rangle$$

Substituting for $\mathcal{H}$ from (6.11), we have

$$a\left(\hbar\omega a^\dagger a + \frac{\hbar\omega_a}{2}\right)|n\rangle = aE_n\,|n\rangle \tag{6.13}$$

However,

$$a(a^\dagger a) = (aa^\dagger)a = (1 + a^\dagger a)a \tag{6.14}$$

where the first equality applies because a and $a^\dagger$ obey the associative law, and the second is obtained by substitution of the commutator relation (6.9). Substituting (6.14) into (6.13), and employing (6.11) we obtain the result

$$\mathcal{H} a\,|n\rangle = (E_n - \hbar\omega)a\,|n\rangle$$

This equality shows that $a\,|n\rangle$ is an eigenvector of $\mathcal{H}$ with eigenvalue $(E_n - \hbar\omega)$. The operator a is termed an *annihilation* operator, because when it operates on an eigenstate it produces another eigenstate whose energy eigenvalue is less than that of the original state. Therefore, operation with a transforms one eigenstate into another with an accompanying annihilation of energy equal to $\hbar\omega$.

Similarly, operation on the eigenvalue equation with $a^\dagger$ yields

$$\mathcal{H} a^\dagger\,|n\rangle = (E_n + \hbar\omega)a^\dagger\,|n\rangle \tag{6.15}$$

Thus the operator $a^\dagger$ transforms one eigenstate into another with an accompanying increase in the energy eigenvalue by an amount $\hbar\omega$. The operator $a^\dagger$ is therefore called a *creation* operator.

If we start with an eigenstate $|n\rangle$, we are able, by the application of a creation or annihilation operator, to generate a new eigenstate removed in energy from the original one by an amount $\hbar\omega$. This process may be repeated, to generate new eigenstates separated from one another by energy $\hbar\omega$.

Now let us determine the range of the eigenvalues E_n to see whether the generation of eigenstates by the application of creation and annihilation operators can be repeated indefinitely in both directions. The answer is negative, since there is a lower limit on the energy eigenvalue because the energy must be greater than or equal to zero. To prove that $E_n \geq 0$, we begin with the eigenvalue equation

$$\mathcal{H}\,|n\rangle = \tfrac{1}{2}(p^2 + \omega^2 q^2)\,|n\rangle = E_n\,|n\rangle$$

Premultiplication of both sides by $\langle n|$ gives

$$\tfrac{1}{2}(\langle n|\,p^2\,|n\rangle + \omega^2\langle n|\,q^2\,|n\rangle) = E_n \tag{6.16}$$

where it is assumed that the $|n\rangle$ make up an orthonormal set. Since p and q are Hermitian, use of the identity operator $I = \sum_m |m\rangle\langle m|$ indicates that

$$\langle n| p^2 |n\rangle = \sum_m \langle n| p |m\rangle\langle m| p |n\rangle$$

$$= \sum_m \langle m| p |n\rangle^* \langle m| p |n\rangle$$

$$= \sum_m |\langle m| p |n\rangle|^2 \geq 0$$

and, similarly, $\langle n| q^2 |n\rangle \geq 0$. Therefore, from (6.16), $E_n \geq 0$.

As a result of the above considerations, we conclude that there is a minimum eigenvalue $E_0 \geq 0$ to which belongs a corresponding eigenstate that we shall label as $|0\rangle$. Since $a|n\rangle$ is an eigenstate with energy $(E_n - \hbar\omega)$, we must have $a|0\rangle = 0$, for if this equality did not hold, there would be an eigenvalue less than E_0. Since $a|0\rangle = 0$, it is also true that

$$\hbar\omega a^\dagger a |0\rangle = 0 \qquad (6.17)$$

From (6.11) we may rewrite (6.17) as

$$\left(\mathcal{H} - \frac{\hbar\omega}{2}\right) |0\rangle = 0$$

Since $\mathcal{H}|0\rangle = E_0|0\rangle$, $E_0 = \hbar\omega/2$. We have thus established that there is a minimum energy eigenstate $|0\rangle$ with a corresponding minimum energy eigenvalue $E_0 = \hbar\omega/2$.

Successive operation by $a^\dagger$ on $|0\rangle$ yields additional energy eigenstates with eigenvalues separated by $\hbar\omega$. This leads to equally spaced energy eigenvalues given by the relationship

$$E_n = (n + \tfrac{1}{2})\hbar\omega \qquad (6.18)$$

where n is a positive integer or zero.

Thus far we have shown that there is a minimum eigenvalue E_0, and an infinite number of equally spaced eigenvalues corresponding to eigenstates obtained by repeated operation with $a^\dagger$. Are there any additional eigenvalues that have not been included by (6.18)? Suppose that there is an energy eigenvalue E_k not included by (6.18) with corresponding eigenstate $|k\rangle$. Successive operation by a on $|k\rangle$ yields eigenvalues given by $(E_k - m\hbar\omega)$, where m is a positive integer. Since the eigenvalue cannot be negative, there will be a minimum energy eigenvalue E_0', to which belongs the eigenvector $|0'\rangle$, where $a|0'\rangle = 0$. From (6.11) we obtain the result as before that $E_0' = E_0 = \hbar\omega/2$. Since E_k is greater than E_0' by an integer times $\hbar\omega$, this implies that E_k is included in (6.18).

The energy level diagram for the quantized harmonic oscillator is shown in Figure 6.1. When the oscillator is excited to its nth eigenstate, it possesses energy $(n + \tfrac{1}{2})\hbar\omega$, which is $n\hbar\omega$ above the lowest energy level. In the case of the

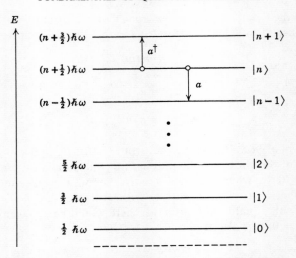

figure 6.1 Energy level diagram for quantized harmonic oscillator.

electromagnetic field we describe the excitation of the normal mode harmonic oscillator to its nth eigenstate by saying that there are n photons in that mode, each of energy $\hbar\omega$. That is, state $|n\rangle$ is a state with n quanta of energy.

State $|0\rangle$, termed the vacuum state, has no quanta but possesses a "zero-point energy" $\hbar\omega/2$. There is nothing sacred about the zero-point energy reference level, since energy measurements are relative rather than absolute, and it is permissible to subtract the quantity $\hbar\omega/2$ from the Hamiltonian (6.11) and thereby shift the energy scale so that the zero-point energy is zero. Such a procedure is allowed because the addition of a constant to a Hamiltonian does not affect either the classical or the quantum mechanical equations of motion. For example, the expectation values of the electric and magnetic fields, their squares, and so on, are not altered by removal of the zero-point energy from the Hamiltonian (see Problem 6.3).

When the creation operator $a^\dagger$ operates on an eigenvector $|n\rangle$, which represents an eigenstate containing n photons, a new eigenvector $|n+1\rangle$ is generated corresponding to an eigenstate containing $n+1$ photons. The creation operator therefore describes the addition of a photon to the field that raises the energy by $\hbar\omega$. Similarly, the annihilation operator corresponds to the annihilation of a photon with a corresponding reduction of $\hbar\omega$ in energy.

In the photon interpretation the operators a and $a^\dagger$ describe the creation and annihilation of photons, hence their names. Combining (6.7), (6.11), and (6.18), we note that the product operator $a^\dagger a$ satisfies the eigenvalue equation

$$a^\dagger a |n\rangle = n |n\rangle \tag{6.19}$$

6 FIELD QUANTIZATION

The operator $a^\dagger a$ corresponds to the number of photons in the mode, in that the expectation value $\langle a^\dagger a \rangle$ for a system in eigenstate $|n\rangle$ is given by n. Therefore, $a^\dagger a$ is referred to as the number operator.

For the general case in which the state of the field is not described by a given eigenstate $|n\rangle$ with n photons, but rather by a statistical mixture of states, the statement "number of photons in the mode" is understood to refer to the expectation value of the number operator,

$$\langle n \rangle = \langle a^\dagger a \rangle = \mathrm{Tr}\,(\rho a^\dagger a) = \sum_{n,m} \rho_{nm}(a^\dagger a)_{mn} = \sum_{n,m} \rho_{nm} n \delta_{mn} = \sum_n \rho_{nn} n \quad (6.20)$$

where the matrix elements $(a^\dagger a)_{mn}$ are obtained by premultiplication of (6.19) by $\langle m|$. The last term in (6.20) is in accord with the standard form for average values, since ρ_{nn} is the probability of occupation of an eigenstate that contains n photons.

To evaluate matrix elements for operators that are expressed in terms of a and $a^\dagger$, it is necessary to determine the normalization constants associated with these operators. Since a operating on $|n\rangle$ produces the eigenstate $|n-1\rangle$, we may write

$$a\,|n\rangle = C_n\,|n-1\rangle$$

where C_n is the proportionality constant. Similarly

$$a^\dagger\,|n\rangle = D_n\,|n+1\rangle$$

where D_n is the proportionality constant for the creation operator. From (6.19), we have

$$a^\dagger a\,|n\rangle = n\,|n\rangle = a^\dagger C_n\,|n-1\rangle = C_n D_{n-1}\,|n\rangle$$

from which it follows that

$$C_n D_{n-1} = n \quad (6.21)$$

In addition, since $a^\dagger$ is the adjoint of a and the $|n\rangle$ are orthonormal, we have

$$\langle n-1|\,a\,|n\rangle = C_n = \langle n|\,a^\dagger\,|n-1\rangle^* = D^*_{n-1} \quad (6.22)$$

Constants C_n and D_n may be chosen to be real without restricting the generality of the solution, and it follows from (6.21) and (6.22) that

$$C_n = \sqrt{n}, \qquad D_n = \sqrt{n+1}$$

The operations with a and $a^\dagger$ are summarized as follows:

$$\begin{aligned}
a\,|0\rangle &= 0 \\
a\,|n\rangle &= \sqrt{n}\,|n-1\rangle, \qquad n \neq 0 \\
a^\dagger\,|n\rangle &= \sqrt{n+1}\,|n+1\rangle \\
a^\dagger a\,|n\rangle &= n\,|n\rangle
\end{aligned} \quad (6.23)$$

TABLE 6.1 MATRIX ELEMENTS OF THE OPERATORS a, $a^\dagger$, AND $a^\dagger a$

$$a \rightarrow \begin{pmatrix} 0 & \sqrt{1} & 0 & 0 & \cdots \\ 0 & 0 & \sqrt{2} & 0 & \cdots \\ 0 & 0 & 0 & \sqrt{3} & \cdots \\ \cdot & \cdot & \cdot & \cdot & \\ \cdot & \cdot & \cdot & \cdot & \\ \cdot & \cdot & \cdot & \cdot & \end{pmatrix}$$

$$a^\dagger \rightarrow \begin{pmatrix} 0 & 0 & 0 & 0 & \cdots \\ \sqrt{1} & 0 & 0 & 0 & \cdots \\ 0 & \sqrt{2} & 0 & 0 & \cdots \\ \cdot & \cdot & \cdot & \cdot & \\ \cdot & \cdot & \cdot & \cdot & \\ \cdot & \cdot & \cdot & \cdot & \end{pmatrix}$$

$$a^\dagger a \rightarrow \begin{pmatrix} 0 & 0 & 0 & 0 & \cdots \\ 0 & 1 & 0 & 0 & \cdots \\ 0 & 0 & 2 & 0 & \cdots \\ \cdot & \cdot & \cdot & \cdot & \\ \cdot & \cdot & \cdot & \cdot & \\ \cdot & \cdot & \cdot & \cdot & \end{pmatrix}$$

Premultiplication of each equation in (6.23) by $\langle m|$ yields the corresponding matrix elements, as given by (6.24) and also presented in Table 6.1:

$$\begin{aligned} a_{mn} &= \sqrt{n}\,\delta_{m,n-1} \\ a^\dagger_{mn} &= \sqrt{n+1}\,\delta_{m,n+1} \\ (a^\dagger a)_{mn} &= n\delta_{mn} \end{aligned} \quad (6.24)$$

With the matrix elements thus determined, we are in a position to apply the results to the solution of problems of interest.

6.2.1 Energy Decay in a Cavity Mode

Let us consider the application of the formalism developed in the preceding section to the simple problem of the growth and decay of electromagnetic energy in a resonator.

The energy operator for the cavity field is given by the Hamiltonian (6.11). The expectation value of the energy is given by

$$\langle \mathcal{H} \rangle = \text{Tr}\,(\rho \mathcal{H})$$

6 FIELD QUANTIZATION

To obtain an equation of motion governing the cavity energy, we follow the procedure developed in Chapter 2 in which a differential equation in the expectation value of a variable of interest is derived. By Table 6.1 we see that all but the diagonal elements of $\mathscr{H} = \hbar\omega(a^\dagger a + \tfrac{1}{2})$ are zero. Therefore the appropriate equation of motion for the first derivative of $\langle\mathscr{H}\rangle$ is given by (1.49),

$$\langle\dot{\mathscr{H}}\rangle + \frac{\langle\mathscr{H}\rangle - \langle\mathscr{H}\rangle^e}{T_1} = 0 \qquad (6.25)$$

where the dot denotes $\partial/\partial t$. It is important at this point to realize that this is the first example in the text in which the derivative-of-expectation-value approach of Section 1.4.5 has been applied to a system with more than two levels (the harmonic oscillator is an infinite-level system). Therefore it is necessary to examine the assumption that all longitudinal relaxation times T_{jk} are equal and given by T_1, which is a requirement that must be satisfied if the formalism is to be applicable. For the case of a resonator with linear wall losses, we observe that the energy in the cavity decays exponentially with a time constant $T_1 = \tau_c$, the cavity lifetime, independently of the level of excitation. Since the taking of the $T_{jk} = T_1$ so that (6.25) holds is consistent with such an exponential decay, we may infer that the equal-relaxation-times assumption is justified in this case.

With the aid of (6.11) and (6.20), we can rewrite (6.25) in the form

$$\langle\dot{\mathscr{H}}\rangle + \frac{\langle\mathscr{H}\rangle}{T_1} = \frac{1}{T_1}\hbar\omega(\langle n\rangle^e + \tfrac{1}{2})$$

The quantity $\langle n\rangle^e$ represents the number of photons in the mode under equilibrium conditions, which, from (6.20), is given by

$$\langle n\rangle^e = \sum_n \rho_{nn}^e n$$

Thus we see that the expectation value of the energy in a cavity normal mode follows an exponential decay of the form e^{-t/T_1}, where T_1 is the cavity relaxation time τ_c associated with loss of energy from the field to its surroundings. The final value to which the energy decays is given by

$$\hbar\omega(\langle n\rangle^e + \tfrac{1}{2})$$

which is the sum of the zero-point energy $\hbar\omega/2$ and the energy corresponding to the number of photons in the mode under equilibrium conditions. Under thermal equilibrium conditions the equilibrium level is established by thermal radiation from surrounding bodies.

6.2.2 Field Equation for a Cavity Mode

A second interesting example of the application of field quantization is the determination of the time dependence of the electric field. From (6.12), the electric field operator for a single cavity normal mode can be expressed in terms of the creation and annihilation operators $a^\dagger$ and a as

$$\mathbf{E} = -i\left(\frac{\hbar\omega}{2\epsilon}\right)^{1/2} \mathbf{E}_a(\mathbf{r})(a^\dagger - a) \qquad (6.26)$$

By Table 6.1 we see that all the diagonal elements of $\mathbf{E}$ are zero. Therefore we apply (1.48) to obtain an equation of motion for the first derivative of the expectation value $\langle \mathbf{E} \rangle = \mathrm{Tr}\,(\rho \mathbf{E})$,

$$\langle \dot{\mathbf{E}} \rangle + \frac{\langle \mathbf{E} \rangle}{T_2} = \frac{1}{i\hbar} \langle [\mathbf{E}, \mathscr{H}] \rangle$$

Evaluation of the commutator $[\mathbf{E}, \mathscr{H}]$ from (6.11) and (6.26) yields

$$[\mathbf{E}, \mathscr{H}] = i\hbar\omega\left(\frac{\hbar\omega}{2\epsilon}\right)^{1/2} \mathbf{E}_a(\mathbf{r})(a + a^\dagger) \qquad (6.27)$$

which we see from (6.12) is proportional to the magnetic field operator $\mathbf{H}$. Let us proceed to the expression for the second derivative of $\langle \mathbf{E} \rangle$, given by (1.51),

$$\langle \ddot{\mathbf{E}} \rangle + \frac{2}{T_2}\langle \dot{\mathbf{E}} \rangle + \frac{1}{T_2^2}\langle \mathbf{E} \rangle = -\frac{1}{\hbar^2}\langle [[\mathbf{E}, \mathscr{H}], \mathscr{H}] \rangle \qquad (6.28)$$

The inside commutator on the right-hand side of (6.28) is given by (6.27). Evaluation of the outside commutator yields

$$[[\mathbf{E}, \mathscr{H}], \mathscr{H}] = \hbar^2\omega^2 \mathbf{E}$$

Substitution of the above into (6.28) yields an expression entirely in terms of $\langle \mathbf{E} \rangle$,

$$\langle \ddot{\mathbf{E}} \rangle + \frac{2}{T_2}\langle \dot{\mathbf{E}} \rangle + \omega^2\langle \mathbf{E} \rangle = 0, \qquad (6.29)$$

where we have taken $\omega \gg 1/T_2$.

Equation 6.29 is the differential equation for the electric field in a cavity of resonant frequency ω with a decay time constant for the field equal to T_2. The solutions to (6.29) are of the form $e^{\pm i\omega t}e^{-t/T_2}$. The decay constant T_2 results from the loss of energy from the field to the surroundings. Since the time constant $T_1 = \tau_c$ in (6.25) for the decay of $\langle \mathscr{H} \rangle$ accounts for the same loss, and since $\langle \mathscr{H} \rangle$ is proportional to the square of the electric field, T_2 is related to T_1 by $1/T_1 = 1/\tau_c = 2/T_2$.

The above examples illustrate how to proceed from the basic results of field quantization to the description of phenomena of interest in terms of the differential equations governing the behavior of the expectation values.

6.3 QUANTIZATION OF PLANE WAVES

In Section 6.2, field quantization was carried out using the orthonormal modes in a cavity as the basis functions. It is sometimes convenient to consider instead an expansion of the field in terms of propagating plane waves.

The wave equation for the electric field $\mathbf{E}$ in a source-free region is

$$\nabla \times (\nabla \times \mathbf{E}) + \frac{\eta^2}{c^2} \frac{\partial^2 \mathbf{E}}{\partial t^2} = 0 \qquad (6.30)$$

where $c = 1/\sqrt{\mu_0 \epsilon_0}$ and $\eta = \sqrt{\mu\epsilon/\mu_0\epsilon_0}$ is the index of refraction. A similar equation can be derived for $\mathbf{H}$.

The solutions to the wave equation for $\mathbf{E}$ can be expressed in terms of plane wave exponentials of the form $\mathbf{1}_\sigma e^{\pm i(\omega_k t - \mathbf{k}\cdot\mathbf{r})}$, where $\mathbf{1}_\sigma$ is a unit vector in the polarization direction and $\mathbf{k}$ is the propagation vector. From the divergence equation $\nabla \cdot \mathbf{D} = 0$, we see that $\mathbf{1}_\sigma \cdot \mathbf{k} = 0$, which shows that the polarization and propagation directions are orthogonal. This is called the transversality condition. Corresponding to each $\mathbf{k}$ there are two independent polarization directions, $\sigma = 1, 2$, in the transverse plane that are orthogonal to each other and orthogonal to $\mathbf{k}$.

Substituting the exponential solution $\mathbf{1}_\sigma e^{\pm i(\omega t - \mathbf{k}\cdot\mathbf{r})}$ into the wave equation (6.30) and applying the identity

$$\nabla \times (\mathbf{1}_\sigma e^{\pm i\mathbf{k}\cdot\mathbf{r}}) = \mp i(\mathbf{1}_\sigma \times \mathbf{k}) e^{\pm i\mathbf{k}\cdot\mathbf{r}}$$

we find that the $\mathbf{k}$ and ω_k are related by

$$k^2 = \frac{\omega_k^2 \eta^2}{c^2}$$

where $k^2 \equiv \mathbf{k} \cdot \mathbf{k}$.

It is convenient to require that the electric and magnetic fields satisfy periodic boundary conditions on opposite faces of a cube with linear dimensions L and volume $V = L^3$. This permits us to express the fields in terms of modes that are discrete and orthogonal. As applied to the x coordinate,

$$e^{i[k_x x + k_y y + k_z z]} = e^{i[k_x(x+L) + k_y y + k_z z]}$$

That is, periodic boundary conditions impose the requirement that translation by a distance L parallel to the edge of the cube does not change the value of the exponential solutions. The requirement is satisfied provided that

$$k_x = \frac{2\pi n_x}{L}, \qquad k_y = \frac{2\pi n_y}{L}, \qquad k_z = \frac{2\pi n_z}{L} \qquad (6.31)$$

where n_x, n_y, and n_z are integers from $-\infty$ to $+\infty$. The propagation constants are thus restricted to a discrete set of values.

If we choose exponential solutions in the form

$$\mathbf{U}_{\mathbf{k}\sigma}(\mathbf{r}) = \mathbf{1}_\sigma \frac{e^{i\mathbf{k}\cdot\mathbf{r}}}{\sqrt{V}} \tag{6.32}$$

then the $\mathbf{U}_{\mathbf{k}\sigma}(\mathbf{r})$ functions satisfy the orthonormality condition

$$\int_V \mathbf{U}_{\mathbf{k}\sigma}^*(\mathbf{r}) \cdot \mathbf{U}_{\mathbf{k}'\sigma'}(\mathbf{r}) \, dV = \delta_{\mathbf{k}\mathbf{k}'}\delta_{\sigma\sigma'} \tag{6.33}$$

where the integral is over the volume of the cube, V.

The electric field may be written as the summation

$$\mathbf{E} = i \sum_{\mathbf{k},\sigma} \left(\frac{\hbar\omega_k}{2\epsilon V}\right)^{1/2} \mathbf{1}_\sigma [a_{\mathbf{k}\sigma} e^{i\mathbf{k}\cdot\mathbf{r}} + b_{\mathbf{k}\sigma} e^{-i\mathbf{k}\cdot\mathbf{r}}] \tag{6.34}$$

where for the classical field $a_{\mathbf{k}\sigma}$ and $b_{\mathbf{k}\sigma}$ are time-varying expansion coefficients, and the constant in front of the bracket is a normalization factor that makes $\mathscr{H}$ for a plane wave expressible in the same form as the $\mathscr{H}$ for fields in a resonator. Classically, $\mathbf{E}$ is a real variable so that $\mathbf{E} = \mathbf{E}^*$, which from (6.34) implies that $b_{\mathbf{k}\sigma} = -a_{\mathbf{k}\sigma}^*$. In general, when we pass to the quantum mechanical picture, time-varying quantities ($a_{\mathbf{k}\sigma}$ and $b_{\mathbf{k}\sigma}$ in this case) become operators. Since $\mathbf{E}$ is an observable, the operator corresponding to $\mathbf{E}$ is Hermitian, and so we have $\mathbf{E} = \mathbf{E}^\dagger$, which from (6.34) indicates that $b_{\mathbf{k}\sigma} = -a_{\mathbf{k}\sigma}^\dagger$. Therefore, for both the classical and quantum mechanical cases we may write

$$\mathbf{E} = i \sum_{\mathbf{k},\sigma} \left(\frac{\hbar\omega_k}{2\epsilon V}\right)^{1/2} \mathbf{1}_\sigma [a_{\mathbf{k}\sigma} e^{i\mathbf{k}\cdot\mathbf{r}} - a_{\mathbf{k}\sigma}^\dagger e^{-i\mathbf{k}\cdot\mathbf{r}}] \tag{6.35}$$

For plane waves the magnetic field is orthogonal to both $\mathbf{k}$ and $\mathbf{1}_\sigma$ and the ratio of $\mathbf{E}$ to $\mathbf{H}$ is given by $\sqrt{\mu/\epsilon}$. Therefore, for the magnetic field we write the summation

$$\mathbf{H} = i \sum_{\mathbf{k},\sigma} \left(\frac{\hbar\omega_k}{2\mu V}\right)^{1/2} (\mathbf{1}_\mathbf{k} \times \mathbf{1}_\sigma)[a_{\mathbf{k}\sigma} e^{i\mathbf{k}\cdot\mathbf{r}} - a_{\mathbf{k}\sigma}^\dagger e^{-i\mathbf{k}\cdot\mathbf{r}}] \tag{6.36}$$

where $\mathbf{1}_\mathbf{k}$ is a unit vector in the $\mathbf{k}$ direction.

With the Hamiltonian chosen to be the electromagnetic energy within the normalization volume V,

$$\mathscr{H} = \frac{1}{2} \int_V (\epsilon \mathbf{E} \cdot \mathbf{E} + \mu \mathbf{H} \cdot \mathbf{H}) \, dV$$

we have, from (6.32)–(6.36),

$$\mathscr{H} = \frac{1}{2} \sum_{\mathbf{k},\sigma} \hbar\omega_k (a_{\mathbf{k}\sigma} a_{\mathbf{k}\sigma}^\dagger + a_{\mathbf{k}\sigma}^\dagger a_{\mathbf{k}\sigma}) \tag{6.37}$$

6 FIELD QUANTIZATION

If the above is to be a valid Hamiltonian, $a_{k\sigma}$ must be expressed in terms of a coordinate q and momentum p so that Hamilton's equations give Maxwell's equations. With $a_{k\sigma}$ defined as in (6.8), $\mathcal{H}$ transforms into the harmonic oscillator form

$$\mathcal{H} = \frac{1}{2} \sum_{k,\sigma} (p_{k\sigma}^2 + \omega_k^2 q_{k\sigma}^2)$$

By proceeding in a manner analogous to the resonator case, we can show that Hamilton's equations yield Maxwell's equations. Then, to quantize the fields we postulate that the $q_{k\sigma}, p_{k\sigma}$ are operators that satisfy the commutation relations

$$[q_{k\sigma}, p_{k'\sigma'}] = i\hbar\, \delta_{kk'}\, \delta_{\sigma\sigma'}$$

The variables $a_{k\sigma}, a_{k\sigma}^\dagger$ are then interpreted as creation and annihilation operators satisfying

$$[a_{k\sigma}, a_{k'\sigma'}^\dagger] = \delta_{kk'}\delta_{\sigma\sigma'}$$

The Hamiltonian (6.37) then becomes

$$\mathcal{H} = \sum_{k,\sigma} \hbar\omega_k (a_{k\sigma}^\dagger a_{k\sigma} + \tfrac{1}{2})$$

and the expressions (6.35) and (6.36) are taken as quantized field expressions in terms of the operators $a_{k\sigma}, a_{k\sigma}^\dagger$.

With the field Hamiltonian reduced to the form above, the solution to the eigenvalue problem and the determination of the matrix elements of a and $a^\dagger$ follow exactly as in the preceding section for the resonator case. The matrix elements are therefore as given in (6.24) and Table 6.1.

If the volume becomes very large, so that $L \to \infty$, the summations in the expressions of this section are replaced by integrals,

$$\frac{1}{\sqrt{V}} \sum_k \to \frac{1}{(2\pi)^{3/2}} \int d\mathbf{k}$$

and the orthogonality relationship (6.33) becomes

$$\frac{1}{(2\pi)^{3/2}} \int \mathbf{U}_{k\sigma}^*(\mathbf{r}) \cdot \mathbf{U}_{k'\sigma'}(\mathbf{r})\, dV = \delta(\mathbf{k'} - \mathbf{k})\, \delta_{\sigma\sigma'}$$

where $\delta(\mathbf{k'} - \mathbf{k})$ is the Dirac delta function.

6.4 INTERACTION OF RADIATION WITH MATTER WHERE BOTH FIELDS AND MATTER ARE QUANTIZED

In Chapters 2 and 3 we examined the interaction between matter and radiation from the semiclassical standpoint in which the medium was quantized but the fields were assumed classical. The semiclassical approach is, in general, quite adequate for the types of problems we have considered. For problems involving spontaneous emission and noise considerations, however, it is necessary to employ a fully quantized approach wherein both matter and fields are quantized.

In this section we shall develop the fully quantized approach to the resonant interaction between radiation and a two-level molecular system coupled by an electric dipole transition. Since this problem is identical with that considered from the semiclassical viewpoint in Chapters 2 and 3, this study will permit us to note the differences between the two approaches. It will also serve as a model for the procedure to be followed in those cases where the fully quantized approach is called for, as in, for example, consideration of spontaneous emission effects and quantum noise.

The system under consideration is assumed to consist of $\mathcal{N}$ atoms or molecules that fill a resonator. The atoms or molecules are assumed to possess a pair of nondegenerate energy eigenstates of opposite parity, $|u_1\rangle$ and $|u_2\rangle$, separated in energy by $E_2 - E_1 = \hbar\Omega$ and coupled by an electric dipole transition. The resonator is assumed to support a single mode of resonant frequency $\omega_a \approx \Omega$. The energy level diagrams for the medium and field are shown in Figure 6.2. The Hamiltonian is given by $\mathcal{H} = \mathcal{H}_0 + \mathcal{H}'$. The term $\mathcal{H}_0$, which is the Hamiltonian excluding the molecule-field interaction, consists of two terms

$$\mathcal{H}_0 = \mathcal{H}_{0m} + \mathcal{H}_{0f}$$

corresponding to the energy of the molecule in the absence of a radiation field and the energy in the field in the absence of a molecule. We shall choose as basis functions the eigenvectors $|u_i, n\rangle = |u_i\rangle |n\rangle$, which satisfy the eigenvalue equation

$$(\mathcal{H}_{0m} + \mathcal{H}_{0f})|u_i, n\rangle = [E_i + (n + \tfrac{1}{2})\hbar\omega]|u_i, n\rangle \quad i = 1, 2; \quad n = 0, 1, 2 \cdots \tag{6.38}$$

where $|u_i\rangle$ is an eigenstate of the medium.

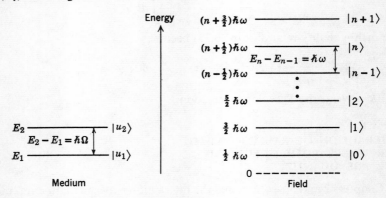

figure 6.2 Energy level diagrams for a two-level quantized medium and quantized electromagnetic field.

6 FIELD QUANTIZATION

The interaction term $\mathcal{H}'$ corresponding to the electric dipole transition is given by

$$\mathcal{H}' = -\boldsymbol{\mu} \cdot \mathbf{E} = -\mu_\alpha E_\alpha$$

where summation over repeated subscripts is understood.

In the semiclassical approach we recall that $\boldsymbol{\mu}$ was an operator whereas $\mathbf{E}$ was a classical variable. In the fully quantized approach, however, both are operators. The matrix elements in this case are given by

$$\mathcal{H}'_{i,n;j,m} = -\langle u_i | \langle n | \mu_\alpha E_\alpha | u_j \rangle | m \rangle = -(\mu_\alpha)_{ij}(E_\alpha)_{nm} \qquad (6.39)$$

Since the states $|u_1\rangle$ and $|u_2\rangle$ are assumed to be of opposite parity, the only nonzero matrix elements of the dipole operator $(\mu_\alpha)_{ij}$ are $(\mu_\alpha)_{12} = (\mu_\alpha)_{21}^* \equiv \mu_\alpha$. For the field a single normal mode is assumed. The matrix elements $\mathbf{E}_{nm}$ are given by (6.24) and (6.26):

$$(E_\alpha)_{nm} = -i\left(\frac{\hbar\omega}{2\epsilon}\right)^{1/2} (E_a)_\alpha (\sqrt{m+1}\,\delta_{n,m+1} - \sqrt{m}\,\delta_{n,m-1})$$

where $\mathbf{E}_a$ is the normal mode field. Therefore, we find that the matrix elements for the interaction Hamiltonian (6.39) are

$$\mathcal{H}'_{1,n;2,n-1} = i\left(\frac{\hbar\omega}{2\epsilon}\right)^{1/2} \mu_\alpha (E_a)_\alpha \sqrt{n}$$

$$\mathcal{H}'_{2,n;1,n-1} = i\left(\frac{\hbar\omega}{2\epsilon}\right)^{1/2} \mu_\alpha^* (E_a)_\alpha \sqrt{n}$$

$$\mathcal{H}'_{1,n;2,n+1} = -i\left(\frac{\hbar\omega}{2\epsilon}\right)^{1/2} \mu_\alpha (E_a)_\alpha \sqrt{n+1} \qquad (6.40)$$

$$\mathcal{H}'_{2,n;1,n+1} = -i\left(\frac{\hbar\omega}{2\epsilon}\right)^{1/2} \mu_\alpha (E_a)_\alpha \sqrt{n+1}$$

6.4.1 Equations of Motion for the Density Operator

The equations of motion for the system are obtained from the density matrix equations (1.30) and (1.36). The subscripting for the elements of the density matrix is slightly more complicated for the fully quantized case than for the semiclassical case because both the state of the medium and the state of the field must be specified. A typical element is of the form $\langle u_i, n | \rho | u_j, m \rangle = \langle u_i | \langle n | \rho | u_j \rangle | m \rangle = \rho_{i,n;j,m}$. The diagonal element $\rho_{i,n;i,n}$ gives the joint probability that the medium is in state $|u_i\rangle$ and that the field is in state $|n\rangle$. From the diagonal elements the total probability that the molecule is in state $|u_i\rangle$ is calculated as

$$\rho_{ii} = \sum_n \rho_{i,n;i,n} \qquad (6.41)$$

Similarly, the total probability that the field is in state $|n\rangle$ is determined by

$$\rho_{nn} = \sum_i \rho_{i,n;i,n} \tag{6.42}$$

The density matrix equations for the system consisting of molecule plus field are generalized from (1.30) and (1.36) and take the form

$$i\hbar\dot{\rho}_{i,n;j,m} - \hbar\omega_{i,n;j,m}\rho_{i,n;j,m} + \frac{i\hbar}{\tau_{i,n;j,m}}\rho_{i,n;j,m} = [\mathscr{H}', \rho]_{i,n;j,m}$$

$$i = 1, 2 \quad j = 1, 2$$
$$n, m = 0, 1, 2, \ldots \tag{6.43}$$
$$i, n \neq j, m$$

and

$$i\hbar\dot{\rho}_{i,n;i,n} + \frac{i\hbar}{T_{i,n;i,n}}(\rho_{i,n;i,n} - \rho^e_{i,n;i,n}) = [\mathscr{H}', \rho]_{i,n;i,n} \tag{6.44}$$

$$i = 1, 2$$
$$n = 0, 1, 2, \ldots$$

where

$$\hbar\omega_{i,n;j,m} = E_i - E_j + (n-m)\hbar\omega \tag{6.45}$$

The appropriate values for the relaxation time constants $\tau_{i,n;j,m}$ and $T_{i,n;i,n}$ will be apparent once equations for the variables of interest are derived.

In most of the text we have not worked directly with the density matrix equations (1.30) and (1.36), but instead have used the time-derivative expectation value equations of Section 1.4.5. The same approach could be used here. However, with product field-matter eigenstates it is convenient to work with the basic density matrix equations, and therefore we shall do so.

6.4.2 Rate Equations

We shall now derive equations of motion for the expectation values of variables of interest. One variable of interest is the number of photons in the mode, $\langle n \rangle$, given by (6.20):

$$\langle n \rangle = \langle a^\dagger a \rangle = \mathrm{Tr}\,(\rho a^\dagger a) = \sum_{i,n;j,m} \rho_{i,n;j,m}(a^\dagger a)_{j,m;i,n} \tag{6.46}$$

The matrix elements of $a^\dagger a$ are determined with the aid of (6.24).

$$(a^\dagger a)_{j,m;i,n} = \langle u_j|\langle m| a^\dagger a |u_i\rangle|n\rangle = (a^\dagger a)_{mn}\delta_{ij} = n\delta_{mn}\delta_{ij} \tag{6.47}$$

Combination of (6.46) and (6.47) then yields

$$\langle n \rangle = \sum_n n\rho_{nn} \tag{6.48}$$

where ρ_{nn} is given by (6.42).

6 FIELD QUANTIZATION

To obtain an equation of motion, we begin by differentiating (6.48) with respect to time

$$\frac{\partial}{\partial t}\langle n\rangle = \sum_n n\dot{\rho}_{nn} = \sum_n n \sum_i \dot{\rho}_{i,n;i,n} = \sum_n n(\dot{\rho}_{1,n;1,n} + \dot{\rho}_{2,n;2,n}) \quad (6.49)$$

In order to evaluate the right-hand side of (6.49), we must carry out some rather lengthy substitutions from the density matrix equations (6.43) and (6.44). This material has been relegated to Appendix 7. The result is

$$\frac{\partial}{\partial t}\langle n\rangle + \frac{\langle n\rangle - \langle n\rangle^e}{\tau_c} = -G\sum_n n(\rho_{1,n;1,n} - \rho_{2,n;2,n}) + G\rho_{22} \quad (6.50)$$

where

$$G = \frac{\pi\omega}{\hbar\epsilon}(\mu_\alpha\mu_\beta^*)(E_a)_\alpha(E_a)_\beta g_L(\omega, \Omega) \quad (6.51)$$

and $g_L(\omega, \Omega)$ is the Lorentzian lineshape function.

In a similar manner an expression may be derived for the difference in occupation probabilities for the upper and lower states, $(\rho_{11} - \rho_{22})$, which from (6.41) is given by

$$\rho_{11} - \rho_{22} = \sum_n (\rho_{1,n;1,n} - \rho_{2,n;2,n}) \quad (6.52)$$

Differentiation of (6.52) with respect to time followed by substitution from (A7.4)–(A7.7) in Appendix 7 and (6.40) yields

$$\frac{\partial}{\partial t}(\rho_{11} - \rho_{22}) + \frac{(\rho_{11} - \rho_{22}) - (\rho_{11} - \rho_{22})^e}{T_1} = -2G\sum_n n(\rho_{1,n;1,n} - \rho_{2,n;2,n}) + 2G\rho_{22} \quad (6.53)$$

where the identification of the time constant on the left-hand side of (6.53) as T_1 is in agreement with the recognition of T_1 as the longitudinal relaxation time of the medium associated with those processes that cause the occupation probabilities ρ_{11} and ρ_{22} to relax toward their equilibrium values.

Equations 6.50 and 6.53 are of the form of rate equations that describe the time evolution of, respectively, the number of photons in the mode and the difference in occupation probabilities of the upper and lower states of the molecule. The rate equations describe how these quantities change as a result of interactions among the field, the molecule, and the surrounding medium.

The significance of the terms in the right-hand side of (6.50) and (6.53) is illustrated by considering two separate cases. First, let us assume that the molecule is in the lower state so that $\rho_{11} = 1$ and $\rho_{2,n;2,n} = 0$. With the aid of (6.42) and (6.48) we see that (6.50) may be written

$$\frac{\partial}{\partial t}\langle n\rangle + \frac{\langle n\rangle - \langle n\rangle^e}{\tau_c} = -G\langle n\rangle \quad (6.54a)$$

Thus when the molecule is in the lower state, the number of photons in the cavity mode decreases at a rate $G\langle n\rangle$ because of interaction with the molecule. This constitutes the process of absorption, and it is seen to progress at a rate proportional to the number of photons present.

If, on the other hand, the molecule is in the upper state so that $\rho_{1,n;1,n} = 0$ and $\rho_{22} = 1$, (6.50) takes the form

$$\frac{\partial}{\partial t}\langle n\rangle + \frac{\langle n\rangle - \langle n\rangle^e}{\tau_c} = G\langle n\rangle + G\rho_{22} = G(\langle n\rangle + 1) \qquad (6.54b)$$

Here we see that when the molecule is in the upper state, the number of photons in the cavity increases as a result of interaction with the molecule. The increase in the number of photons results from two processes. One is proportional to the number of photons in the mode and proceeds at a rate $G\langle n\rangle$. This process, the inverse of the absorption process, is called induced or stimulated emission. It is responsible for the buildup of fields in the case of a laser. The second term on the right-hand side of (6.54b) accounts for the emission of photons into the cavity mode because of a molecule in the upper state by a process that is independent of the number of photons in the mode. This process occurs even in the absence of fields and is therefore known as spontaneous emission. Since the term is of the same form as that for stimulated emission with $\langle n\rangle = 1$, it is sometimes referred to as the "extra photon" emission process.

When the molecule is not completely in either the lower or the upper state, the separation of the interaction into absorption and emission processes proportional to $\langle n\rangle$ is not always possible. The cause of the difficulty is the possibility of what are known as statistically dependent states. The density matrix element $\rho_{i,n;i,n}$ is the probability that the molecule is in state $|u_i\rangle$ and that the field is in state $|n\rangle$. If the probability of finding the molecule in state $|u_i\rangle$ is influenced by which field state $|n\rangle$ the system is in, then

$$\rho_{i,n;i,n} \neq \rho_{ii}\rho_{nn}$$

where ρ_{ii} is the probability that the system is in state $|u_i\rangle$ and ρ_{nn} is the probability that the system is in state $|n\rangle$. When the above inequality applies, the molecular and field states are said to be statistically dependent, and the rate equations cannot be written in terms of ρ_{11}, ρ_{22}, and $\langle n\rangle$. However, if

$$\rho_{i,n;i,n} = \rho_{ii}\rho_{nn}, \qquad (6.55)$$

the molecular and field states are said to be statistically independent. In what follows we shall assume that the states are statistically independent and therefore that (6.55) holds. Mashkevich[4] has considered the conditions under which (6.55) could be violated, and he found that statistical dependence can be significant only for high-power, Q-switched lasers.

With (6.55) assumed, we find with the aid of (6.41) and (6.42) that the rate equations (6.50) and (6.53) reduce to

$$\frac{\partial}{\partial t}\langle n\rangle + \frac{\langle n\rangle - \langle n\rangle^e}{\tau_c} = G(\langle n\rangle + 1)\rho_{22} - G\langle n\rangle\rho_{11} \qquad (6.56)$$

$$\frac{\partial}{\partial t}(\rho_{11} - \rho_{22}) + \frac{(\rho_{11} - \rho_{22}) - (\rho_{11} - \rho_{22})^e}{T_1} = 2G(\langle n\rangle + 1)\rho_{22} - 2G\langle n\rangle\rho_{11} \qquad (6.57)$$

The first and second terms on the right-hand side of (6.56) and (6.57) can be identified as the emission and absorption processes, respectively.

The coefficient G is the coupling constant given by (6.51). For an isotropic medium, G reduces to

$$G = \frac{\pi\omega}{\hbar\epsilon}\frac{|\mu_{12}|^2}{3} g_L(\omega, \Omega) E_a^2 \qquad (6.58)$$

Consider a cavity for which the normal mode function is proportional to $\cos kx$, that is, $E_a \propto \cos kx$. The coupling constant G given by (6.58) is then proportional to $\cos^2 kx$, and therefore has nodes and crests. For an atom located in the region of a node, the coupling constant G is small and, therefore, according to (6.57), transitions take place at a very low rate. It is tempting for one to describe this by saying that there are more photons at the crests of the normal mode field function than at the nodes, and therefore transitions take place at higher rates at the crests. This is not correct, however. Examination of (6.56) and consideration of its physical significance shows that $\langle n\rangle$ is a numeric which stands for the number of photons in the mode but does not vary with position. The rate of change of photons due to an atom located at a particular position depends on the coupling constant G which is a function of coordinate. The variable $\langle n\rangle$ keeps track of the quantized level of excitation of the normal-mode harmonic oscillator.

Let us also consider a proof based on the uncertainty principle that it is incorrect to think of photons as being located in certain regions of space smaller than a wavelength. Let Δx and Δp be the uncertainties in position and momentum. Since $p = \hbar k = 2\pi\hbar/\lambda$ and $\Delta x \Delta p \geq \hbar/2$, we have that $\Delta x \geq \lambda^2/4\pi \Delta\lambda$. The uncertainty in the wavelength $\Delta\lambda$ will always be less than λ. From this we observe that it is impossible to localize the position of a photon within a region Δx which is smaller than a dimension on the order of the wavelength of the photon.

The rate equations (6.56) and (6.57) were obtained for a single molecule. Since the coupling constant G is a function of the normal mode field E_a, and E_a is a function of the coordinates, then the population difference also depends on the coordinates. Let us now consider the case where there are many molecules inside the volume V. The dependence of $\langle n\rangle$ given by (6.56) on all the molecules is obtained by integration of the right hand side of (6.56) over the resonator volume.

If we neglect the spatial variations of E_a and ρ_{ii} and consider only spatial average values, then for $\mathcal{N}$ molecules uniformly distributed throughout the cavity of volume V we have from (6.56)–(6.58),

$$\frac{\partial}{\partial t}\langle n\rangle + \frac{\langle n\rangle - \langle n\rangle^e}{\tau_c} = K(\langle n\rangle + 1)\mathcal{N}_2 - K\langle n\rangle \mathcal{N}_1 \tag{6.59}$$

$$\frac{\partial}{\partial t}(\mathcal{N}_1 - \mathcal{N}_2) + \frac{(\mathcal{N}_1 - \mathcal{N}_2) - (\mathcal{N}_1 - \mathcal{N}_2)^e}{T_1} = 2K(\langle n\rangle + 1)\mathcal{N}_2 - 2K\langle n\rangle \mathcal{N}_1 \tag{6.60}$$

where $\mathcal{N}_1 = \mathcal{N}\rho_{11}$ and $\mathcal{N}_2 = \mathcal{N}\rho_{22}$ are the total number of molecules in the lower and upper states, respectively. The coupling constant K is given by

$$K = \frac{\pi\omega}{\hbar\epsilon} \frac{|\mathbf{\mu}_{12}|^2}{3} g_L(\omega, \Omega) \frac{1}{V}. \tag{6.61}$$

In the evaluation of K we have used the fact that the normalization condition, $\int E_a^2\, dV = 1$, reduces to $E_a^2 V = 1$ when spatial variations in the normal mode field are neglected.

Equations (6.59) and (6.60) are also valid including spatial variations if $\overline{G\rho_{ii}} = \bar{G}\,\bar{\rho}_{ii}$ is a good approximation, where the overbar indicates a spatial average. Otherwise, integration of the right-hand side of (6.56) over the resonator volume is required.

The rate equations (6.59) and (6.60) describe the interaction between a cavity field and a two-level electric dipole transition as derived on the basis of quantization of both the field and the medium. The first term on the right-hand side of (6.59) represents stimulated and spontaneous emission of photons into the cavity mode resulting from $\mathcal{N}_2$ molecules in the upper state, and the second term corresponds to absorption of photons by $\mathcal{N}_1$ molecules in the lower state. Equation 6.60 exhibits the change in population difference as a result of such transitions. The factor 2, which appears in (6.60) but not in (6.59), indicates that for every absorption or emission process that results in the transition of a molecule from one state to another, the population difference changes by 2.

If the populations are inverted by a pumping process that maintains $(\mathcal{N}_2 - \mathcal{N}_1)^e > 0$, then (6.59) and (6.60) are applicable to the description of a laser. Since the equations for this case were derived previously (Section 3.4 and Chapter 4) on the basis of a semiclassical analysis, we can now compare the results of the two approaches, semiclassical and fully quantized, and note the differences between them.

The equations resulting from the semiclassical development are given by (3.78) and (3.79). In comparing (3.78) and (3.79) with (6.59) and (6.60), we note that the semiclassical variables N and φ correspond, respectively, to the inverted population difference *per unit volume* and energy/$\hbar\omega$ *per unit volume*, and must therefore be multiplied by the cavity volume V before a comparison can be made. When this is done, we find that the major difference is the following: *When field quantization is taken into account, the transition rate for emission processes, rather than being proportional to $\langle n \rangle$ as in the semiclassical case, is instead proportional to $(\langle n \rangle + 1)$, the extra factor of unity corresponding to spontaneous emission.* Therefore we conclude that equations derived on a semiclassical basis can be corrected to take into account the results of field quantization by the replacement for emission processes

$$\langle n \rangle \rightarrow \langle n \rangle + 1$$

where the additional factor of unity corresponds to the inclusion of spontaneous emission effects.

The semiclassical approach used in previous chapters, according to which the field is not quantized, is valid when $\langle n \rangle \gg 1$. In a resonator this means that the average energy stored should be much greater than $\hbar\omega$. At optical frequencies ($\lambda = 5000$ Å) the semiclassical approach is a good approximation when the energy is much greater than 4×10^{-19} J. Unless one is interested in laser operation very close to threshold or in the noise properties of a laser, the semiclassical analysis presented in Chapter 4 is an excellent approximation.

6.4.3 Mode Density

In the previous sections we considered only the interaction between a molecular transition and a single-mode radiation field. Now these results must be extended to the transition that couples to many radiation modes simultaneously. Because photons are emitted spontaneously into all possible modes the spontaneous emission term must involve a multimode analysis. Furthermore, when the molecular transition interacts with radiation that has a broadband frequency spectrum, the absorption and stimulated emission terms may also involve many modes.

We shall consider only the case where the cavity dimensions are large compared with the wavelength of interest. This assumption almost always is satisfied at optical frequencies. When the cavity is large compared to the wavelength, the normal modes of the cavity become closely spaced with respect to the optical frequency and are relatively independent of the precise shape of the cavity.

In order to examine interaction with many modes, it is necessary to evaluate the mode density, which is the number of modes per unit volume per unit frequency interval. Let us take as our cavity a cube of linear dimensions L and volume $V = L^3$. The cavity standing-wave normal mode solutions each consist of a triple product of sine and cosine terms with arguments $k_x x$, $k_y y$, and $k_z z$. (We shall consider the traveling-wave representation later.)

The boundary conditions at the perfectly conducting walls impose the requirement

$$k_x = \frac{n_x \pi}{L}, \quad k_y = \frac{n_y \pi}{L}, \quad k_z = \frac{n_z \pi}{L}$$

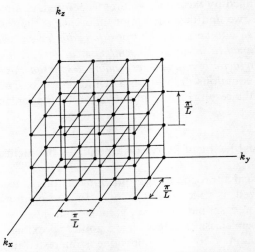

figure 6.3 Distribution of points in **k** space that correspond to allowed cavity standing-wave normal mode solutions.

where n_x, n_y, and n_z take on integer values from 0 to ∞. The allowed values of $\mathbf{k} = \mathbf{1}_x k_x + \mathbf{1}_y k_y + \mathbf{1}_z k_z$ can be represented by a series of points in **k** space, as shown in Figure 6.3. Each point corresponds to an allowed **k** value designated by the integer triplet (n_x, n_y, n_z). We shall use the notation $k \equiv |\mathbf{k}| = (k_x^2 + k_y^2 + k_z^2)^{1/2}$.

Associated with each k value, that is, with each point, there are two cavity modes, a $TE_{n_x n_y n_z}$ and a $TM_{n_x n_y n_z}$ mode. Therefore there are two modes in a **k**-space volume $(\pi/L)^3$ leading to a mode density of $2(L/\pi)^3$ modes per unit volume in **k** space. The number of modes, $d\mathcal{N}$, in an incremental range dk is given by the product of the mode density and the **k**-space volume of a spherical shell, where the volume element is $4\pi k^2 \, dk$. The volume is restricted to the first quadrant, which introduces a $\tfrac{1}{8}$ factor, since the integers n_x, n_y, and n_z take on positive values only. The result is

$$d\mathcal{N} = 2 \frac{V}{\pi^3} \times \frac{1}{8} \times 4\pi k^2 \, dk \qquad (6.62)$$

where $V = L^3$. In passing from a set of discrete points to a continuum, we are assuming that the range dk is large in comparison with the mode spacing (π/L). Since the mode frequency ω and mode **k** value are related by $k = \omega \eta / c$, where η is the refractive index of the medium filling the cavity, we have $dk = (\eta/c) \, d\omega$. Therefore, from (6.62), for the number of modes in a frequency range $d\omega$ we have

$$d\mathcal{N} = \frac{\omega^2 \eta^3 V}{\pi^2 c^3} \, d\omega$$

6 FIELD QUANTIZATION

The corresponding number of modes per unit volume in the frequency range $d\omega$ is therefore given by

$$p(\omega)\, d\omega = \frac{d\mathcal{N}}{V} = \frac{\omega^2 \eta^3}{\pi^2 c^3} d\omega \qquad (6.63)$$

Since $\omega = 2\pi\nu$, (6.63) may be rewritten as

$$p(\nu)\, d\nu = \frac{8\pi\nu^2 \eta^3}{c^3} d\nu \qquad (6.64)$$

We have just considered the mode density problem for the case in which the cavity normal mode solutions are in the form of standing waves. When the normal mode solutions are assumed to be in the form of traveling waves, as in Section 6.3, the results are identical although some of the intermediate steps differ. In the traveling-wave case the periodic boundary conditions impose the requirement (6.31). As in the standing-wave case, the allowed values of **k** can be represented by a series of points in **k** space. Whereas in the standing-wave case the points are separated by (π/L) and are restricted to one quadrant, as shown in Figure 6.3, in the traveling-wave case the points are separated by $(2\pi/L)$ and are permitted in all quadrants. Since for each **k** value two polarizations may exist, there are two modes in a **k**-space volume $(2\pi/L)^3$ leading to a mode density of $2(L/2\pi)^3$ modes per unit volume in **k** space. The number of modes, $d\mathcal{N}$, in an incremental volume in **k** space can be written

$$d\mathcal{N} = 2\, \frac{V}{(2\pi)^3} dk_x\, dk_y\, dk_z$$

In the spherical coordinate system illustrated in Figure 6.4, the volume element $dk_x\, dk_y\, dk_z$ becomes $k^2\, dk\, \sin\theta_k\, d\theta_k\, d\varphi_k$. Therefore, $d\mathcal{N}$ may be rewritten as

$$d\mathcal{N} = \frac{2V}{(2\pi)^3} k^2\, dk\, \sin\theta_k\, d\theta_k\, d\varphi_k$$

$$= \frac{2V}{(2\pi)^3} k^2\, dk\, d\zeta$$

where $d\zeta$ corresponds to an incremental solid angle in **k** space. Hence, $d\mathcal{N}$ is the number of modes in a solid angle $d\zeta$ and **k** space increment dk. With $k = \omega\eta/c$, $dk = (\eta/c)\, d\omega$, for the number of modes per unit solid angle per unit volume in a frequency range $d\omega$ we have

$$\frac{dp(\omega)}{d\zeta} d\omega = \frac{1}{V} \frac{d\mathcal{N}}{d\zeta} = 2\, \frac{\omega^2 \eta^3}{(2\pi c)^3} d\omega \qquad (6.65)$$

For processes that are angular-dependent, it is often necessary to express the mode density in the above form before integration over solid angle occurs. If the process

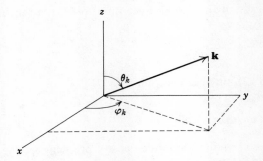

figure 6.4 The **k** vector in cylindrical coordinates.

is not angular dependent, then the number of modes per unit volume in the frequency range $d\omega$ and solid angle $\zeta = 4\pi$, (6.65) reduces to the result (6.63) obtained for the standing-wave case.

In the following sections the expressions for the mode density will be used to determine the spontaneous emission rate to extend previously derived single-mode results to the multimode or broadband frequency case.

6.4.4 Spontaneous Emission

According to (6.59), the rate at which photons are emitted into a given cavity mode as a result of spontaneous emission is given by

$$\frac{\partial \langle n \rangle_a}{\partial t} = K \mathcal{N}_2 \tag{6.66}$$

where the subscript a designates the particular cavity mode under consideration, $\mathcal{N}_2$ is the number of atoms or molecules in the cavity in the upper state, and K as given by (6.61) is the probability per unit time that an atom in the upper state will emit a photon into the cavity mode. As a result of the spontaneous emission into a given mode, the population difference given by (6.60) changes according to

$$\frac{\partial}{\partial t}(\mathcal{N}_1 - \mathcal{N}_2) = 2K\mathcal{N}_2 \tag{6.67}$$

In Section 6.4.3, we saw that there are generally a multitude of cavity modes in a given frequency increment. The range of cavity modes to which the transition couples is determined by the frequency dependence of the coupling constant K, which contains the lineshape factor of the transition. In general, many cavity modes fall within this linewidth. The total photon output—due to spontaneous emission into all cavity modes,—known as fluorescence, is therefore given by a generalization of (6.66),

$$\text{fluorescence photon output} = \sum_a \frac{\partial \langle n \rangle_a}{\partial t} = \sum_a K(\omega_a) \mathcal{N}_2 \tag{6.68}$$

where the major contribution to the sum over all modes are the modes that lie within the molecular linewidth. This leads to a spontaneous emission rate, $A = 1/\tau_{sp}$, defined as the probability per unit time that an atom in the upper state will emit a photon,

$$A = \frac{1}{\tau_{sp}} = \sum_a K(\omega_a) \rightarrow V \int_0^\infty K(\omega) p(\omega) \, d\omega \tag{6.69}$$

where $p(\omega) \, d\omega$ is the number of modes per unit volume in a frequency range $d\omega$ given by (6.63), V is the cavity volume, $K(\omega)$ is the single-mode spontaneous emission rate given by (6.61), and τ_{sp} is defined as the spontaneous emission time.

The frequency dependence of the integrand in (6.69) is determined primarily by the rapidly varying lineshape factor contained in $K(\omega)$. Therefore the linear dependence of $K(\omega)$ on ω and quadratic dependence of $p(\omega)$ on ω may be removed from the integral, with ω replaced by the transition frequency Ω. The lineshape factor can then be integrated ($\int g(\omega) \, d\omega = 1$) to yield the spontaneous emission rate

$$A = \frac{1}{\tau_{sp}} = \frac{\Omega^3 \eta^3 |\mu_{12}|^2}{3\pi\hbar\epsilon c^3} \tag{6.70}$$

Equation 6.70 relates $|\mu_{12}|^2$ to the reciprocal of the spontaneous emission time. Therefore a measurement of τ_{sp} gives a value for the matrix element, and this is the technique that is often used to determine $|\mu_{12}|^2$. The change in population difference as a result of spontaneous emission into all modes is given by the generalization of (6.67),

$$\frac{\partial}{\partial t}(\mathcal{N}_1 - \mathcal{N}_2) = 2A\mathcal{N}_2 \tag{6.71}$$

6.4.5 Modified Rate Equations for the Single-Mode Case

Since there is spontaneous emission into many modes, the change in population difference resulting from spontaneous emission is determined by A rather than K, that is, follows (6.71) rather than (6.67). As a result, it is necessary to modify the rate equations, (6.59) and (6.60), to read

$$\frac{\partial}{\partial t}\langle n \rangle + \frac{\langle n \rangle - \langle n \rangle^e}{\tau_c} = -K\langle n \rangle(\mathcal{N}_1 - \mathcal{N}_2) + K\mathcal{N}_2 \tag{6.72}$$

$$\frac{\partial}{\partial t}(\mathcal{N}_1 - \mathcal{N}_2) + \frac{(\mathcal{N}_1 - \mathcal{N}_2) - (\mathcal{N}_1 - \mathcal{N}_2)^e}{T_1} = 2K\langle n \rangle(\mathcal{N}_2 - \mathcal{N}_1) + 2A\mathcal{N}_2 \tag{6.73}$$

That is, the population difference is depleted at a rate $2A = 2/\tau_{sp}$ because of spontaneous emission into all possible modes. We see from the above that the growth rate of photons in the single mode of interest, $\langle n \rangle$, is, however, still given by K. The number of photons spontaneously emitted into the mode of interest therefore make up K/A of the total spontaneous emission output.

Equations 6.72 and 6.73 can be reduced to more compact forms which for $\langle n \rangle \gg 1$ are identical to equations 3.78 and 3.79 derived on a semiclassical basis:

$$\frac{\partial}{\partial t} \langle n \rangle + \frac{\langle n \rangle - \langle n \rangle^{e'}}{\tau_c} = -K(\langle n \rangle + \tfrac{1}{2}) \Delta \mathcal{N} \tag{6.74}$$

$$\frac{\partial}{\partial t} \Delta \mathcal{N} + \frac{\Delta \mathcal{N} - \Delta \mathcal{N}^{e'}}{T_1'} = -2K \langle n \rangle \Delta \mathcal{N} \tag{6.75}$$

where $\Delta \mathcal{N} = \mathcal{N}_1 - \mathcal{N}_2$, $\mathcal{N} = \mathcal{N}_1 + \mathcal{N}_2$, $\langle n \rangle^{e'} = \langle n \rangle^e + K \mathcal{N} \tau_c / 2$

$$\Delta \mathcal{N}^{e'} = \left(\Delta \mathcal{N}^e + \mathcal{N} \frac{T_1}{\tau_{sp}} \right) \left(\frac{T_1'}{T_1} \right)$$

and

$$(1/T_1') = (1/T_1 + 1/\tau_{sp})$$

Equations 6.74 and 6.75 are of the same form as (3.78) and (3.79) if we make the substitutions

$$\frac{\langle n \rangle}{V} \rightarrow \varphi \qquad \frac{\Delta \mathcal{N}}{V} \rightarrow -N$$

In (6.74) and (6.75) the Lorentz correction factor L was taken to be unity, and we allowed for a nonzero equilibrium photon density $\langle n \rangle^e$. Aside from these differences, for $\langle n \rangle \gg 1$ the corresponding equations are identical.

The longitudinal relaxation rate $1/T_1'$ is made up of two separate contributions, one resulting from the usual thermalizing processes such as collisions or excitation of lattice vibrations, and the other from spontaneous emission. As temperature is reduced, longitudinal relaxation caused by thermalizing processes decreases, bringing about a reduction in the rate $1/T_1$. For the 6943 Å laser transition in ruby, $T_1' = 3.0$ msec at room temperature where both contributions are significant, but increases to 4.3 msec at 77°K where the relaxation is essentially caused entirely by spontaneous emission.[5]

6.4.6 Interaction of Matter and A Broadband Radiation Field

In Section 6.4.5 we considered a single monochromatic radiation mode that interacts with an atomic or molecular transition. We shall here extend those results to the molecular system that interacts with a broadband continuum of radiation modes whose total bandwidth exceeds that of the molecular transition.

6 FIELD QUANTIZATION

From the right-hand side of (6.72), we can define a transition rate from level 1 to level 2 in the presence of a single-mode field as

$$W_{12} = K\langle n\rangle \qquad (6.76)$$

whereas the transition rate from level 2 to level 1 is given by

$$W_{21} = K\langle n\rangle + A \qquad (6.77)$$

In terms of the transition rates, the equation for the population difference, (6.73), can be written

$$\frac{\partial}{\partial t}(\mathcal{N}_1 - \mathcal{N}_2) + \frac{(\mathcal{N}_1 - \mathcal{N}_2) - (\mathcal{N}_1 - \mathcal{N}_2)^e}{T_1} = 2W_{21}\mathcal{N}_2 - 2W_{12}\mathcal{N}_1 \qquad (6.78)$$

As we pass from the single-mode radiation case to the broadband radiation case, (6.78) still holds, but now the transition rates W_{12} and W_{21} given by (6.76) and (6.77) must be generalized by summation or integration over frequency,

$$W_{12} = \sum_a K(\nu_a)\langle n(\nu_a)\rangle \to V \int_0^\infty K(\nu)p(\nu)\langle n(\nu)\rangle\, d\nu \qquad (6.79)$$

$$W_{21} \to V \int_0^\infty K(\nu)p(\nu)\langle n(\nu)\rangle\, d\nu + A \qquad (6.80)$$

where $p(\nu)\, d\nu$ is the number of modes per unit volume in a frequency range $d\nu$, V is the cavity volume, and $K(\nu)$ is the single-mode rate constant given by (6.61). If the frequency distribution of the radiation field represented by $\langle n(\nu)\rangle$ and the mode density $p(\nu)$ are each assumed to vary slowly with respect to the lineshape factor contained in $K(\nu)$, all frequency dependence other than the lineshape factor may be removed from the integral. With this assumption, (6.79) and (6.80) reduce to

$$W_{12} = B_{12}\rho(\nu) \qquad (6.81)$$

$$W_{21} = B_{21}\rho(\nu) + A \qquad (6.82)$$

where $\rho(\nu)$ is the energy per unit volume per unit frequency interval, defined by

$$\rho(\nu) = p(\nu)h\nu\langle n(\nu)\rangle \qquad (6.83)$$

and $B_{12} = B_{21} = B = (1/h\nu)V\int_0^\infty K\, d\nu$. With A given by (6.69), we see that the ratio of A to B is as follows:

$$\frac{A}{B} = h\nu p(\nu) = \frac{8\pi h\nu^3 \eta^3}{c^3} \qquad (6.84a)$$

Alternatively, from (6.83) and (6.84a) we may write

$$\frac{A}{B\rho} = \frac{1}{\langle n\rangle} \qquad (6.84b)$$

where $\langle n \rangle$ is the expectation value of the number of photons in a single mode. In deriving the above we make use of the mode density relationship (6.64). From (6.81) and (6.82) it is clear that A is associated with spontaneous emission whereas B is associated with the processes of induced or stimulated emission and absorption. The coefficients A and B are referred to as the Einstein coefficients since they were derived originally by Einstein, not on the basis of field quantization, but rather by the use of classical arguments and thermodynamic considerations. This alternative approach is presented in Section 6.4.8. From (6.84b) we see, that the induced transition rate is $\langle n \rangle$ times that for spontaneous transitions.

6.4.7 Blackbody Radiation

From field quantization we may predict the spectral distribution of blackbody radiation, which is the radiation from a body in thermal equilibrium with its environment.[6] We take as our blackbody radiator a cavity filled with a medium. Under thermal equilibrium conditions, radiation modes are excited as a result of energy exchange between transitions in the medium and the radiation field. Under steady-state conditions the number of molecules in each energy level remains constant at the equilibrium values as a result of the balance between absorption and emission processes.

Now let us consider a particular pair of levels separated in energy by $h\nu$. For the steady-state we have, from (6.78),

$$W_{12}\mathcal{N}_1 = W_{21}\mathcal{N}_2$$

which, with the aid of (6.81)–(6.83), reduces to

$$\langle n(\nu) \rangle \mathcal{N}_1 = (\langle n(\nu) \rangle + 1) \mathcal{N}_2 \tag{6.85}$$

Since the total system is in thermal equilibrium, the level populations conform to the Boltzmann distribution so that $\mathcal{N}_2 = \mathcal{N}_1 e^{-h\nu/\kappa T}$. Substitution into (6.85) leads to an expression for the expectation value of the number of photons in a mode,

$$\langle n(\nu) \rangle = \frac{1}{e^{h\nu/\kappa T} - 1} \tag{6.86}$$

Since the number of modes per unit volume in a frequency interval $d\nu$, assuming a continuum, is given by $(8\pi\nu^2\eta^3/c^3)\,d\nu$, and the energy per mode is given by $h\nu\langle n \rangle$, from (6.83), for the energy per unit volume in a frequency range $d\nu$, we have

$$\rho(\nu)\,d\nu = \frac{8\pi h\nu^3\eta^3}{c^3} \frac{1}{e^{h\nu/\kappa T} - 1}\,d\nu \tag{6.87}$$

The above distribution is the blackbody radiation curve known as Planck's law.

6.4.8 Einstein Treatment of Induced and Spontaneous Transitions

We have derived expressions for the spontaneous and induced transition rate coefficients A and B, (6.70) and (6.84), on the basis of field quantization. Even before the introduction of field quantization, Einstein hypothesized that in the steady-state an expression of the form[7,8]

$$\rho(\nu)B_{12}\mathcal{N}_1 = \rho(\nu)B_{21}\mathcal{N}_2 + A\mathcal{N}_2 \tag{6.88}$$

must be true to account for the transitions that take place under the influence of a broadband radiation field, where the quantities are those discussed in Section 6.4.6. We shall now show that certain results of the previous work follow directly from the simple hypothesis (6.88). In order to find relationships among B_{12}, B_{21}, and A, it is sufficient to choose an experiment in which $\mathcal{N}_1$, $\mathcal{N}_2$, and $\rho(\nu)$ are known *a priori*. An example that comes immediately to mind is blackbody radiation. Here the level populations follow Boltzmann's law, and the radiation distribution is given by Planck's law.

Setting $\mathcal{N}_2 = \mathcal{N}_1 e^{-h\nu/\kappa T}$ in (6.88) and solving for $\rho(\nu)$, we obtain

$$\rho(\nu) = \frac{A/B_{21}}{(B_{12}/B_{21})e^{h\nu/\kappa T} - 1}$$

Comparison of the above with (6.87) yields the results

$$B_{21} = B_{12}$$

$$\frac{A}{B_{12}} = \frac{8\pi h\nu^3 \eta^3}{c^3}$$

These expressions are found to be in agreement with the results obtained in Section 6.4.6 from equations derived on the basis of field quantization.

REFERENCES

1. A. Einstein, "Über einen die Erzeugung und Verwandlung des Lichtes betreffenden heuristischen Gesichtspunkt," *Ann. d. Physik*, **17**, 132, 1905.
2. A. Einstein, "Zur Theorie des Lichterzeugung und Lichtabsorption," *Ann. d. Physik*, **20**, 199, 1906.
3. C. Kittel, *Quantum Theory of Solids*, Wiley, New York, 1963, p. 12.
4. V. S. Mashkevich, *Laser Kinetics*, American Elsevier Publishing Co., New York, 1967, Chapter 2.
5. B. A. Lengyel, *Introduction to Laser Physics*, Wiley, New York, 1966, pp. 99ff.
6. D. Bohm, *Quantum Theory*, Prentice-Hall, Englewood Cliffs, N.J., 1951, pp. 18ff.
7. A. Einstein, "Zur Quantentheorie der Strahlung," *Phys. Zeit.*, **18**, 121, 1917.

8. A. Einstein, "Strahlungs-Emission und Absorption nach der Quantentheorie," *Verh. d. Deutsch. Phys. Ges.*, **18**, 318, 1916.

For additional discussion of field quantization see:

(a) W. Louisell, *Radiation and Noise in Quantum Electronics*, McGraw-Hill, New York, 1964, Chapter 4.

(b) W. Heitler, *The Quantum Theory of Radiation*, 2nd ed., Oxford University Press, Fair Lawn, N.J., 1944, Chapters 1 and 2.

PROBLEMS

6.1 Show that Hamilton's equations based on the Hamiltonian (6.5) lead to the Maxwell equation $\nabla \times H = \epsilon\, \partial E/\partial t$.

6.2 Determine $\langle \mathbf{E} \rangle$, $\langle \mathbf{H} \rangle$, $\langle \mathbf{E}^2 \rangle$, $\langle \mathbf{H}^2 \rangle$, and $\langle \mathscr{H} \rangle$ for a single-mode cavity field in the vacuum state $|0\rangle$. Use the results to express the variance of the zero-point fluctuations of **E** and **H**, given by expressions of the form

$$(\Delta \mathbf{E})^2 = \langle \mathbf{E}^2 \rangle - \langle \mathbf{E} \rangle^2$$

6.3 Start with the field Hamiltonian with zero-point energy removed,

$$\mathscr{H} = (\tfrac{1}{2})(p^2 + \omega^2 q^2) - \hbar\omega/2$$

(a) Determine the energy eigenvalues.

(b) What are the results for Problem 6.2 with the zero-point energy removed?

(c) What statement can be made about the significance of the presence of the zero-point energy in $\mathscr{H}$?

6.4 (a) For a single mode quantized plane wave, show that $\langle \mathbf{E} \rangle$ satisfies the wave equation

$$\nabla \times (\nabla \times \langle \mathbf{E} \rangle) + \frac{\eta^2}{c^2} \frac{\partial^2 \langle \mathbf{E} \rangle}{\partial t^2} = 0$$

(b) For quantized plane waves, obtain an expression for the Poynting vector in terms of a and $a^\dagger$.

(c) Obtain an expression for the matrix elements of the Poynting vector.

6.5 (a) For two-photon absorption we found that the transition probability was proportional to the intensity. Therefore, we may write

$$\frac{dI}{dt} = CI^2$$

where C is a proportionality constant and I is intensity. The above expression was derived semiclassically for absorption from $|1\rangle$ to $|2\rangle$ as shown in Figure P6.5a. Using the $n \to n+1$ rule for emission and $n \to n$ for absorption, write the rate equation that applies for the fully quantized approach for both two-photon absorption and emission. $\mathscr{N}_1$ and $\mathscr{N}_2$ are the level populations.

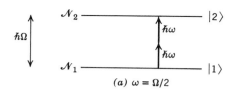

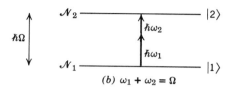

figure P6.5 Two-photon absorption and emission: (a) photon energies identical; (b) photon energies not identical.

(b) Consider two-photon absorption and emission when the two photons have different energies, as shown in Figure 6.5b. Write rate equations, fully quantized for dn_1/dt, dn_2/dt. Under what conditions is it possible to obtain gain at ω_1 without population inversion?

6.6 In Compton scattering, energy is absorbed by electrons, where the incident photon energy is $\hbar\omega_1$ and the radiation is re-emitted with photon energy $\hbar\omega_2$. For classical fields the appropriate expression is

$$\frac{dI_2}{dt} = C \mathcal{N} I_1 I_2$$

where C is a proportionality constant, I_1 and I_2 are intensities at frequencies ω_1 and ω_2, respectively, and $\mathcal{N}$ is the number of electrons in the interaction volume. Write the equation for quantized fields, allowing for the reverse process (absorption at $\hbar\omega_2$) to occur.

6.7 (a) How many modes exist in a 1-cm³ ruby crystal ($\eta = 1.76$) within the linewidth ($\Delta\nu = 330$ GHz) of the 6943 Å laser transition?
(b) How many TEM$_{00n}$ modes exist within the laser linewidth if we consider those modes for which k exists only along one direction of the crystal, as might be excited if only two faces are silvered and the crystal pumped above threshold for laser action (that is, k_z allowed, $k_x = k_y = 0$)?
(c) What is the frequency spacing between these latter modes?

6.8 Show that the absorption constant $\Gamma(\omega)$ for an isotropic medium with a Lorentzian absorption line can be expressed in terms of the spontaneous emission time τ_{sp} as

$$\Gamma = \frac{\pi^2 c^2}{\Omega^2 \eta^2 \tau_{sp}} (N_1 - N_2) g_L(\omega, \Omega)$$

where N_1 and N_2 are the number of molecules per unit volume in the lower and upper states, respectively.

INTERACTIONS BETWEEN RADIATION AND PHONONS

7

7.1 INTRODUCTION

In this chapter we consider the interaction of electromagnetic radiation and the vibrational modes of a medium. In particular, we concentrate on the vibrational modes of crystals, although the results obtained carry over to continuous media as a limiting case.

Section 7.2, devoted to a description of crystal lattice vibrations, shows that crystal vibrations can be divided into two categories, acoustic and optical vibrations. The acoustic vibrations are sound waves that propagate through the crystal in which adjacent atoms vibrate with a small relative phase displacement from one unit cell to the next. Optical vibrations, on the other hand, constitute a set of modes in which adjacent atoms vibrate against each other in a seesaw fashion. Acoustic and optical modes are distinguished from one another in their propagation characteristics, in their light-interaction properties, and in their resonant frequencies.

In Section 7.3 the lattice vibrational modes are quantized. The quantization procedure is similar to that applied to electromagnetic radiation in Chapter 6. The results are also similar: the vibrations act as a collection of harmonic oscillators, with the corresponding interpretation that the mode properties can be described in terms of discrete quanta called phonons.

Since we are interested primarily in the interaction of electromagnetic radiation and vibrational modes, Section 7.4 considers the general problem of wave interaction and energy transfer. Some very general energy and momentum conservation conditions are seen to apply.

Sections 7.5, 7.6, and 7.7 consider specific cases of the interaction of electromagnetic radiation with phonons. In Section 7.5 direct linear absorption by the optical modes is considered. The nonlinear light-scattering processes known as the Raman and Brillouin effects are considered in Sections 7.6 and 7.7, respectively. Both effects are a consequence of the modulation of the dielectric constant by vibrational modes present in the medium. As a result of such fluctuations, an incident light wave is scattered from the medium with a shift in frequency corresponding to that of the vibrational mode from which the scattering takes place. With the Raman effect, scattering from the optical modes is involved, and the frequency shifts are on the order of several hundred Angstroms. Brillouin scattering is similar to Raman scattering except that acoustic rather than optical modes are involved, and thus the frequency shifts are smaller, on the order of several GHz. With a laser source providing the incident beam, both effects have been observed above a certain threshold as stimulated effects, with the result that the scattered light beams have an intensity and collimation comparable to that of the exciting laser. Such nonlinear effects thus provide an efficient means for generating new frequencies and studying material properties.

Section 7.8 concludes the chapter with a description of the phenomenon of self-trapping and self-focusing of light beams. This phenomenon is of interest because it markedly affects the threshold properties of nonlinear processes such as the stimulated Raman and Brillouin effects.

7.2 CRYSTAL LATTICE VIBRATIONS

A discussion of the interaction between radiation and the vibrational modes of a crystal requires an examination of the properties of the vibrational modes themselves. In our analysis we shall, for simplicity, idealize our model of a crystal by assuming that the forces that act on given atoms are caused by nearest neighbors only and are central, that is, act along the lines joining the atoms. In a real crystal the forces are not necessarily central if covalent bonds exist, and interactions between other than nearest neighbors must be taken into account.[1,2] Nevertheless, analysis of the idealized model enables us easily to obtain results that are descriptive of the behavior of real crystals.

7.2.1 Modes of a Monatomic Crystal Lattice

We shall begin our discussion of crystal modes by considering a monatomic crystal lattice in which all the atoms are identical and for which the primitive cell† contains only one atom. An example is the face-centered-cubic structure of aluminum. Furthermore, we shall restrict our consideration to waves propagating in directions for which the waves are purely transverse or purely longitudinal. In a cubic crystal these are the [100], [111], and [110] directions.‡ For propagation in

---

† A primitive cell is a minimum-volume unit-cell from which the entire crystal lattice can be reproduced by a series of suitable translations.

‡ The indices of a direction in a crystal [hkl] are the smallest integers which have the same ratios as the components of a vector in the desired direction referred to the crystal axis vectors.

7 INTERACTIONS BETWEEN RADIATION AND PHONONS 193

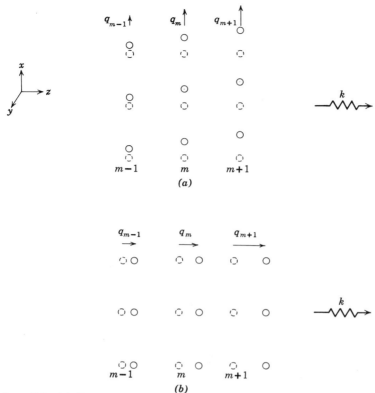

figure 7.1 (a) Propagation of a transverse acoustic wave. The dashed circles indicate equilibrium positions of the atoms and the solid circles show the positions during a vibration. (b) Propagation of a longitudinal acoustic wave.

these directions, entire planes of atoms in the crystal move in phase. For propagation in an arbitrary direction, on the other hand, the waves are no longer purely longitudinal or purely transverse, but are rather a mixture, and the motion of the atoms is more complex.

First, let us consider the propagation of a wave in the z direction that causes successive planes of atoms to be displaced in the x direction as shown in Figure 7.1a. Such a wave is purely transverse since the displacements of the atoms are perpendicular to the direction of the propagation vector, denoted by **k**.

As one plane of atoms is displaced in relation to another, there will be restoring forces between the planes because of the difference in their displacements. The first-order term in a Taylor's series expansion for the force on an atom in a given plane resulting from those in another in terms of the relative displacements is directly proportional to the difference of their displacements. This is equivalent to a Hooke's law assumption, which is valid for small displacements. For nearest-neighbor plane interactions,

$$M \frac{d^2 q_m}{dt^2} = C_T(q_{m+1} - q_m) - C_T(q_m - q_{m-1}) = C_T(q_{m+1} - 2q_m + q_{m-1}) \quad (7.1)$$

where M is the atomic mass, C_T is the force constant applicable to transverse wave motion, the subscript m is an integer that denotes a given plane of atoms, and q_m is the displacement of an atom at site m from its equilibrium position.

We look for solutions to (7.1) in the form of a traveling wave,

$$q_m = \tfrac{1}{2} \tilde{q} e^{i(\omega_k t - kma)} + \text{c.c.} \quad (7.2)$$

where $\tilde{q}$ is a complex amplitude, a is the spacing between planes, k is the wave vector, and ω_k is the frequency of vibration for a given value of k. Substitution of (7.2) into (7.1) then yields what is termed the dispersion relationship connecting ω_k and k:

$$\omega_k = 2 \left(\frac{C_T}{M} \right)^{1/2} \left| \sin \frac{ka}{2} \right| \quad (7.3)$$

which is plotted in Figure 7.2 for $-\pi/a \leq k \leq \pi/a$. Positive values of k correspond to transverse waves propagating in the $+z$ direction, and negative values of k to waves propagating in the $-z$ direction. In addition, we note that although there is nothing in the dispersion relationship (7.3) to limit the range of k values to $\pm \pi/a$, we find that values outside this range merely reproduce solutions that correspond to exactly the same particle motions as solutions with k values inside the range. This is easily verified by considering the substitution $k = k_0 + 2p\pi/a$, where k_0 lies within the range $\pm \pi/a$ and p is any positive or negative integer. It is found that the solutions for various values of p are exactly equivalent to the

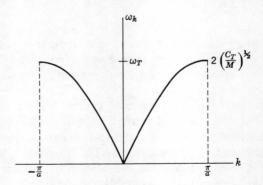

figure 7.2 Dispersion characteristic for transverse modes of a monatomic crystal.

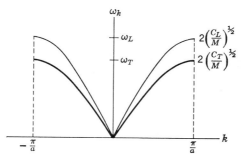

figure 7.3 Dispersion characteristics for transverse and longitudinal modes of a monatomic crystal.

solution for $k = k_0$. The restricted range of k values lying within the range $\pm \pi/a$ is referred to as the first Brillouin zone.

At the zone boundary $k = \pi/a$, the frequency has its maximum value, and $\omega_T = 2\sqrt{C_T/M}$, which is the cutoff frequency for the transverse modes. The wavelength λ of the propagating wave, defined by $k = 2\pi/\lambda$, equals $2a$, and the solution (7.2) becomes

$$q_m = \tfrac{1}{2}\tilde{q}e^{i(\omega_k t - m\pi)} + \text{c.c.} = |\tilde{q}| \cos(\omega_k t - m\pi + \varphi)$$

where $\tilde{q} = |\tilde{q}|\, e^{i\varphi}$. Therefore, for a k value at the zone boundary, neighboring planes oscillate 180 deg out of phase with each other. The motion of the particles in this case does not correspond to progressive wave motion along the crystal, but rather to a standing wave, sometimes called the π mode.

In summary, then, we find that transverse lattice waves propagate with k values lying in the range $\pm \pi/a$. As the k value goes from zero to π/a, the frequency of the wave increases to the cutoff value, the wavelength of the wave decreases to where the half-wavelength corresponds to the lattice spacing, and the traveling wave becomes a standing wave at the zone boundary.

Before leaving the discussion of transverse waves, we note that a transverse wave with displacements in the y direction, rather than in the x direction as shown in Figure 7.1a, is also possible. This leads to a second dispersion curve, which may or may not be degenerate with respect to the first, depending on the relative values of C_T for the two cases.

Now consider the propagation of a wave in the z direction, which causes successive planes of atoms to be displaced in the z direction as shown in Figure 7.1b Such a wave is purely longitudinal. The equation of motion and its solution are of the same form as (7.1)–(7.3) except that C_T, the spring constant for the transverse case, is replaced by C_L, the spring constant for the longitudinal case. As a result, the dispersion curve for the longitudinal case differs from that of the transverse case in the manner indicated by Figure 7.3. This lack of degeneracy between

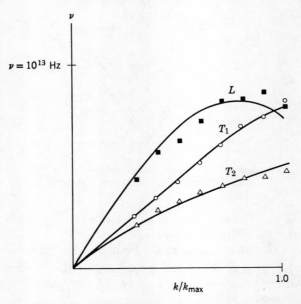

figure 7.4 Dispersion curves for elastic waves propagating along the [110] axis in aluminum. The measured data for the longitudinal wave are shown by the solid circles; the data for the transverse wave T_1, whose polarization is parallel to the [100] axis, are given by the open circles; and the data for the transverse wave T_2, whose polarization is parallel to the [110] axis, are given by the solid triangles. The smooth curves represent the fitted solutions where noncentral forces and interactions up to third neighbors are allowed.[4]

longitudinal and transverse modes holds even for a medium that is completely isotropic elastically.[3]

Dispersion curves for elastic waves propagating along the [110] axis in aluminum as determined by the inelastic scattering of X rays is shown in Figure 7.4. These curves are in qualitative agreement with the above considerations. To obtain quantitative agreement it is necessary to include noncentral forces and other than nearest-neighbor interactions.[4]

Thus far we have considered waves that propagate along certain select axes. If propagation in a general direction is considered, it is found that the dispersion curve still has three branches, but the corresponding modes in general, are not, purely longitudinal or purely transverse.

7.2.2 Modes of a Diatomic Crystal Lattice

In this section we shall determine the vibrational modes of a diatomic crystal structure, that is, a structure that contains two atoms per primitive cell, such as

7 INTERACTIONS BETWEEN RADIATION AND PHONONS

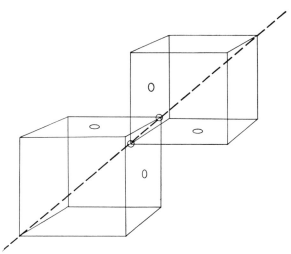

figure 7.5 Diamond lattice structure. It consists of two interpenetrating face-centered-cubic lattices, one of which is displaced along the body diagonal of the other by one-fourth its length.

NaCl or diamond. We note that a diatomic crystal is not necessarily composed of two different atoms, but may consist entirely of one element as in the case of diamond, which contains only carbon atoms. The diamond structure is diatomic because it consists of two face-centered-cubic lattices, one of which is displaced along the body diagonal of the other by one-fourth its length, as shown in Figure 7.5. Such a structure cannot be reproduced by translation of a primitive cell containing only one atom, but rather requires a cell containing two atoms.

Analysis of the diatomic lattice proceeds in a manner analogous to the analysis of the monatomic lattice. Once again we shall restrict our considerations to waves that propagate in such directions that the waves are purely transverse or purely longitudinal. For example, pure transverse or pure longitudinal waves propagate in the [111] direction of the NaCl structure.

Let atoms of mass M_1 lie on the odd-numbered planes, and atoms of mass M_2 lie on the even-numbered planes, as indicated in Figure 7.6. We shall assume that $M_1 \geqq M_2$. If the force constant between all pairs of neighboring planes is identical and nearest-neighbor plane interactions only are considered, then for transverse wave propagation in the two-atom case, we have

$$M_1 \frac{d^2 q_{2m+1}}{dt^2} = C_T(q_{2m+2} - 2q_{2m+1} + q_{2m})$$

$$M_2 \frac{d^2 q_{2m}}{dt^2} = C_T(q_{2m+1} - 2q_{2m} + q_{2m-1})$$

(7.4)

where m denotes a given plane of atoms.

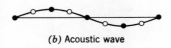

(a) Diatomic lattice

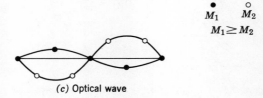

(b) Acoustic wave

$M_1 \quad M_2$
$M_1 \geq M_2$

(c) Optical wave

figure 7.6 Transverse wave motion in a diatomic crystal. In the acoustic wave the two sublattices vibrate together; in the optical wave they vibrate against one another.

We look for solutions to (7.4) in the form of traveling waves with separate solutions for odd- and even-numbered planes,

$$q_{2m+1} = \tfrac{1}{2}\tilde{q}_1 e^{i[\omega_k t - k(2m+1)a]} + \text{c.c.}$$
$$q_{2m} = \tfrac{1}{2}\tilde{q}_2 e^{i[\omega_k t - k2ma]} + \text{c.c.} \tag{7.5}$$

Substitution of (7.5) into (7.4) yields

$$\begin{pmatrix} 2C_T - \omega_k^2 M_1 & -2C_T \cos ka \\ -2C_T \cos ka & 2C_T - \omega_k^2 M_2 \end{pmatrix} \begin{pmatrix} \tilde{q}_1 \\ \tilde{q}_2 \end{pmatrix} = 0 \tag{7.6}$$

The above has a nontrivial solution only if the determinant of coefficients of $\tilde{q}_1$ and $\tilde{q}_2$ vanishes, or

$$\omega_k = \left\{ C_T\left(\frac{1}{M_1} + \frac{1}{M_2}\right) \pm C_T\left[\left(\frac{1}{M_1} + \frac{1}{M_2}\right)^2 - \frac{4\sin^2 ka}{M_1 M_2}\right]^{1/2} \right\}^{1/2} \tag{7.7}$$

The above expression is plotted in Figure 7.7 for $-\pi/2a < k < \pi/2a$. We see that in the case of the diatomic crystal we have, in addition to the acoustic branch, a second branch whose frequency is nonzero at $k = 0$. The nature of the vibrational modes that correspond to this branch is revealed by the form of the solution near $k = 0$, where (7.7) gives

$$\omega_k = \left[2C_T\left(\frac{1}{M_1} + \frac{1}{M_2}\right)\right]^{1/2} \tag{7.8}$$

Substitution of the above into (7.6) yields, for the amplitudes of the vibrations of alternate planes,

$$\frac{\tilde{q}_1}{\tilde{q}_2} = -\frac{M_2}{M_1} \tag{7.9}$$

That is, alternate planes vibrate against each other with the center of mass remaining fixed, as shown in Figure 7.6c. If the two atoms carry opposite charges, this type of motion can be excited by the electric field of an electromagnetic wave, since such a field would drive the atomic motion in this mode. This would not occur for the acoustic mode, where atoms of both charges vibrate together rather than against one another. For typical crystal parameters the frequency of oscillation when adjacent planes vibrate against each other is in the infrared part of the optical spectrum, and therefore this branch of the dispersion curve is called the optical branch. The optical and acoustic wave motions are shown in Figure 7.6.

The foregoing analysis can be repeated for longitudinal waves. In this case we obtain a set of dispersion curves which differ only in the value of the force constant C. Thus, for the diatomic crystal there are six branches on the dispersion diagram: two transverse acoustic, one longitudinal acoustic, two transverse optical, and one longitudinal optical. Dispersion curves for the elastic waves propagating in the [111] direction of KBr, which has the NaCl diatomic structure, are shown in Figure 7.8.

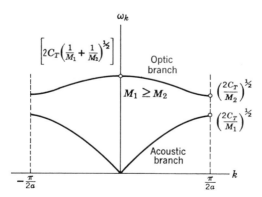

figure 7.7 Dispersion characteristics for transverse optical and acoustic modes of a diatomic crystal.

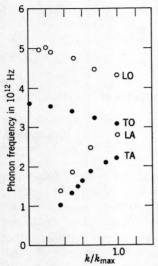

figure 7.8 Optic and acoustic branches of the dispersion characteristic for waves propagating in the [111] direction of potassium bromide (KBr) as determined by inelastic scattering of neutrons. The open circles correspond to the longitudinal modes; the closed circles correspond to the transverse modes. The transverse branches are doubly degenerate.[5] The designations LO, TA, etc., stand for longitudinal optical, transverse acoustic, etc.

The approach taken for the monatomic and diatomic lattices can be generalized so that it applies when there are p atoms in the primitive cell with the result that there are $3p$ branches to the phonon dispersion characteristic, 3 acoustic, and $3p - 3$ optical.

7.3 QUANTIZATION OF LATTICE VIBRATIONS

In Section 7.2 we saw that atomic motion in a vibrating crystal lattice can be described in terms of traveling-wave modes that can be classified as either acoustic or optical modes. In direct analogy with the case of electromagnetic wave propagation we pass from the classical to the quantum situation by quantizing such modes. As in the electromagnetic case, we find upon quantization that the modes of interest act as a collection of harmonic oscillators. Again, the existence of a set of discrete evenly spaced energy levels lends itself to the interpretation that the excitation of a given harmonic oscillator to its nth energy eigenstate can be described by saying that there are n quanta in the mode, each of energy $\hbar\omega$, where ω is the radian frequency of the harmonic oscillator in question. For acoustic and optical vibrational modes the quanta are referred to as phonons. We shall first consider the quantization of the longitudinal acoustic branch of the monatomic crystal lattice discussed in Section 7.2.1. This is followed by the quantization of the transverse optical branch of the diatomic crystal lattice discussed in Section 7.2.2. The development in these sections parallels closely the development for the corresponding electromagnetic case given in Section 6.2.

7.3.1 Periodic Boundary Conditions and Normal Modes

We saw in Section 7.2.1 that the displacement q_m of an atom in plane m of a vibrating monatomic crystal lattice can be described in terms of traveling-wave solutions of the form, in one dimension,

$$q_m = \tfrac{1}{2}\tilde{q}e^{i(\omega_k t - kma)} + \text{c.c.} \qquad (7.10)$$

The first step in the quantization procedure requires classification of the possible solutions into a set of normal modes that satisfy orthogonality conditions. This is accomplished by the introduction of appropriate boundary conditions. As in the case of electromagnetic radiation in a cavity, we have a choice regarding the form of the normal mode solutions. We may combine traveling-wave solutions of the form (7.10) into standing waves that satisfy boundary conditions at, for example, the crystal boundaries. Alternatively, we may employ the traveling-wave solutions directly. In this case we introduce the required normalization of the traveling-wave functions by imposing periodic boundary conditions. For a system that includes a large number of lattice sites the results obtained by either approach do not differ in any essential respect. In the development presented here we shall employ the traveling-wave approach with periodic boundary conditions.

To apply periodic boundary conditions we require that the displacement of atoms in the $m + \mathcal{N}$ plane be identical to the displacement in plane m. That is, we require that

$$q_{m+\mathcal{N}} = q_m \tag{7.11}$$

There are thus $\mathcal{N}$ independent variables in the displacements of the $\mathcal{N}$ planes. We assume that $\mathcal{N}$ is a very large number corresponding, for example, to the number of planes in a macroscopic section of crystal.

The periodic boundary conditions are shown in Figure 7.9, where we have taken $\mathcal{N} = 8$ for illustrative purposes. As is shown in Figure 7.9b, imposition of the periodic boundary conditions can be thought of as a bending of the crystal to form a circular ring. We shall return to this analogy when we consider the form of the Hamiltonian.

Application of the periodic boundary condition (7.11) to the traveling-wave solution (7.10) results in the requirement

$$e^{-ik(m+\mathcal{N})a} = e^{-ikma}$$

or

$$e^{-ik\mathcal{N}a} = 1$$

which is satisfied providing that

$$k = \frac{2\pi l}{\mathcal{N}a} \tag{7.12}$$

where l is any positive or negative integer or zero. However, we recall that in the discussion of acoustic waves in a monatomic crystal all possible particle motions could be accounted for by restricting the k values to the range $\pm \pi/a$. Therefore, in (7.12) we can restrict l to the range $\pm \mathcal{N}/2$, where for convenience we assume that $\mathcal{N}$ is even. Since the particle motion described by $k = -\pi/a$ is identical to that described by $k = +\pi/a$, the independent modes are described by k and l values given by

$$k = \frac{2\pi l}{\mathcal{N}a}$$

$$-\frac{\mathcal{N}}{2} + 1 \leq l \leq \frac{\mathcal{N}}{2} \qquad l = 0, \pm 1, \pm 2, \ldots \tag{7.13}$$

$$-\frac{\pi}{a} < k \leq \frac{\pi}{a}$$

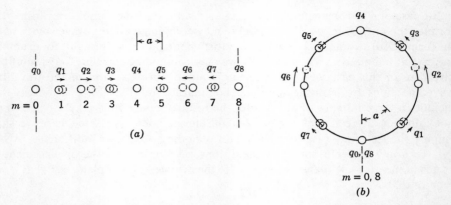

figure 7.9 The application of periodic boundary conditions (b.c.) to the longitudinal acoustic mode case. To apply periodic b.c., we set $q_{m+\mathcal{N}} = q_m$, shown above for $\mathcal{N} = 8$. An array of linear displacements that satisfies periodic b.c. is shown in (a). An equivalent representation is shown in (b), where the atoms are assumed to be constrained to move on a circular ring in such a way that the periodic boundary conditions are satisfied automatically.

The restriction of allowed k values to a finite range is one point where quantization of acoustic waves differs from quantization of the electromagnetic field. In the latter, the k values are allowed to extend to $\pm \infty$. This difference arises because of the inability to distinguish between short wavelength waves (that is, waves with half-wavelength shorter than the interplane spacing) and long wavelength waves (that is, those with half-wavelength longer than the interplane spacing) as far as the displacement of the atoms are concerned. Therefore, the shorter wavelength waves are eliminated from consideration in the case of vibrational modes of a crystal lattice.

From the above discussion we conclude that there are $\mathcal{N}$ independent modes designated by k values evenly distributed in k space as shown in Figure 7.10. We see that the number of independent modes corresponds as it should, to the number of independent variables, namely, the displacements of the $\mathcal{N}$ planes. Therefore the dispersion curve is quasicontinuous, approaching a continuum as $\mathcal{N}$ becomes large.

We now wish to prove the orthogonality of the traveling-wave modes that correspond to the above selected set of k values. That is, we wish to prove that if two traveling-wave solutions of the form (7.10) with specified k values are multiplied together and summed over the range of lattice planes $m = 1$ to $m = \mathcal{N}$, the result is zero unless the k values are identical.

We obtain the required orthogonality condition by showing that

$$\sum_{m=1}^{\mathcal{N}} e^{i(k-k')ma} = \mathcal{N} \delta_{kk'} \tag{7.14}$$

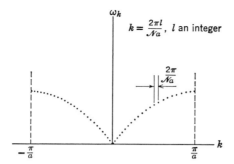

figure 7.10 Quasicontinuous dispersion characteristic for the longitudinal acoustic branch. The discrete allowed k values each correspond to one of a set of orthogonal traveling-wave modes obtained from the application of periodic boundary conditions. The distribution of k values approaches a continuum as $\mathcal{N}$ (the number of planes over which periodic boundary conditions are applied) approaches infinity.

For $k = k'$, the result follows immediately. For $k \neq k'$, from (7.12) we have

$$\sum_{m=1}^{\mathcal{N}} e^{i(k-k')ma} = \sum_{m=1}^{\mathcal{N}} \exp\left(i\,\frac{2\pi(l-l')m}{\mathcal{N}}\right)$$

Let $l - l' = p$, an integer less than $\mathcal{N}$. Then

$$\sum_{m=1}^{\mathcal{N}} e^{i(k-k')ma} = \sum_{m=1}^{\mathcal{N}} \exp\left(i\,\frac{2\pi pm}{\mathcal{N}}\right) = \sum_{m=0}^{\mathcal{N}-1} \exp\left(i\,\frac{2\pi pm}{\mathcal{N}}\right)$$

Carrying out the sum, the terms of which form a geometric progression, we have

$$\sum_{m=1}^{\mathcal{N}} e^{i(k-k')ma} = \frac{e^{i2\pi p} - 1}{e^{i2\pi p/\mathcal{N}} - 1}$$

which is zero for $p \neq 0$ ($k \neq k'$). This completes the proof.

It has been proved that with the k values restricted as required by periodic boundary conditions, traveling-wave solutions of the form (7.10) make up a set of orthogonal basis functions. Let us expand an arbitrary displacement in terms of these functions. Thus we let

$$q_m = \sum_k \left(\frac{\hbar}{2\mathcal{N}M\omega_k}\right)^{1/2} [a_k(0)^* e^{i(\omega_k t - kma)} + \text{c.c.}]$$

which may be rewritten

$$q_m = \sum_k \left(\frac{\hbar}{2\mathcal{N} M \omega_k}\right)^{1/2} [a_k e^{ikma} + a_k^\dagger e^{-ikma}] \tag{7.15a}$$

For the classical case considered here, a_k and $a_k^\dagger$ are simply the time-varying expansion coefficients defined as

$$a_k = a_k(0) e^{-i\omega_k t}, \qquad a_k^\dagger = a_k(0)^* e^{i\omega_k t} \tag{7.15b}$$

The square root factor in front of the bracket is at this point an arbitrary normalization factor in terms of the mass M of the atoms forming the monatomic crystal, and the number of planes $\mathcal{N}$ over which periodic boundary conditions are applied. In the following section we shall verify that, written in this particular form, Equation 7.15a for q_m holds also for the quantized case with $a^\dagger$ and a interpreted there as the creation and annihilation operators for the acoustic wave.

7.3.2 Quantization of the Acoustic Modes of a Monatomic Crystal

The first step in the quantization procedure is to write down the Hamiltonian as a function of the coordinates q_m and momenta p_m. For the set of vibrating atoms, the configuration shown in Figure 7.9, we are led to write for the longitudinal modes

$$\mathcal{H} = \frac{1}{2} \sum_{m=1}^{\mathcal{N}} \left[\frac{p_m^2}{M} + C_L(q_{m+1} - q_m)^2\right] \tag{7.16}$$

where for the $m = \mathcal{N}$ term, $q_{\mathcal{N}+1} \equiv q_1$ because of the periodic boundary conditions. The first term in the brackets corresponds to the kinetic energy of the oscillating atoms, the second to the potential energy associated with the interplane Hooke's law forces. It is easily verified that with the above choice for the Hamiltonian, Hamilton's equations,

$$\dot{q}_i = \frac{\partial \mathcal{H}}{\partial p_i} = \frac{p_i}{M}$$

$$\dot{p}_i = -\frac{\partial \mathcal{H}}{\partial q_i} = C_L(q_{i+1} - q_i) - C_L(q_i - q_{i-1}) \tag{7.17}$$

lead to the appropriate equation of motion, (7.1).

We shall now proceed to rewrite $\mathcal{H}$ so that it has the same form as the Hamiltonian for the unit-mass harmonic oscillator, whereupon the results obtained from the quantization of the electromagnetic field in Chapter 6 may be applied directly to the quantization of lattice vibrations.

Let us express the coordinates q_m and momenta p_m in terms of the traveling-wave normal mode expansion (7.15a),

$$q_m = \sum_k \left(\frac{\hbar}{2\mathcal{N}M\omega_k}\right)^{1/2} [a_k e^{ikma} + a_k^\dagger e^{-ikma}]$$

$$p_m = M\dot{q}_m = -i\sum_k \left(\frac{\hbar M\omega_k}{2\mathcal{N}}\right)^{1/2} [a_k e^{ikma} - a_k^\dagger e^{-ikma}]$$

(7.18)

where, from (7.15b), we have used the relations

$$\dot{a}_k = -i\omega_k a_k, \qquad \dot{a}_k^\dagger = i\omega_k a_k^\dagger$$

From (7.18) we find that

$$\sum_{m=1}^{\mathcal{N}} p_m^2 = \sum_{m=1}^{\mathcal{N}} \left\{(-i)\sum_k \left(\frac{\hbar M\omega_k}{2\mathcal{N}}\right)^{1/2} [a_k e^{ikma} - a_k^\dagger e^{-ikma}](-i)\sum_{k'} \left(\frac{\hbar M\omega_{k'}}{2\mathcal{N}}\right)^{1/2} \right.$$
$$\left. \times [a_{k'} e^{ik'ma} - a_{k'}^\dagger e^{-ik'ma}]\right\}$$

which, with the application of (7.14), reduces to

$$\sum_{m=1}^{\mathcal{N}} p_m^2 = -\sum_k \frac{\hbar M\omega_k}{2} [a_k a_{-k} - a_k a_k^\dagger - a_k^\dagger a_k + a_k^\dagger a_{-k}^\dagger] \tag{7.19}$$

In deriving (7.19) we have been careful to preserve the order of the $a^\dagger$ and a coefficients even though for the classical case it is not necessary. However, we wish the development to apply also to the quantum case, in which $a^\dagger$ and a are non-commuting operators.

By a development similar to that which led to (7.19) we find that

$$\sum_{m=1}^{\mathcal{N}} (q_{m+1} - q_m)^2 = \sum_k \frac{\hbar\omega_k}{2C_L} [a_k a_{-k} + a_k a_k^\dagger + a_k^\dagger a_k + a_k^\dagger a_{-k}^\dagger] \tag{7.20}$$

In the derivation of (7.20) it was necessary to use the dispersion equation (7.3) that relates ω_k to k.

Substituting (7.19) and (7.20) into (7.16), we obtain

$$\mathcal{H} = \frac{1}{2}\sum_k \hbar\omega_k(a_k a_k^\dagger + a_k^\dagger a_k) \tag{7.21}$$

If a_k and $a_k^\dagger$ are now defined in terms of the new variables Q_k and P_k as follows:

$$a_k = \frac{1}{\sqrt{2\hbar\omega_k}}(\omega_k Q_k + iP_k)$$

$$a_k^\dagger = \frac{1}{\sqrt{2\hbar\omega_k}}(\omega_k Q_k - iP_k)$$

(7.22)

then (7.21) reduces to the harmonic oscillator form

$$\mathcal{H} = \frac{1}{2}\sum_k (P_k^2 + \omega_k^2 Q_k^2) \tag{7.23}$$

With P_k taken to be the momentum and Q_k taken to be the coordinate, Hamilton's equations based on this Hamiltonian lead to the correct equations of motion, (7.17). (See Problem 7.3.)

With the Hamiltonian in the form (7.23) we proceed to quantize the acoustic vibrations by postulating that the Q_k, P_k are operators that satisfy the commutation relations

$$[Q_k, P_{k'}] = i\hbar \delta_{kk'} \tag{7.24}$$

as we did for the electromagnetic field traveling-wave case in Chapter 6. The variables $a_k^\dagger$, a_k, which are related to Q_k, P_k by (7.22), are then interpreted as creation and annihilation operators, which on the basis of (7.22) and (7.24), satisfy

$$[a_k, a_{k'}^\dagger] = \delta_{kk'}$$

If the commutator relationship for a_k and $a^\dagger$ is used in (7.21), the Hamiltonian can be rewritten as

$$\mathcal{H} = \sum_k \hbar\omega_k (a_k^\dagger a_k + \tfrac{1}{2}) \tag{7.25}$$

It is easily verified that q_m and p_m as given by (7.18) are, in the quantum mechanical case, Hermitian operators, as they must be since they correspond to observables.

With the acoustic field Hamiltonian reduced to the form (7.25), the solution to the eigenvalue problem follows the development for the electromagnetic field case in Section 6.2.

A given term in the Hamiltonian (7.25) labeled by a particular value of k corresponds to the energy in the k traveling-wave mode which is represented by one of the dots in Figure 7.10. The frequency ω_k of this particular mode is read from Figure 7.10 as the ordinate. Associated with each k mode is an energy level diagram as shown in Figure 7.11. This diagram portrays the solution to the eigenvalue equation

$$\mathcal{H}|n_k\rangle = E_{n_k}|n_k\rangle$$

which is given by

$$E_{n_k} = (n_k + \tfrac{1}{2})\hbar\omega_k$$

where $\mathcal{H}$ is a single term in the series on the right hand side of (7.25). When the k traveling-wave mode is excited to its nth eigenstate, we say that there are n_k phonons in that mode, each of energy $\hbar\omega_k$.

In summary, then, the total energy in the longitudinal acoustic modes of a monatomic crystal lattice is given by (7.25). This energy consists of a sum of contributions from individual traveling-wave modes, each of which is labeled by k and corresponds to a particular dot on the dispersion characteristic Figure 7.10.

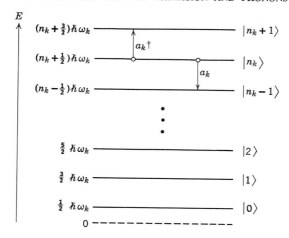

figure 7.11 Energy level diagram for a longitudinal acoustic traveling-wave mode with a given k value. When the k mode is excited to its nth energy eigenstate, the energy contained in the mode is given by

$$E_{n_k} = (n_k + \tfrac{1}{2})\hbar\omega_k$$

which corresponds to the sum of the zero-point energy $\hbar\omega_k/2$ and the energy resulting from n_k photons. The operators $a_k^\dagger$ and a_k are the operators that correspond, respectively, to the creation and annihilation of phonons in the mode.

Each of these individual modes, in turn, has its own energy level diagram, as in Figure 7.11, that indicates the energy, that is, the number of phonons, in that particular mode.

7.3.3 Quantization of the Optical Modes of a Diatomic Crystal

In Section 7.2.2 we saw that the diatomic crystal possesses, in addition to the acoustic modes, optical modes in which adjacent atoms vibrate against one another. For long-wavelength vibrations ($\lambda \gg a$) the frequency for transverse optical waves is given by (7.8),

$$\Omega = \left(\frac{2C_T}{M}\right)^{1/2}, \quad \frac{1}{M} = \frac{1}{M_1} + \frac{1}{M_2} \qquad (7.26)$$

The corresponding amplitudes of the displacements of alternate planes of atoms are given by (7.9),

$$\frac{\tilde{q}_1}{\tilde{q}_2} = -\frac{M_2}{M_1}$$

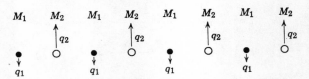

figure 7.12 Long-wavelength transverse optic mode in a diatomic crystal. Alternate planes of atoms move in phase as the two sublattices vibrate against one another.

Figure 7.12 is an illustration of a long-wavelength transverse optical mode of vibration.

For the analysis of effects in which light interacts with the optical modes, it is sufficient to consider the problem from the standpoint of the long-wavelength approximation since the wavelength of light for infrared radiation is always much greater than the lattice spacing. Infrared absorption and Raman scattering, for example, can be handled in the long-wavelength approximation. Under the long-wavelength assumption the equations of motion for the vibration of the two sublattices assume the form

$$M_2 \frac{d^2 q_2}{dt^2} = C_T(q_1 - q_2) - C_T(q_2 - q_1) = -2C_T(q_2 - q_1)$$

$$M_1 \frac{d^2 q_1}{dt^2} = C_T(q_2 - q_1) - C_T(q_1 - q_2) = 2C_T(q_2 - q_1)$$

If we divide the first equation by M_2 and the second by M_1, and subtract the second from the first, we obtain an equation in terms of the difference variable $Q = q_1 - q_2$,

$$M \frac{d^2 Q}{dt^2} + \Omega^2 M Q = 0 \tag{7.27}$$

where M and Ω are given by (7.26). Thus we see that the optical-mode vibration in the long-wavelength approximation assumes the form of a simple harmonic oscillator of effective mass M and frequency Ω.

Since the quantization procedure for a harmonic oscillator is now familiar (Section 6.2), we know that introduction of the transformation

$$Q = \left(\frac{\hbar}{2M\Omega}\right)^{1/2}(a + a^\dagger) \qquad P = M\dot{Q} = -i\left(\frac{M\hbar\Omega}{2}\right)^{1/2}(a - a^\dagger) \tag{7.28}$$

into the harmonic oscillator Hamiltonian

$$\mathcal{H} = \frac{P^2}{2M} + \tfrac{1}{2}\Omega^2 M Q^2 \tag{7.29}$$

leads to

$$\mathcal{H} = \hbar\Omega(a^\dagger a + \tfrac{1}{2}) \tag{7.30}$$

Therefore the analysis of effects involving long-wavelength optical-mode phonons can proceed with the use of the operators in (7.28) and the Hamiltonian (7.30). The summary in Section 6.2 for the results of the quantum mechanical analysis of the harmonic oscillator is directly applicable.

7.4 CONSERVATION OF ENERGY AND MOMENTUM IN PROCESSES INVOLVING PHONONS

7.4.1 Conservation Conditions

When radiation interacts with the vibrational modes of a medium, certain conservation conditions apply. In the classical picture the conservation conditions take the form that the sum of the frequencies $\Sigma_j \omega_j$ and the sum of the wave vectors $\Sigma_j \mathbf{k}_j$ of the vibrational and electromagnetic waves gaining energy must match the sums corresponding to those losing energy. In the language of mode quantization, the matching of frequencies and wave vectors corresponds to the conservation of energy and momentum. In this section we shall examine the conservation conditions from a very general standpoint. Of special interest are certain geometrical construction techniques based on the dispersion characteristic, which are useful in the application and interpretation of the conservation conditions. Although the discussion is carried out on a classical basis, it has a counterpart in quantum language that justifies the application of the results to the description of quantum processes. This is because the conservation of energy and momentum applies equally well to both cases. The results that we derive are applicable whether the vibrational modes are those of a lattice or those of a continuum.

When an electromagnetic field exchanges energy with the vibrational modes of a medium, energy is delivered to the medium at the rate

$$\frac{\partial \mathcal{W}}{\partial t} = \int_{\Delta V} dV \mathbf{J} \cdot \mathbf{E} = \int_{\Delta V} dV \dot{\mathbf{P}} \cdot \mathbf{E} \tag{7.31}$$

where ΔV is the interaction volume, and $\mathbf{J}$ and $\mathbf{P}$ are those parts of the current density and polarization, respectively, resulting from vibrational motion. In linear absorption the polarization is associated directly with the vibration of charged masses or sublattices against each other. In nonlinear processes, such as Brillouin and Raman scattering, which are discussed later in the section, vibrational motion serves to modulate the dielectric constant or polarizability of the medium and thereby gives rise to nonlinear polarization terms.

According to (7.31) the energy delivered to the medium in a time Δt is given by

$$\mathcal{W} = \int_{\Delta t} dt \int_{\Delta V} dV \dot{\mathbf{P}} \cdot \mathbf{E} \tag{7.32}$$

In order for a significant amount of energy to be exchanged, it is necessary that the integrand contain terms that yield a finite contribution when integrated over time and space. It is the necessity of meeting this requirement that leads to the conservation of frequency and wavevector.

To derive the conservation conditions, let us begin by assuming that a polarization **P** is being generated in the presence of a field **E** as a result of vibrations in the medium. Next, consider the polarization and field to be expressed in terms of plane waves,

$$\mathbf{P} = \frac{1}{2} \sum_{\omega_P} \tilde{\mathbf{P}}(\omega_P) e^{i(\omega_P t - \mathbf{k}_P \cdot \mathbf{r})} + \text{c.c.} \tag{7.33}$$

$$\mathbf{E} = \frac{1}{2} \sum_{\omega_E} \tilde{\mathbf{E}}(\omega_E) e^{i(\omega_E t - \mathbf{k}_E \cdot \mathbf{r})} + \text{c.c.} \tag{7.34}$$

The subscript E refers to the electric field, and the subscript P refers to the polarization.

If the expansions (7.33) and (7.34) are substituted into (7.32), we obtain

$$\mathscr{W} = \frac{1}{4} \sum_{\omega_P, \omega_E} i\omega_P \tilde{\mathbf{P}}(\omega_P) \cdot \tilde{\mathbf{E}}^*(\omega_E) \int_{\Delta t} dt \int_{\Delta V} dV \, e^{i[(\omega_P - \omega_E)t - (\mathbf{k}_P - \mathbf{k}_E) \cdot \mathbf{r}]} + \text{c.c.}$$

$$= \frac{1}{2} \sum_{\omega_P, \omega_E} \text{Re}\left\{ i\omega_P \tilde{\mathbf{P}}(\omega_P) \cdot \tilde{\mathbf{E}}^*(\omega_E) \int_{\Delta t} dt \, e^{i(\omega_P - \omega_E)t} \int_{\Delta V} dV \, e^{-i[(\mathbf{k}_P - \mathbf{k}_E) \cdot \mathbf{r}]} \right\}$$

where Re indicates "the real part of." The only terms in the above expression that can be expected to yield a significant contribution when integration over a large number of cycles and a large number of wavelengths is considered are those that result in slowly varying exponentials for the integrands.

For example, for the integration over time we have

$$I = \int_{-\Delta t/2}^{\Delta t/2} dt \, e^{i(\omega_P - \omega_E)t} = \frac{2 \sin \left[(\omega_P - \omega_E) \Delta t/2 \right]}{\omega_P - \omega_E} \tag{7.35}$$

which is plotted in Figure 7.13. We see that the contribution is largest for $\omega_P = \omega_E$. Furthermore, if the integral is to exceed a given finite value, the amount by which ω_P and ω_E can differ becomes smaller as the integration time Δt becomes larger. In fact we have in the limit

$$\lim_{\Delta t \to \infty} I = 2\pi \, \delta(\omega_P - \omega_E)$$

where $\delta(\omega_P - \omega_E)$ is the Dirac delta function.

The conclusion to be drawn from the above is that for long times $[\Delta t \gg 2\pi/(\omega_P - \omega_E)]$ a significant amount of energy $\mathscr{W}$ will be exchanged only if

$$\omega_P = \omega_E \tag{7.36}$$

7 INTERACTIONS BETWEEN RADIATION AND PHONONS

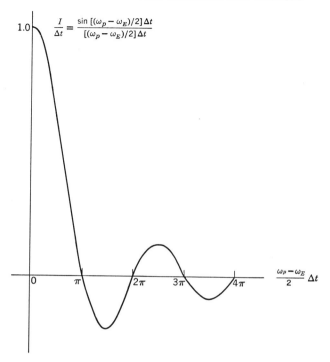

figure 7.13 Plot of integral $I/\Delta t$ in Equation 7.35. Integral I is a factor in the expression for the average energy transferred in time Δt between a field and polarization wave whose frequencies ω_E and ω_P, respectively, differ.

The integral in space leads to the conclusion that for $\Delta x \gg 2\pi/(\mathbf{k}_P - \mathbf{k}_E)_x$, and similarly for other coordinate directions, only for

$$\mathbf{k}_P = \mathbf{k}_E \qquad (7.37)$$

is there appreciable energy exchange between **P** and **E**. That is, energy is exchanged only if the polarization generated as a result of vibrational activity has the same frequency and wave vector as the field. Physically, we may say that (7.36) guarantees that the polarization and field remain in time phase and (7.37) insures that the polarization and field remain in phase in space so that power generated at a given instant in time and at a particular location in space adds in phase to power generated elsewhere in time and space. We shall now consider some specific applications of the above results. It will become apparent in the applications that (7.36) and (7.37) correspond, respectively, to the conservation of energy and momentum.

7.4.2. Conservation Conditions for Brillouin Scattering

As a first example of the conservation conditions let us consider the process of Brillouin scattering, which will be discussed in detail in Section 7.7. Brillouin scattering is a nonlinear process in which light incident on a medium is scattered from an acoustic vibration with a shift in frequency corresponding to that of the acoustic mode. The fractional frequency shift is small, approximately one part in 10^5. For scattering of visible light, shifts on the order of several GHz are typical. If an intense laser is used as the exciting source, *stimulated* Brillouin scattering can take place, and if it does, the scattered light wave at the Brillouin-shifted frequency has an intensity and spectral purity comparable to that of the laser. A detailed description of stimulated scattering is presented in Section 7.7.

The Brillouin effect occurs as a result of modulation of the optical index of refraction or polarizability of a medium by a sound wave. Since polarization is proportional to the product of polarizability and electric field, the fluctuations in the medium serve to produce beat note polarization terms at frequencies corresponding to the sum and difference of the applied optical frequency and the vibrational frequency. Let us confine our attention to the difference term.

In light of the above discussion, the description of the salient features of Brillouin scattering is based on a polarization term that is proportional to the product of a vibrational coordinate and the electric field. Therefore, we have

$$P = bqE \tag{7.38}$$

where b, q, and E are the proportionality constant, vibrational coordinate, and electric field, respectively. We have assumed that the polarization, vibration, and electric field are polarized in the same direction, in which case the vector notation may be dropped.

An incident optical wave

$$E_i = \tfrac{1}{2}\tilde{E}_i e^{i(\omega_i t - \mathbf{k}_i \cdot \mathbf{r})} + \text{c.c.}$$

interacting with a vibrational wave

$$q = \tfrac{1}{2}\tilde{q} e^{i(\omega_k t - \mathbf{k} \cdot \mathbf{r})} + \text{c.c.}$$

produces a difference frequency polarization term

$$P = \tfrac{1}{4}b\tilde{q}^*\tilde{E}_i e^{i[(\omega_i - \omega_k)t - (\mathbf{k}_i - \mathbf{k}) \cdot \mathbf{r}]} + \text{c.c.}$$

If the above polarization is to deliver appreciable energy to a scattered wave, E_s, given by

$$E_s = \tfrac{1}{2}\tilde{E}_s e^{(i\omega_s t - \mathbf{k}_s \cdot \mathbf{r})} + \text{c.c.}$$

then the conservation conditions (7.36) and (7.37) must apply:

$$\omega_s = \omega_i - \omega_k$$
$$\mathbf{k}_s = \mathbf{k}_i - \mathbf{k}$$

or

$$\hbar\omega_i = \hbar\omega_s + \hbar\omega_k \tag{7.39}$$
$$\hbar\mathbf{k}_i = \hbar\mathbf{k}_s + \hbar\mathbf{k} \tag{7.40}$$

The conservation conditions (7.39) and (7.40) correspond to the conservation of energy and momentum, respectively. The association of ω with energy and $\mathbf{k}$ with momentum is a result of the De Broglie relations that connect the wave and particle concepts inherent in the wave-particle duality of nature. The De Broglie relations state that with a free particle of energy E and momentum p there is associated a wave of frequency ω and wavelength λ given by[6]

$$E = \hbar\omega \tag{7.41}$$

$$p = \frac{h}{\lambda} = \frac{2\pi\hbar}{\lambda} = \hbar k \tag{7.42}$$

The Brillouin effect can therefore be described as the annihilation of an incident photon that results in the production of an acoustic phonon and a scattered photon with energy and momentum being conserved in the process. The scattering event is shown in Figure 7.14.

In the case of lattice vibrations, the vibrational wave vector $\mathbf{k}$ that appears in the conservation condition (7.40) may be replaced by the quantity $\mathbf{k} + \mathbf{G}$, where $\mathbf{G}$ is any vector in $\mathbf{k}$ space that connects a $\mathbf{k}$ vector for which $|\mathbf{k}| < \pi/a$ with those $\mathbf{k}$ vectors that correspond to identical particle motions for which $|\mathbf{k}| > \pi/a$, where a is the lattice spacing. (Recall the discussion following Equation 7.3.) For example, in one dimension we may replace k_x by $k_x + 2\pi n/a$, where n is an integer.

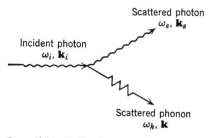

figure 7.14 Brillouin scattering consists of the annihilation of an incident photon, with the production of a phonon and scattered photon.

The vector **G** is termed a reciprocal lattice vector. The addition of **G** is permitted in all wave vector conservation laws applicable to crystal lattices. Processes for which **G** $\neq 0$ are called umklapp processes, which are important, for example, in determining the thermal resistivity of a lattice.[7]

One further statement is appropriate at this point concerning the labeling of $\hbar \mathbf{k}$ as a momentum. It can be shown that phonons other than the one for which $\mathbf{k} = 0$ do not actually carry physical momentum, because phonon coordinates involve the *relative* motion of all the atoms. (See Problem 7.2.) However, because of wave vector conservation laws, $\hbar \mathbf{k}$ plays the role of momentum and is therefore referred to as such.[8]

Since ω and **k** are the ordinate and abscissa, respectively, of a dispersion characteristic, we see that a photon or phonon can each be represented by a single vector on the appropriate dispersion characteristic, as shown in Figure 7.15a. We refer to such a vector as a dispersion vector, defined as a vector in $\omega \mathbf{k}$ space that originates at the origin and terminates at a point on a dispersion curve that corresponds to an allowed ω, **k** value of a photon or phonon under consideration. The conservation laws (7.39) and (7.40) can be expressed as the requirement that the dispersion vectors form a closed vector diagram as in Figures 7.15b and 7.15c.

Figure 7.15b shows collinear back-scattering of an incident light wave (ω_i, k_i) from an acoustic mode to produce a scattered light wave (ω_s, k_s). Note that the dispersion vector (ω_s, k_s), when translated parallel to itself so as to pass through the origin, lies on the dispersion curve for light, as it must (that is, vectors $0s$ and $0's'$ are the same length and parallel).

To consider scattering at an angle, as in Figure 7.14, we must replace the one-dimensional diagram of Figure 7.15b with a two-dimensional figure as in Figure 7.15c. These figures have been drawn for an isotropic medium in which the dispersion characteristics for light and acoustic waves are surfaces of revolution. The conservation condition can also be expressed by subtracting the acoustic dispersion curve from the point on the dispersion curve for light that corresponds to the input photon. The dispersion vectors corresponding to the scattered light then originate at the origin and terminate on the intersection of the two surfaces, as in Figure 7.15d. From this diagram we observe that the frequency of the scattered light is a function of the angle between the incident light, assumed propagating in the z direction, and the scattered light propagating in the yz plane. This frequency dependence on the angle of scattering is discussed in Section 7.7.

7.4.3 Raman Scattering

The phenomenon of Raman scattering offers a second example of conservation of energy and momentum in processes involving phonons. Raman scattering is similar to Brillouin scattering, differing primarily in that it involves scattering of light from optical rather than acoustic vibrational modes. The scattering of visible light from optical modes yields frequency shifts typically on the order of several hundred Angstroms.

If a powerful laser source is used to excite the Raman effect, *stimulated Raman emission* can occur—a process in which the shifted frequencies exhibit the spectral purity and high-intensity characteristic of laser emission. The stimulated

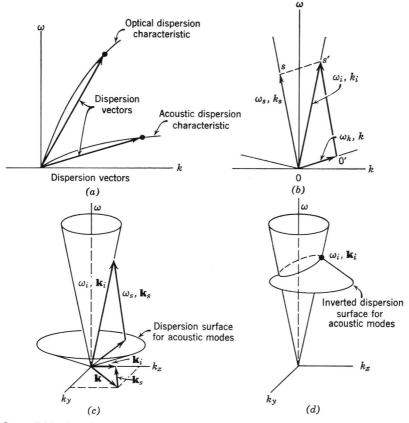

figure 7.15 (a) Dispersion vectors. (b) Generation of backscattered light, illustrating energy and momentum conservation for Brillouin scattering. (c) Dispersion vector diagram for scattering in a plane. The projection of a dispersion vector onto the vertical axis yields the frequency of the wave under consideration; the projection onto the horizontal plane yields the wave vector. Wave vector conservation is displayed by the projection of the closed vector diagram onto the horizontal plane. (d) In accordance with the conservation conditions, the locus of allowed scattered-light dispersion vectors is given by the intersection of two surfaces obtained by subtracting the $(\omega_k, \mathbf{k})$ surface corresponding to acoustic dispersion from the point on the dispersion characteristic for light corresponding to the input photon $(\omega_i, \mathbf{k}_i)$. The frequency of the scattered light is seen to be dependent on the angle in the yz plane.

Raman effect therefore provides a useful technique for frequency translation of coherent optical signals. A detailed discussion of the stimulated Raman effect is presented in Section 7.6.

The nonlinear polarization source term generated by the Raman process is of the same form as that applicable in the Brillouin process, and is therefore given by (7.38). The conservation conditions are given by (7.39) and (7.40). The corresponding closed dispersion vector diagrams are shown in Figure 7.16. They

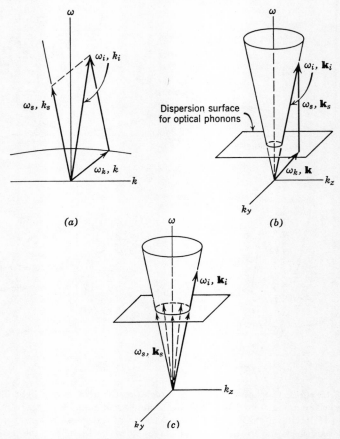

figure 7.16 Conservation of energy and momentum in Raman scattering as displayed on dispersion diagrams: (a) backward Raman scattering; (b) Raman scattering in a plane; (c) locus of allowed scattered-light vectors, given by the intersection of two surfaces. The frequency of the scattered light is seen to be essentially independent of the angle in the yz plane because the optical mode has very little dispersion in the region of interest.

are analogous to those of Figure 7.15 except that the dispersion characteristic for the acoustic mode is replaced by that of an optical mode. In Figure 7.16c the locus of allowed dispersion vectors for the scattered light is obtained by subtraction of the optical-mode dispersion characteristic from the dispersion vector of the incident light. Since the optical mode shown is relatively flat in the region of interest, which is typically the case, the frequency of the scattered light is essentially independent of the angle between incident and scattered light. For the case where the optical vibrational mode interacts strongly with an electromagnetic wave (that is, the mode is infrared-absorbing as discussed in Section 7.5) the optical mode does have dispersion. In this case the frequency of the mode and hence of the scattered radiation is a function of angle.

The technique that combines the conservation laws and dispersion diagrams, applied in this section to Brillouin and Raman scattering, can be applied to scattering processes in general. (See Problem 7.4.)

7.5 INFRARED PROPERTIES OF OPTICAL PHONONS

When optical phonons propagate in a diatomic crystal, charge separation may or may not take place, depending on the type of bonding. A crystal formed by ionic bonding, such as NaCl, is made up of positive and negative ions. In this case charge separation takes place during optical phonon propagation as the charged sublattice structures vibrate against one another. For a crystal such as diamond which is formed by covalent bonding, on the other hand, the sublattice structures are essentially neutral and little or no charge separation takes place as the sublattices vibrate against one another. A given crystal may fall anywhere in the range between pure ionic and pure covalent bonding. Crystals formed with atoms with nearly filled shells (alkali halides) tend to be ionic, whereas crystals formed with atoms in columns II, III, and IV of the periodic table (C, Ge, Si, Te) tend to be covalent.

As a result of the separation of charge that accompanies optical phonon propagation in ionically bonded crystals, optical phonons interact strongly with an electromagnetic field that oscillates at the vibrational frequency, typically in the infrared. This interaction leads to marked infrared absorption and reflection properties, which we shall now examine. An optical mode that couples strongly to an electromagnetic field is said to be *infrared active*.

We shall employ the semiclassical approach, by which the fields are treated classically and the optical vibrations of the medium are treated quantum mechanically.

The total Hamiltonian for a unit cell is given by

$$\mathcal{H} = \mathcal{H}_0 + \mathcal{H}' \tag{7.43}$$

where $\mathcal{H}_0$ is the unperturbed part as given by (7.29)

$$\mathcal{H}_0 = \frac{1}{2M}(P^2 + M^2\Omega^2 Q^2) = \hbar\Omega(a^\dagger a + \tfrac{1}{2}) \tag{7.44}$$

and $\mathcal{H}'$ is the electric dipole interaction Hamiltonian

$$\mathcal{H}' = -\mathbf{\mu}\cdot\mathbf{E} = -ZEQ = -ZE\left(\frac{\hbar}{2M\Omega}\right)^{1/2}(a + a^\dagger) \tag{7.45}$$

The interactions Hamiltonian $\mathcal{H}'$ corresponds to the separation of a charge Z as a result of the relative displacement Q of the two sublattices, $Q = q_2 - q_1$. We consider here an electric field and phonon displacement that are parallel.

The equation of motion for the expectation value of the vibrational mode amplitude, $\langle Q \rangle$, is found in the manner now familiar to the reader by use of (1.51):

$$\langle \ddot{Q} \rangle + \frac{2}{T_2}\langle \dot{Q} \rangle + \frac{1}{T_2^2}\langle Q \rangle = -\frac{1}{\hbar^2}\langle [[Q,\mathcal{H}],\mathcal{H}]\rangle \tag{7.46}$$

Evaluation of the inside commutator in (7.46) with the aid of (7.44) and (7.45) yields

$$[Q, \mathcal{H}] = [Q, \mathcal{H}_0 + \mathcal{H}'] = [Q, \mathcal{H}_0] + [Q, \mathcal{H}']$$
$$= [Q, \mathcal{H}_0] = \hbar\Omega\left(\frac{\hbar}{2M\Omega}\right)^{1/2}(a - a^\dagger)$$

since $\mathcal{H}'$ commutes with Q. Evaluation of the outside commutator proceeds similarly, yielding

$$[[Q,\mathcal{H}],\mathcal{H}] = \hbar^2\Omega^2 Q - \frac{\hbar^2 ZE}{M}$$

The resulting equation of motion is, therefore,

$$\langle \ddot{Q} \rangle + \Delta\omega_L \langle \dot{Q} \rangle + \Omega^2 \langle Q \rangle = \frac{ZE}{M} \tag{7.47}$$

where $\Delta\omega_L = 2/T_2$ is the linewidth and we have taken $\Omega^2 \gg 1/T_2^2$.

If there are $N_V = \mathcal{N}/V$ identical unit cells per unit volume, the polarization associated with the phonon wave is given by

$$P = \frac{1}{V}\sum_{i=1}^{\mathcal{N}}\langle \mu \rangle_i = N_V \langle \mu \rangle = N_V Z \langle Q \rangle \tag{7.48}$$

where Z is the charge separation which takes place in the unit cell during the relative displacement Q of the sublattices.

Equations 7.47 and 7.48 describe the behavior of the medium. To this we add the equation of propagation for the electromagnetic field, (2.50),

$$\nabla \times (\nabla \times \mathbf{E}) + \frac{\eta_\infty^2}{c^2}\frac{\partial^2 E}{\partial t^2} = -\mu_0 \frac{\partial^2 P}{\partial t^2} \tag{7.49}$$

We shall show by (7.52) which follows that η_∞ is the refractive index measured on the high-frequency side of the transition of interest. In (7.47) and (7.48) the fields are local fields and the polarization P is the actual polarization defined by (7.48), which, for dense media, are different from the macroscopic electric field and polarization source term that appear in the wave equation, (7.49). As discussed in Section 2.4.7, the differences between the local and macroscopic fields and between the actual polarization and the polarization source term that drives the field equation result from the influence of nearby polarizable matter on the local electric field seen by the unit cell. Consideration of this effect leads to Lorentz correction factors.

If (7.47) and (7.48) are rewritten with the Lorentz factors included by use of the technique discussed in Section 2.4.7, (7.47) and (7.48) retain the same form with a relabeling of the fields and polarization as the macroscopic variables that appear in the field equation, (7.49). The introduction of Lorentz correction factors alters the resonant frequency Ω slightly and introduces a constant factor as a multiplier of the effective charge Z. Therefore we can proceed with the understanding that Equations 7.47 through 7.49 are compatible as they stand, where P and E are now macroscopic variables, and the Lorentz correction factor is absorbed into Z.

To solve (7.47)–(7.49), we assume solutions for P, $\langle Q \rangle$, and E of the form

$$E = \frac{\tilde{E}}{2} e^{i(\omega t - kz)} + \text{c.c.} \tag{7.50}$$

that correspond to traveling waves propagating in the z direction. Equations 7.47–7.49 yield two types of solutions, transverse and longitudinal, that is, solutions in which P, $\langle Q \rangle$, and E are either perpendicular to or parallel with the direction of propagation. We shall consider the transverse solutions first. In the transverse solutions, $\nabla \times (\nabla \times E) = k^2 E$, and the simultaneous solution of (7.47)–(7.49) yields

$$k^2 = \frac{\omega^2 \kappa(\omega)}{c^2} \tag{7.51}$$

where κ is the complex dielectric constant given by

$$\kappa(\omega) = \kappa_\infty + \frac{\Omega^2 \Delta\kappa}{\Omega^2 - \omega^2 + i\Delta\omega_L \omega} \tag{7.52}$$

The parameter $\Delta\kappa = N_V Z^2 / \Omega^2 M \epsilon_0$ is the contribution to the dielectric constant resulting from the ionic vibration as one passes through the transition frequency, and is therefore a measure of the strength of the transition. The parameter $\kappa_\infty = \eta_\infty^2$ is the dielectric constant on the high-frequency side of the transition of interest, as may be seen from (7.52) by letting $\omega \to \infty$.

Let us take as an example the cubic crystal GaP of point group $\bar{4}3m$. The appropriate parameter values are

$$\lambda_t = \frac{2\pi c}{\Omega} = 27.3\mu \quad \text{(the wavelength of the transverse vibration)}$$

$$\kappa_\infty = 8.457$$

$$\Delta\kappa = 1.725$$

$$\frac{\Delta\omega_L}{\Omega} = 0.003$$

Since $\Delta\omega_L/\Omega$ is so small, we shall ignore the effect of this term (that is, we shall ignore the effect of damping) in the discussion that follows unless we specifically state to the contrary. With $\Delta\omega_L/\Omega$ set equal to zero, the dielectric function $\kappa(\omega)$ as given by (7.52) is then purely real.

A plot of $\kappa(\omega)$ for GaP is shown in Figure 7.17. At $\omega = 0$ the dielectric constant starts out at its dc value, $\kappa_{dc} = \kappa_\infty + \Delta\kappa = 10.182$, and rises sharply as resonance is approached. On the high-frequency side of resonance, the dielectric constant climbs from a large negative value, passes through zero at a frequency ω_l

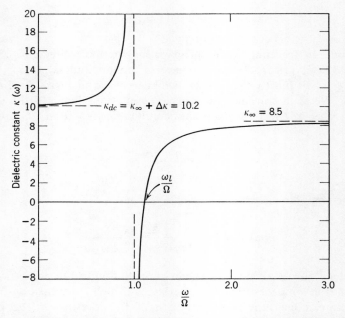

figure 7.17 Frequency dependence of dielectric constant $\kappa(\omega)$ of GaP.

given by $(\omega_l/\Omega) = 1.1$, and asymptotically approaches the high-frequency value $\kappa_\infty = 8.457$. As we shall see from the longitudinal solution given later in this section, the frequency ω_l is the longitudinal mode frequency and corresponds in this case to a wavelength $\lambda_l = 24.9\ \mu$. We note from Figure 7.17 that in the region $\Omega < \omega < \omega_l$ the dielectric constant $\kappa(\omega)$ is negative. This implies that in this region the index of refraction $\eta(\omega) = \sqrt{\kappa(\omega)}$ is imaginary. The surface reflectivity $\mathscr{R}$, which measures the fractional part of incident power reflected from a surface of refractive index $\eta(\omega)$, is given by[9]

$$\mathscr{R} = \left[\frac{\eta(\omega) - 1}{\eta(\omega) + 1}\right]^2$$

Therefore, $\mathscr{R}$ has a magnitude of unity in the region where $\eta(\omega)$ is imaginary, implying that in this frequency range there is a stopband in which electromagnetic radiation is totally reflected from the surface. This is known as *Reststrahl* or *residual ray* reflection.† A plot of the reflectivity for GaP is given in Figure 7.18. The deviation of the maximum reflectivity from unity can be accounted for when the effects of damping are considered. The stopband is observed to lie between $\lambda_t = 27.3\ \mu$ and $\lambda_l = 24.9\ \mu$.

The dispersion characteristic (ωk diagram) is shown in Figure 7.19. At frequencies well below the vibrational frequency, radiation propagates at a velocity $c/\sqrt{\kappa_{dc}}$. In this region the lattice can follow the slowly varying fields, and this motion contributes to the dielectric constant. Since the frequency is far below resonance, however, the mechanical displacements are small and therefore most

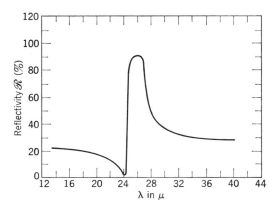

figure 7.18 Reflectivity of GaP.

† This nomenclature results from the experimental use of crystals of this type as multiple reflection surfaces from which a relatively monochromatic residual ray can be extracted from an initially broadband light source.

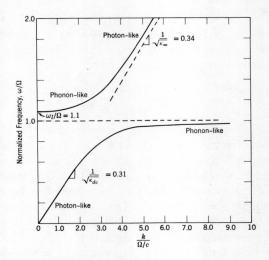

figure 7.19 Dispersion characteristic of GaP.

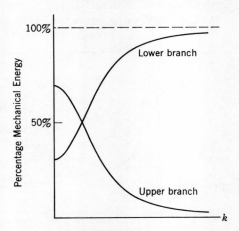

figure 7.20 Percentage mechanical energy in the transverse modes.[10] The upper and lower branches correspond to frequencies above and below the transition frequency Ω, respectively, as shown in Figure 7.19.

of the energy is in the electromagnetic wave. In this region the coupled-mode system can be said to be photon-like.

As the frequency approaches the transverse lattice vibrational frequency Ω, the lattice vibrations couple strongly to the infrared radiation and the dispersion curve departs radically from the photon-like characteristic. In this region most of the energy is in the form of mechanical energy associated with the lattice displacements. The percentage of mechanical energy as a function of k is shown in Figure 7.20.[10]

In the stopband region $\Omega < \omega < \omega_l$ the coupled mode does not propagate. On the high-frequency side of the stopband the mode begins again as a predominantly mechanical vibration. Away from resonance, however, the dispersion characteristic of the coupled-mode system again becomes photon-like, although with a reduced dielectric constant because of the inability of the lattice vibrations to follow the rapidly varying field. The contributions to the dielectric constant in this region are from higher frequency resonances such as electronic resonances. For some crystals there may be higher-lying ionic vibrations of the type under discussion, and if so, their contributions must also be included. The parameter values for several common crystals that possess a single ionic vibration are listed in Table 7.1.

The dispersion curve just described is that of a coupled-mode system, partly electromagnetic, partly mechanical in nature. As the coupling between the radiation and the lattice vibration approaches zero (that is, $Z \to 0$), the dispersion characteristic displayed in Figure 7.19 approaches the form of two intersecting straight lines; one is horizontal, corresponding to a flat phonon dispersion

TABLE 7.1 CRYSTAL PARAMETERS FOR INFRARED-ACTIVE OPTICAL MODES[11]

Crystal	Structure	Static dielectric constant κ_{dc}	Optical dielectric constant κ_∞	Ω in 10^{13} sec^{-1} (experimental)
LiF	NaCl	8.9	1.9	5.8
NaF	NaCl	5.1	1.7	4.5
NaCl	NaCl	5.9	2.25	3.1
NaBr	NaCl	6.4	2.6	2.5
KCl	NaCl	4.85	2.1	2.7
KI	NaCl	5.1	2.7	1.9
RbI	NaCl	5.5	2.6	1.4
CsCl	CsCl	7.2	2.6	1.9
TlCl	CsCl	31.9	5.1	1.2
TlBr	CsCl	29.8	5.4	0.81
AgCl	NaCl	12.3	4.0	1.9
AgBr	NaCl	13.1	4.6	1.5
MgO	NaCl	9.8	2.95	7.5

characteristic that would exist in the absence of coupling to radiation, and the other is diagonal, corresponding to radiation dispersion in the absence of vibrations. Viewed from this standpoint, the dispersion characteristic for finite coupling is seen to be a splitting of the uncoupled dispersion characteristics in the region where the frequencies and wave vectors are nearly identical. Such behavior is typical of coupled-mode systems.

When damping is taken into account, the dielectric constant given by (7.52) has both real and imaginary parts because of the linewidth factor $\Delta\omega_L$. The effect of the inclusion of damping on the curves for the dielectric constant and dispersion is to connect the separate parts in a continuous fashion as shown in Figure 7.22c and 7.22d. To discuss the effects of damping further, we may introduce a complex index of refraction $\sqrt{\kappa(\omega)} = \eta(\omega) - iK(\omega)$, where $\eta(\omega)$ is the real part and $K(\omega)$, known as the extinction coefficient, accounts for loss. From (7.50) and (7.51) we see that the extinction coefficient K is related to the energy absorption constant Γ by $K = \Gamma c/2\omega$. The quantities η and K for GaP are plotted in Figure 7.21.

In addition to the transverse solutions to (7.47)–(7.49) considered above, longitudinal solutions also exist. For this case P, $\langle Q \rangle$, and E are parallel to the direction of propagation. Here $\nabla \times (\nabla \times E) = 0$ and the solution of (7.47)–(7.49) yields $\kappa = 0$, where κ is again given by (7.52). If we can neglect the damping term and solve (7.52) for ω, the expression for the longitudinal mode frequency, ω_l, is

$$\frac{\omega_l}{\Omega} = \left(\frac{\kappa_{dc}}{\kappa_\infty}\right)^{1/2} \tag{7.53}$$

where $\kappa_{dc} = \kappa_\infty + \Delta\kappa$. The above expression (7.53) is known as the Lyddane-Sachs-Teller relation. The frequency of the longitudinal mode is seen to be independent of the wavevector k, and therefore the corresponding dispersion curve is flat.

For the longitudinal modes P, $\langle Q \rangle$, and E are polarized parallel to the direction of propagation. Therefore, the longitudinal lattice vibrations are not coupled to optical radiation fields, for in the latter case the fields are perpendicular to the direction of propagation.

From (7.51) we see that the longitudinal condition $\kappa = 0$ also corresponds to the $k = 0$ point for the transverse solution with finite ω. Thus the upper branch of the transverse mode solution begins at $\omega = \omega_l$, so that ω_l is also the upper frequency of the stopband associated with the transverse modes.

The treatment presented in this section can be extended to cover several vibrational modes existing between $\omega = 0$ and the frequencies of the electronic resonances.[12] In this case (7.52) generalizes to

$$\kappa(\omega) = \kappa_\infty + \sum_j \frac{\Omega_j^2 \Delta\kappa_j}{\Omega_j^2 - \omega^2 + i(\Delta\omega_L)_j \omega}$$

7 INTERACTIONS BETWEEN RADIATION AND PHONONS

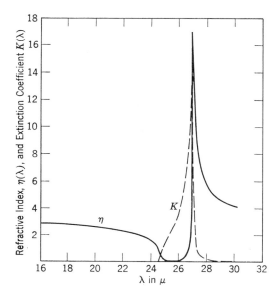

figure 7.21 Index of refraction and extinction coefficient for GaP.

The static dielectric constant κ_{dc} is then given by

$$\kappa_{dc} = \kappa_\infty + \sum_j \Delta\kappa_j$$

The parameter values, reflectivity, dielectric constant, and the dispersion curves for the transverse modes in MgF_2 with E perpendicular to the c axis are shown in Figure 7.22.

Before leaving the subject of the absorption and dispersion properties of optical phonons, we observe that the results describe the absorption and dispersion by electronic resonances as well. This was considered in Chapter 3 in terms of the interaction of radiation with a two-level electric dipole transition. The analysis in Chapter 3 led to an equation for the propagation constant, (3.11a), which is of the same form as (7.51) and (7.52). Therefore it is apparent that the results based on (7.51) and (7.52) apply also to the electronic case.

7.6 RAMAN EFFECT

In 1928, Raman and Krishnan, in a series of light-scattering experiments with liquids and gaseous vapors, observed that a certain small fraction of the light intensity (10^{-6}–10^{-7}) experienced a substantial frequency shift ($\sim$ several hundred Angstroms) during the scattering process.[13] In the same year the effect was

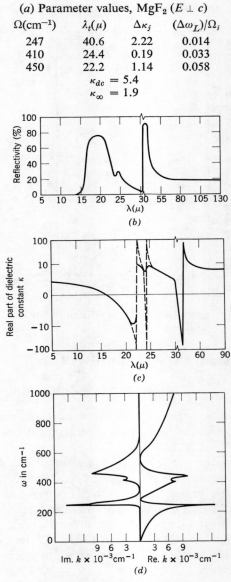

figure 7.22 Transverse optic modes in MgF$_2$ for $E \perp c$ axis: (a) parameter values; (b) reflectivity; (c) real part of dielectric constant $\kappa(\omega)$ as determined from reflection data. The dashed curve corresponds to damping set equal to zero. (d) Dispersion curves.

observed independently by Landsberg and Mandelstam in work with crystals.[14] In their experiments the frequency shifts were found to correspond to optical vibrational modes of the scattering substance under consideration. From these observations it was inferred that the dielectric constant or polarizability of the medium was being modulated by the vibrational modes, resulting in a mixing action that beat together the input radiation and optical modes to produce sum and difference frequencies. These processes are illustrated in Figure 7.23. The phenomenon has become known as the Raman effect and is now a basic tool in spectroscopic work for determining the optical mode structure. As a spectroscopic tool it is a useful adjunct to infrared absorption spectroscopy, in that modes that are not infrared-absorbing because of parity considerations may be active in Raman scattering.

Interest in the Raman effect has increased sharply as intense laser sources have become available. One reason is that the effect is now more readily observable

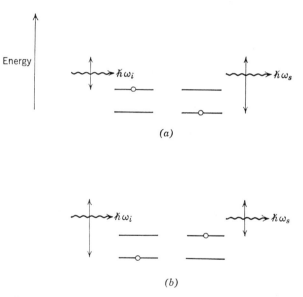

figure 7.23 The Raman effect is a nonlinear light-scattering process in which an incident light wave is scattered from an optic mode of the medium to produce photons at the sum and difference frequencies. This figure indicates how photons in the incident beam are converted into frequency-shifted scattered photons with an accompanying change of state of the medium. The lengths of the vertical arrows correspond to the photon energies. (a) Sum frequency production (anti-Stokes scattering). (b) Difference frequency production (Stokes scattering).

and therefore more useful as a tool in the study of material properties. Another reason is that the shifted lines can themselves participate in stimulated processes characteristic of lasers and thus act as new sources of coherent optical radiation. This effect is called the *stimulated* Raman effect.

The essential features of the stimulated Raman effect are best described through a comparison with ordinary or spontaneous Raman scattering. In spontaneous Raman scattering, light that is reradiated at the Raman frequency (incident frequency minus vibrational mode frequency) is emitted randomly in all directions. Since each radiator acts independently of the others, there is no phase coherence between the radiation from separate radiators. The scattering efficiency is low, on the order of 10^{-6} scattered photons per incident photon. The spectral width of the output radiation corresponds to the width of the vibrational level of the medium.

In stimulated Raman scattering, on the other hand, the output radiation is not emitted in all directions, but in a well-defined direction. Thus the radiators do not act independently, but rather are locked in phase and emission direction because of the effects of high intensity radiation at the Raman frequency. In the stimulated effect, contrary to the spontaneous effect, the emission at the Raman frequency is influenced by the radiation already present at the Raman frequency. The high intensity, which may approach that of the incident beam, results from a gain mechanism at the Raman frequency. Since the gain mechanism is sensitively dependent on frequency, the stimulated output occurs primarily at that frequency where the gain is highest. As a result, the spectral width of the output is narrow compared with that of the spontaneous emission. The difference between spontaneous and stimulated emission is discussed more fully in Section 7.6.2.

The stimulated Raman effect was discovered in 1962 in an experiment in which an unexpected emission line at 7670 Å in the output of a Q-switched ruby laser (6943 Å) was observed.[15] From its intensity and collimation it was apparent that the new line resulted from stimulated emission, although the wavelength corresponded to no known transition in ruby. The effect was interpreted as resulting from stimulated Raman emission from molecular vibrations in the nitrobenzene-filled Kerr cell used for Q-switching.[16,17] Soon after the original discovery, stimulated Raman emission was observed from several other materials, including both organic[18] and inorganic[19] liquids, crystals,[20] and gases.[21] Concurrent with the early experimental work, theory was developed to explain the stimulated effect.[15,23–25]

Whereas ordinary Raman scattering efficiencies are of the order of 10^{-6} scattered photons per incident photon, the stimulated effect has yielded laser-to-Raman conversion efficiencies as high as 30%, and efficiencies approaching unity are theoretically possible. The stimulated Raman effect thus provides an efficient means for producing coherent optical signals throughout the visible and infrared spectrum in a selectively controlled manner using only a single laser source as

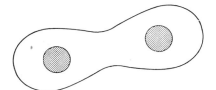

figure 7.24 Nonvibrating molecule consisting of nuclei surrounded by electron cloud.

exciter, and either a crystal or a cell containing a liquid or gas as a scattering medium. More than 100 coherent lines have been observed,[26,27] and the range of frequencies covered extends considerably beyond the simple Raman shifts because of iterative and parametric processes in which the initially generated Raman line itself acts as a pump source for further shifts. A list of materials from which stimulated Raman scattering has been observed as of this writing is given in Appendix 8.

7.6.1 General Considerations—Classical Approach

The basic properties of Raman scattering can be understood with the aid of a relatively simple classical picture. An individual molecule or unit cell of a crystal from which scattering takes place typically consists of two or more nuclei bonded together and surrounded by an electron cloud, as shown in Figure 7.24.

If an incident radiation field in the visible or near infrared portion of the spectrum is allowed to interact with the molecule or unit cell, an electric dipole moment μ is induced. If the frequency of the radiation is well above the vibrational resonances and well below the electronic resonances, then, in accordance with (7.47)–(7.48) and (2.35), the dipole moment is directly proportional to the electric field,

$$\mu_i = \alpha_{ij} E_j = \alpha_{ij} \tilde{E}_j \cos \omega t \tag{7.54}$$

where α_{ij} is the electronic polarizability tensor of the molecule. Summation over repeated indices is understood. The polarizability α_{ij} results from a sum over expressions of the form (2.35) when the frequency of the field is well below the transition frequency. This represents a sum over electronic transitions. For a physical description we say that the dipole moment results from the displacement of the electron charge cloud with respect to the nuclei. In this interaction the relatively light electron cloud follows the incident field, and to a first approximation the heavier nuclei remain stationary. Assuming that the nuclei remain stationary, the electronic polarizability α_{ij} at any frequency is simply a constant that has a value characteristic of the molecule or unit cell. It is this polarizability that accounts for the index of refraction of a medium at frequencies above the ionic vibrational resonances.

The assumption that the nuclei remain stationary is not quite correct, however, for some of the energy absorbed by the electrons is transferred to the nuclei because of the motion of the electron cloud. As a result, the nuclei begin to oscillate about their equilibrium positions as shown in Figure 7.25.

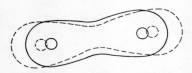

figure 7.25 Vibrating molecule. The vibrations are induced by energy transferred to the nuclei by the motion of the electron cloud. The frequency of oscillation of the nuclei corresponds not to the frequency of the oscillating electron cloud that follows the applied field, but rather to a molecular vibrational frequency. For typical structures (organic liquids such as benzene or chloroform, or crystals such as lithium niobate or diamond) the molecular vibrations lie in the 3–30 μ portion of the infrared spectrum.

As the nuclei oscillate, the electronic polarizability α_{ij} changes as the nuclear configuration changes resulting from the distortion of the environment seen by the electrons. (See Figure 7.26.) The situation can be expressed mathematically by expanding the electronic polarizability α_{ij} in a Taylor's series in terms of normal mode vibrational coordinates such as $Q = q_2 - q_1$, applicable to the diatomic crystal discussed in Section 7.3.3. This approach, due originally to Placzek, takes the form[28]

$$\alpha_{ij} = \alpha_{ij}^0 + \left(\frac{\partial \alpha_{ij}}{\partial Q}\right)_0 Q + \cdots$$
$$= \alpha_{ij}^0 + \left(\frac{\partial \alpha_{ij}}{\partial Q}\right)_0 \tilde{Q} \cos \omega_v t + \cdots \quad (7.55)$$

where ω_v is the vibrational frequency and $(\partial \alpha_{ij}/\partial Q)_0$ is a derivative evaluated at the equilibrium positions of the nuclei.

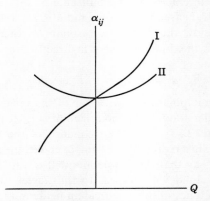

figure 7.26 Polarizability as a function of a vibrational coordinate Q. If the polarizability changes as in I, then a first-order Raman effect exists; if the polarizability changes as in II, then the first-order effect vanishes.

Substitution of (7.55) into (7.54) leads to

$$\mu_i = \alpha_{ij}^0 \tilde{E}_j \cos \omega t + \frac{1}{2}\left(\frac{\partial \alpha_{ij}}{\partial Q}\right)_0 \tilde{Q}\tilde{E}_j \left[\cos(\omega_i + \omega_v)t + \cos(\omega_i - \omega_v)t\right] + \cdots \quad (7.56)$$

From this equation we see that as the molecule vibrates, changes in the electronic polarizability lead to dipole moments that give rise to pairs of emission lines, one above and one below the excitation frequency. Each is displaced by an amount corresponding to the vibrational frequency. It is the generation of these displaced lines that constitutes the Raman effect. The lower-frequency emission lines are called Stokes lines, the higher the anti-Stokes. The terms "Stokes" and "anti-Stokes" follow from Stokes' law in fluorescence studies, which states that emission lines generally are lower in frequency than excitation lines.

As is evident in (7.56), the Raman effect is proportional to the factor $(\partial \alpha_{ij}/\partial Q)_0$. Therefore, if the polarizability changes with vibrational coordinate as shown in curve I, Figure 7.26, the Raman effect is present; if it changes as in curve II, the first-order Raman effect vanishes and the next higher order term in the expression (7.55) must be considered. The Raman effect therefore depends on how the polarizability changes as a function of vibrational displacements. Whether the Raman effect is present for a particular symmetry vibration of a particular symmetry molecule or crystal can be determined by reference to Herzberg,[29] where the Raman properties of vibration of the 32 point groups are listed.

The intensities of the Raman lines depend strongly on the type of bonding, and are usually much higher for covalent than for ionic bonds. The Raman intensity depends on how much the polarizability changes with displacement during a vibration. For a covalent bond, valence electrons are shared by the atoms, and a change in the distance between the nuclei strongly affects the polarizability. In ionic bonding, however, each electron is essentially under the influence of only one nucleus and the change in polarizability during the vibration is small.

We may now obtain an expression for the term in the Hamiltonian that gives rise to the Raman process. From (7.54) and (7.55) we see that the dipole moment associated with the Raman effect is given by

$$\mu_i = \alpha'_{ij}(Q) E_j \quad (7.57)$$

where $\alpha'_{ij}(Q)$ is given by

$$\alpha'_{ij}(Q) = \left(\frac{\partial \alpha_{ij}}{\partial Q}\right)_0 Q \quad (7.58)$$

The interaction energy term in the presence of an electric field takes the form characteristic of the potential energy stored in a polarizable medium,[30,31]

$$\mathcal{H}' = -\tfrac{1}{2} \alpha'_{ij}(Q) E_j E_i \quad (7.59)$$

7.6.2 Quantum Mechanical Treatment

In this section we derive the equations of motion that describe the Raman effect. We employ a semiclassical approach, treating the fields classically and the medium quantum mechanically. Just as with resonance absorption and emission, a semiclassical treatment of the fields results in a description of the *stimulated* Raman effect only. A description of the spontaneous Raman effect, which follows from field quantization, is, however, included on the basis of the "extra photon" approach. This approach, we recall from Chapter 6, provides that spontaneous emission terms can be introduced into semiclassically derived formulas for emission processes provided that we let $n \to n + 1$, where n is the number of photons in a radiation mode at the emission frequency.

Since the Raman effect depends on optical vibrational modes, the properties of the medium can be described in terms of the quantized optical-mode development derived for the diatomic crystal in Section 7.3.3. That derivation for the oscillation of two sublattices against each other is equally applicable to other harmonic oscillations (for example, vibrational modes of a molecule) provided that the effective mass is suitably defined.

The total Hamiltonian for the problem is given by $\mathcal{H} = \mathcal{H}_0 + \mathcal{H}'$. The unperturbed part $\mathcal{H}_0$ is, from (7.30),

$$\mathcal{H}_0 = \hbar\Omega(a^\dagger a + \tfrac{1}{2}) \tag{7.60}$$

On the basis of the discussion in Section 7.6.1, the interaction term for a given vibrational mode takes the form

$$\mathcal{H}' = -\tfrac{1}{2}\alpha'(Q)E^2$$

$$= -\tfrac{1}{2}\left(\frac{\partial\alpha}{\partial Q}\right)_0 Q E^2 \tag{7.61}$$

where from (7.28) the operator Q corresponding to the normal mode vibration is given by

$$Q = \left(\frac{\hbar}{2M\Omega}\right)^{1/2}(a + a^\dagger) \tag{7.62}$$

For the sake of simplicity, we are considering a one-dimensional problem in which the polarization of the fields, vibration, and induced dipole moments are all parallel, which is often the case in experimental work.

The equation of motion for the expectation value of the vibrational mode amplitude, $\langle Q \rangle$, is found by application of (1.51):

$$\langle \ddot{Q} \rangle + \frac{2}{T_2}\langle \dot{Q} \rangle + \frac{1}{T_2^2}\langle Q \rangle = -\frac{1}{\hbar^2}\langle [[Q, \mathcal{H}], \mathcal{H}]\rangle \tag{7.63}$$

Evaluation of the inside commutator in (7.63) with the aid of (7.60)–(7.62) yields

$$[Q, \mathcal{H}] = [Q, \mathcal{H}_0 + \mathcal{H}'] = [Q, \mathcal{H}_0] + [Q, \mathcal{H}'] = [Q, \mathcal{H}_0]$$

$$= \hbar\Omega\left(\frac{\hbar}{2M\Omega}\right)^{1/2}(a - a^\dagger)$$

since $\mathcal{H}'$ commutes with $\langle Q \rangle$.

Evaluation of the outside commutator proceeds similarly, yielding

$$[[Q, \mathcal{H}], \mathcal{H}] = \hbar^2\Omega^2 Q - \frac{\hbar^2}{2M}\left(\frac{\partial\alpha}{\partial Q}\right)_0 E^2(aa^\dagger - a^\dagger a)$$

$$= \hbar^2\Omega^2 Q - \frac{\hbar^2}{2M}\left(\frac{\partial\alpha}{\partial Q}\right)_0 E^2 \qquad (7.64)$$

where elimination of the operator $(aa^\dagger - a^\dagger a)$ on the right-hand side of (7.64) results from the application of the commutator relationship $[a, a^\dagger] = 1$. We shall see that it is sometimes useful to retain the longer operator form to facilitate physical interpretation of the results.

Substitution of (7.64) into (7.63) yields these equivalent alternative expressions for $\langle Q \rangle$:

$$\langle \ddot{Q} \rangle + \Delta\omega_L \langle Q \rangle + \Omega^2 \langle Q \rangle = \frac{1}{2M}\left(\frac{\partial\alpha}{\partial Q}\right)_0 E^2(\langle n_p \rangle + 1) - \frac{1}{2M}\left(\frac{\partial\alpha}{\partial Q}\right)_0 E^2 \langle n_p \rangle \quad (7.65a)$$

$$= \frac{1}{2M}\left(\frac{\partial\alpha}{\partial Q}\right)_0 E^2 \qquad (7.65b)$$

where $\Delta\omega_L = 2/T_2$ is the linewidth of the vibration, Ω is the resonant frequency of the vibration, and we have taken $\Omega^2 \gg 1/T_2^2$. The quantity $\langle n_p \rangle$ is the expectation value of the number of phonons in the vibration. The coefficients written in terms of $\langle n_p \rangle$ result from the application of (6.20) to the operator in parentheses in (7.64), that is,

$$\langle a^\dagger a \rangle = \langle n_p \rangle$$
$$\langle a a^\dagger \rangle = \langle a^\dagger a + 1 \rangle = \langle n_p \rangle + 1$$

We shall examine the physical significance of the longer form, (7.65a), later in this section.

If there are $N_V = \mathcal{N}/V$ identical unit cells or molecules per unit volume, the nonlinear polarization associated with the phonons is, from (7.57) and (7.58),

$$P = \frac{1}{V}\sum_{i=1}^{\mathcal{N}} \langle\mu\rangle_i = N_V \overline{\langle\mu\rangle} = N_V \overline{\left(\frac{\partial\alpha}{\partial Q}\right)_0 \langle Q\rangle E} \qquad (7.66)$$

where the overbar indicates a spatial average.

Equations 7.65 and 7.66 describe the behavior of the medium in the presence of an electromagnetic field. In order to close the system of equations, we add the wave equation (2.50) for the propagation of the field, ignoring loss,

$$\nabla \times (\nabla \times E) + \frac{\eta^2}{c^2} \frac{\partial^2 E}{\partial t^2} = -\mu_0 \frac{\partial^2 P}{\partial t^2} \qquad (7.67)$$

In (7.65) and (7.66) the fields are local fields and the polarization P is the microscopic polarization, given by (7.65), whereas in the wave equation (7.67) the field and polarization are macroscopic variables. As we saw in Section 2.4.7, the influence of nearby polarizable matter on the local electric field seen by a given molecule or unit cell leads to Lorentz local field correction factors. Such calculations, when carried out, simply introduce constant factors as multipliers of $(\partial \alpha / \partial Q)_0$ in (7.65) and (7.66) with a relabeling of the fields and polarization as macroscopic variables compatible with the field equation (7.67). Therefore, Equations 7.65–7.67 are compatible provided that Lorentz correction factors are absorbed into the $(\partial \alpha / \partial Q)_0$ coefficient.

Equations 7.65–7.67 comprise a self-consistent descriptive approach that describes the Raman effect resulting from the interaction of an electromagnetic field and an optical phonon mode. The phonon mode described by (7.65) acts as a harmonic oscillator of frequency Ω that can be driven near resonance by a product of an incident field and a scattered Stokes field whose frequencies differ by the vibrational frequency. The polarization at the Stokes frequency required to generate the Stokes field in accordance with (7.67) is, in turn, provided by the mixing of the vibrational mode and incident field as indicated in (7.66).

7.6.3 Raman Susceptibility

We now consider the solution of the equations of motion (7.65) and (7.67). We assume that the vibrational, incident, and scattered Stokes waves propagate collinearly in the z direction. The vibrational wave is assumed to have a solution of the form

$$\langle Q \rangle = \frac{\tilde{Q}_v}{2} e^{i(\omega_v t - k_v z)} + \text{c.c.} \qquad (7.68)$$

and the other variables are defined in a similar manner, for example,

$$E = \tfrac{1}{2} \tilde{E}_i e^{i(\omega_i t - k_i z)} + \tfrac{1}{2} \tilde{E}_s e^{i(\omega_s t - k_s z)} + \text{c.c.}$$

where v, i, and s denote the vibrational, incident, and scattered Stokes waves, respectively. Factors of like exponentials are equated to obtain relationships between the magnitudes of the various waves, and we assume energy and wave vector conservation, $\omega_s = \omega_i - \omega_v$ and $k_s = k_i - k_v^*$, as discussed in Section 7.4.3. The real part of the k equation is the usual wave vector conservation, and the imaginary part shows that the Stokes radiation and vibration grow or decay

7 INTERACTIONS BETWEEN RADIATION AND PHONONS

at the same rate. The incident wave vector k_i is taken to be real, corresponding to an assumed negligible depletion of the incident beam. The polarization at the Stokes frequency is found from (7.65) and (7.66) to be of the form

$$\tilde{P}_s = \epsilon_0 \chi_R(\omega_v) |\tilde{E}_i|^2 \tilde{E}_s \tag{7.69}$$

where $\chi_R(\omega_v)$ is the Raman susceptibility defined by

$$\chi_R(\omega_v) = \frac{N_V (\partial \alpha/\partial Q)_0^2}{4M\epsilon_0} \frac{1}{\Omega^2 - \omega_v^2 - i(\Delta \omega_L)\omega_v} \tag{7.70}$$

The Raman scattering process can therefore be described in terms of a Raman susceptibility $\chi_R(\omega_v) = \chi_R'(\omega_v) + i\chi_R''(\omega_v)$, which has a Lorentzian lineshape and is pure imaginary on resonance. Since $\chi_R(\omega_v)$ relates the polarization to a cubic product of fields, the Raman susceptibility for the general three-dimensional case is a fourth-rank tensor property of the scattering medium.

The field equation (7.67) becomes

$$\left(k_s^2 - \frac{\eta_s^2 \omega_s^2}{c^2}\right) \tilde{E}_s = \mu_0 \omega_s^2 \tilde{P}_s$$

which when combined with (7.69) becomes

$$\left(k_s^2 - \frac{\eta_s^2 \omega_s^2}{c^2}\right) \tilde{E}_s = \mu_0 \omega_s^2 \epsilon_0 \chi_R(\omega_v) |\tilde{E}_i| \tilde{E}_s^2 \tag{7.71}$$

7.6.4 Raman Gain

The Stokes propagation constant is, from (7.71),

$$k_s^2 = \frac{\omega_s^2 \eta_s^2}{c^2} \left[1 + \frac{\chi_R |\tilde{E}_i|^2}{\eta_s^2}\right] \tag{7.72}$$

where $c^2 \mu_0 = 1/\epsilon_0$ has been used. Since the second term in brackets resulting from the Raman nonlinearity is small, the square root approximation $\sqrt{1 + x} \approx 1 + x/2$ may be employed with the result

$$k_s = k_s' + ik_s'' = \frac{\omega_s \eta_s}{c}\left[1 + \frac{\chi_R' |\tilde{E}_i|^2}{2\eta_s^2}\right] + i\frac{\omega_s \chi_R'' |\tilde{E}_i|^2}{2c\eta_s} \tag{7.73}$$

From the exponential form of the solutions we observe that the imaginary term in k_s leads to gain at the Stokes frequency. The Stokes power gain g_s defined by the growth in the Stokes intensity,

$$|\tilde{E}_s|^2 e^{2k_s'' z} = |\tilde{E}_s|^2 e^{g_s z} \tag{7.74}$$

can be written

$$g_s = \frac{2\pi \chi_R'' |\tilde{E}_i^2|}{\lambda_s \eta_s} = \frac{4\pi \chi_R'' I_i}{\lambda_s \eta_s \eta_i \epsilon_0 c} \quad m^{-1} \tag{7.75}$$

where λ_s is the free-space wavelength of the Stokes radiation and $I_i = (\eta_i\epsilon_0 c\,|\tilde{E}_i|^2)/2$ is the power per unit area carried by the incident wave. The Stokes wave solution can be summarized by noting from (7.74) that the Stokes intensity $I_s = [(\eta_s\epsilon_0 c\,|\tilde{E}_s|^2)/2]\,e^{g_s z}$ satisfies the growth equation

$$\frac{\partial I_s}{\partial z} = g_s I_s \tag{7.76}$$

where g_s is given by (7.75). Since the growth in the Stokes intensity is proportional to the Stokes intensity itself, the effect is an induced or stimulated effect. The phenomenon described by (7.76) is therefore the stimulated Raman effect. This definition is consistent with the definition of a stimulated process given in Chapter 6 since the intensity grows at a rate proportional to the intensity itself. The distinction between stimulated and spontaneous effects in the Raman process is identical to the distinction encountered in single-photon emission processes as discussed in Chapter 6. The spontaneous Raman effect is discussed in Section 7.6.5.

As an example of typical values, we consider the 992 cm^{-1} vibration of the organic liquid molecule benzene (C_6H_6). With a ruby laser as pump source ($\lambda_i = 6943$ Å), the following parameters apply:

$$\chi''_R(\omega_v = \Omega) = 8 \times 10^{-21} \text{ MKS units}\dagger$$
$$\lambda_s = 7450 \text{ Å} = 7.45 \times 10^{-7} \text{ m} \tag{7.77}$$
$$\eta_i = \eta_s = 1.49$$

For a pump power $I_i = 400$ MW/cm^2, (7.75) yields a gain ~ 1 cm^{-1}. Such gains are not uncommon, even with lasers of modest power, for beam intensities are often enchanced by one or two orders of magnitude because of the self-focusing phenomenon discussed in Section 7.8. Gains of this magnitude readily exceed typical optical losses, and therefore the Stokes intensity can build up to a large value in a cell of several centimeters' length.

Reflection of the Stokes beam by mirrors or other surfaces, for example, cell windows, so that it traverses the pump beam more than once, serves to enhance the effect. For example, consider a cell of length L that contains a Raman scatterer such as benzene and that has mirrors at each end to form an optical cavity. For a single pass through the liquid, (7.76) indicates that the Stokes signal will grow from an initial value $I_s(0)$ to $I_s(0)e^{g_s L} \approx I_s(0)(1 + g_s L)$ for small gain. If we assume that the fractional single-pass loss is δ, caused, for example, by transmission through the end mirrors, then the threshold condition for buildup of stimulated emission is that the fractional single-pass gain $g_s L$ equal the single-pass loss δ:

$$g_s L = \delta$$

† The value for χ''_R is obtained from (7.93), a relationship developed in Section 7.6.5 which relates χ''_R to the spontaneous scattering cross-section.

7 INTERACTIONS BETWEEN RADIATION AND PHONONS

This requirement leads from (7.75) to a threshold intensity expression

$$(I_i)_{\text{th}} = \frac{\lambda_s \eta_s \eta_i \epsilon_0 c\, \delta}{4\pi \chi_R'' L}$$

Threshold for the 992 cm^{-1} line in benzene for a 10-cm cell and 10% loss per pass is found to be $(I_i)_{\text{th}} = 4$ MW/cm^2.

Equation 7.76 can be written in terms of the expectation value of the number of Stokes photons in a mode, $\langle n_s \rangle$. In the discussion of field quantization of traveling-wave modes in Chapter 6, the modes were defined in a volume $V = L^3$ over which periodic boundary conditions were applied. We assume here that the dimension L, although large with respect to wavelength, is small compared with the macroscopic spatial variations considered in (7.76), that is, small compared with the distance over which the intensity changes significantly as a result of growth. At some point z the number of Stokes photons in a mode contained in a small volume $V = L^3$ is $\langle n_s \rangle$. Since they propagate at velocity c/η_s, in $L/(c/\eta_s)$ seconds the $\langle n_s \rangle$ photons will pass through an area L^2 at one end of the volume carrying $\hbar \omega_s \langle n_s \rangle$ joules. This implies an intensity

$$I_s = \frac{\hbar \omega_s \langle n_s \rangle}{L/(c/\eta_s)} \frac{1}{L^2} = \frac{\hbar \omega_s c \langle n_s \rangle}{\eta_s V} \frac{\text{watts}}{m^2} \tag{7.78}$$

Similar considerations apply to the incident beam. Therefore, (7.76) can be rewritten in terms of the expectation values of the number of photons in a mode. We obtain

$$\frac{\partial \langle n_s \rangle}{\partial z} = K_s \langle n_i \rangle \langle n_s \rangle \tag{7.79}$$

where, with the aid of (7.75) we have that

$$K_s = \frac{8\pi^2 \hbar c \chi_R''}{\epsilon_0 \eta_i^2 \eta_s \lambda_i \lambda_s V} \tag{7.80}$$

It is important for us to note that the replacement of intensity by photon number in the substitution (7.78) does not alter the fact that (7.79) remains a semiclassically derived expression in which the fields have not been quantized. Therefore, (7.78) is to be considered simply as a substitution of variables. We shall return to this point later when we consider generalizing (7.79) to the case where the fields are quantized.

To understand the physical significance of (7.79), we return to (7.65), the basis of our work to this point. In the longer form (7.65a) coefficients appear that are functions of $\langle n_p \rangle$, the expectation value of the number of phonons in the vibrational mode. Although we have carried through the derivation based on the

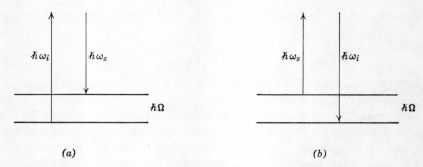

figure 7.27 Emission and absorption of Stokes photons: (a) conversion of photon of energy $\hbar\omega_i$ into a Stokes photon of energy $\hbar\omega_s$, accompanied by excitation of the vibrational mode (creation of a phonon of energy $\hbar\Omega$); (b) conversion of Stokes photon of energy $\hbar\omega_s$ into a photon of energy $\hbar\omega_i$, accompanied by deexcitation of the vibrational mode (phonon absorption).

second shorter form of (7.65b), we may at this point reinsert the first form of (7.65a), so that (7.79) becomes

$$\frac{\partial \langle n_s \rangle}{\partial z} = K_s(\langle n_p \rangle + 1)\langle n_i \rangle \langle n_s \rangle - K_s \langle n_p \rangle \langle n_i \rangle \langle n_s \rangle \tag{7.81}$$

The first term on the right-hand side is positive, and therefore corresponds to growth in Stokes photons with distance. In order to conserve energy, the annihilation of an incident photon and generation of a Stokes photon must be accompanied by the excitation of the vibration that can be described as the emission of a phonon. (See Figure 7.27a.) This is precisely the process described by the first term in (7.81). As we have seen in Chapter 6, the emission of energy into a quantized harmonic oscillator mode is proportional to $(\langle n \rangle + 1)$, where $\langle n \rangle$ is the number of quanta in the mode. This holds for the phonon case, since the process involving the generation of phonons is proportional to $(\langle n_p \rangle + 1)$.

The second term on the right-hand side of (7.81) corresponds to a decrease in Stokes photons, which, accompanied by annihilation of phonons, results in the generation of photons at the frequency of the incident wave. The absorption of acoustic quanta is proportional to $\langle n_p \rangle$ as is characteristic of the absorption of energy in a quantized harmonic oscillator mode. The net result of these two processes is that the Stokes wave grows as indicated by (7.79).

As we discussed following (7.80), the field variables $\langle n_i \rangle$ and $\langle n_s \rangle$ that appear in (7.81) resulted from the substitution of (7.78) into the semiclassically derived expression (7.76). That is, the fields themselves have not been quantized, and therefore (7.78) is simply a substitution of variables. Chapter 6 showed that when field quantization was considered, the results concerning the absorption and emission of radiation were identical, with one exception, to those obtained on a

7 INTERACTIONS BETWEEN RADIATION AND PHONONS

semiclassical basis. In semiclassically derived expressions describing emission processes, the number of photons in the mode at the emission frequency, $\langle n \rangle$, is to be replaced by $\langle n \rangle + 1$. The additional factor of unity corresponds to a spontaneous emission process. Consideration of the field quantization problem for different processes indicates that such a procedure is generally applicable. Therefore, in (7.81) we have only to replace $\langle n \rangle$ by $\langle n \rangle + 1$ in the field emission terms in order to account for spontaneous emission processes. Thus (7.81) becomes

$$\frac{\partial \langle n_s \rangle}{\partial z} = K_s(\langle n_p \rangle + 1)\langle n_i \rangle(\langle n_s \rangle + 1) - K_s\langle n_p \rangle\langle n_s \rangle(\langle n_i \rangle + 1) \qquad (7.82)$$

Equation 7.82 describes the result that is obtained when the field as well as the medium is quantized. The acoustic, incident, and scattered waves are all assumed to be single modes.

7.6.5 Spontaneous Raman Scattering

Let us now to consider spontaneous Stokes emission, in which an incident photon is scattered from the vibrational mode and a Stokes photon is emitted spontaneously. Measurement of the spontaneous scattering yields information from which the Raman susceptibility χ_R and stimulated gain constant g_s can be determined.

We begin by taking $\langle n_p \rangle$, $\langle n_s \rangle \ll 1$, $\langle n_i \rangle \gg 1$, whereupon (7.82) reduces to

$$\frac{\partial \langle n_s \rangle}{\partial z} = K_s \langle n_i \rangle$$

We see that for the spontaneous Stokes emission process the participation of Stokes photons is not required as it is in the stimulated emission case described by (7.79). Solving the above, we find that the number of Stokes photons in a mode volume V caused by spontaneous emission alone grows with distance l as

$$\langle n_s \rangle = K_s \langle n_i \rangle l$$

The number of photons generated spontaneously is therefore directly proportional to the laser intensity and the length of the scattering material.

The coefficient K_s defined in (7.80) and the gain constant g_s given by (7.75) can be determined if χ_R'' is known. This parameter, in turn, can be determined from a measurement of the spontaneous Raman scattering cross-section σ_c. The relationship between χ_R'' and σ_c is determined as follows.

First, we return to the beginning of Section 7.6.3 and determine the polarization at the incident instead of the scattered frequency. We find that the polarization is given by

$$\tilde{P}_i = \epsilon_0 \chi_R^* |\tilde{E}_s|^2 \tilde{E}_i$$

Following through the same development (which simply means interchanging subscripts i and s and replacing χ_R with χ_R^*), we find that for the incident beam, (7.79) and (7.81) are replaced by

$$\frac{\partial \langle n_i \rangle}{\partial z} = -K_i \langle n_s \rangle \langle n_i \rangle$$
$$= -K_i(\langle n_p \rangle + 1)\langle n_s \rangle \langle n_i \rangle + K_i \langle n_p \rangle \langle n_s \rangle \langle n_i \rangle \quad (7.83)$$

where

$$K_i = \frac{8\pi^2 \hbar c \chi_R''}{\epsilon_0 \eta_s^2 \eta_i \lambda_s \lambda_i V} \quad (7.84)$$

Again, in order to account for the effects of spontaneous emission that would be predicted on the basis of a fully quantized approach, we replace $\langle n \rangle$ by $\langle n \rangle + 1$ in the field emission terms, and consequently (7.83) becomes

$$\frac{\partial \langle n_i \rangle}{\partial z} = -K_i(\langle n_p \rangle + 1)(\langle n_s \rangle + 1)\langle n_i \rangle + K_i \langle n_p \rangle \langle n_s \rangle (\langle n_i \rangle + 1) \quad (7.85)$$

In the generation of spontaneous Stokes emission, taking $\langle n_p \rangle$, $\langle n_s \rangle \ll 1$, $\langle n_i \rangle \gg 1$, the incident beam decreases as

$$\frac{\partial \langle n_i \rangle}{\partial z} = -K_i \langle n_i \rangle$$

or

$$\langle n_i \rangle = \langle n_i \rangle_{z=0} e^{-K_i z} \quad (7.86)$$

For the single-mode case under consideration we now define a Raman scattering cross-section per scatterer, σ_1, by writing (7.86) in the form

$$\langle n_i \rangle(z) = \langle n_i \rangle(0) e^{-N_V \sigma_1 z} \quad (7.87)$$

where N_V is the number of scatterers per unit volume. This definition can be shown to be consistent with the concept that each scatterer has an effective area σ_1 as far as its ability to remove incident photons due to forward scattering is concerned. (See Problem 3.1.) A similar definition was encountered in connection with the cross-section for absorption by a resonant transition as discussed in Chapter 3. From (7.84), (7.86), and (7.87) we obtain

$$\sigma_1 = \frac{8\pi^2 \hbar c \chi_R''}{\epsilon_0 \eta_s^2 \eta_i \lambda_s \lambda_i V N_V} \quad (7.88)$$

which has the units of area. The above value for σ_1 is for collinear forward scattering into a single mode.

Thus far we have considered spontaneous emission into a single mode at the Stokes frequency. This yields a relationship between σ_1 and χ_R''. What is generally measured, however, is the total cross section σ_c that accounts for spontaneous

7 INTERACTIONS BETWEEN RADIATION AND PHONONS

emission into all possible modes. Therefore, parallel to the single-photon emission process (Sections 6.4.3 and 6.4.4), it is necessary to sum over all radiation modes in which spontaneous emission may occur in order to describe the total spontaneous emission process. We see that in (7.88) the cross-section that measures the removal of incident photons as a result of spontaneous emission into a single mode is proportional to χ_R'', which has a Lorentzian lineshape. Therefore the total Raman scattering cross-section for spontaneous emission into all modes is to be found by summing over those modes that lie within the profile of the Lorentzian line.

In Chapter 6 a mode density expression was derived for the number of traveling-wave modes per unit solid angle per unit volume in a frequency range $d\omega$ for a single polarization,† (6.65),

$$\frac{dp(\omega_s)}{d\zeta} d\omega_s = \frac{\omega_s^2 \eta_s^3}{(2\pi c)^3} d\omega_s \qquad (7.89)$$

where $dp(\omega_s)/d\zeta$ is the number of modes per unit solid angle per unit volume per unit frequency interval, and $d\zeta$ is the differential solid angle. Therefore the spontaneous Raman scattering cross-section per unit solid angle, $d\sigma_c/d\zeta$, for scattering into all modes within the differential solid angle is found by combining (7.88) with the above mode density expression and integrating over frequency,

$$\frac{d\sigma_c}{d\zeta} = \int \sigma_1 \frac{dp}{d\zeta} d\omega_s = \int_{-\infty}^{\infty} \sigma_1 \frac{V}{(2\pi)^3} \frac{\omega_s^2 \eta_s^3}{c^3} d\omega_s \qquad (7.90)$$

where $d\sigma_c/d\zeta$ and σ_1 are functions of the angle between the incident and scattered beams. The total scattering cross-section is then found by integrating over solid angle,

$$\sigma_c = \int_0^{4\pi} \left(\frac{d\sigma_c}{d\zeta} \right) d\zeta = \int_0^{2\pi} d\varphi \int_0^\pi d\theta \frac{d\sigma_c}{d\zeta} \sin\theta \qquad (7.91)$$

As an example consider the case of the 992 cm^{-1} line of liquid benzene (C_6H_6). Liquid benzene is an isotropic substance. For an isotropic scatterer the angular dependence of $d\sigma_c/d\zeta$ is the same as for ordinary dipole radiation. Namely, the following two conditions hold: (a) when the plane of observation is perpendicular to the electric field vector of the incident wave, the intensity at the scattered wave (and therefore cross section) is constant with the observation angle φ shown in Figure 7.28; (b) when the plane of observation is parallel to the incident electric field vector, the intensity distribution follows a $\cos^2 \psi$ law, where ψ is the angle between the propagation directions of the incident and scattered waves, also shown

---

† Raman lines are typically highly polarized when driven by a polarized source, and therefore only one polarization needs to be included in the calculation.[26] The mode density given in (7.89) is one-half the value given by (6.74) for two polarizations.

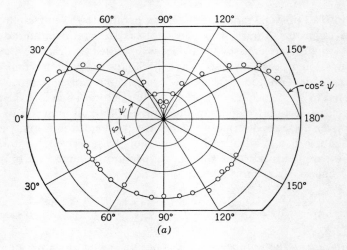

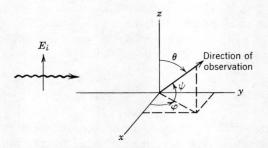

figure 7.28 Angular dependence of the intensity of the Raman-scattered 992 cm^{-1} line from benzene.[32] (a) The curve in the upper half-plane is obtained when the plane of observation is parallel to the incident electric field vector, and the lower curve is obtained when the two are perpendicular. (b) Geometry indicating the direction of observation of the scattered wave. The incident wave propagates in the y direction and is polarized in the z direction.

in Figure 7.25. It follows from the above discussion that (7.91) reduces to

$$\sigma_c = 2\pi \left(\frac{d\sigma_c}{d\zeta}\right)_{\max} \int_0^\pi d\theta \cos^2\psi \sin\theta = 2\pi \left(\frac{d\sigma_c}{d\zeta}\right)_{\max} \int_0^\pi d\theta \sin^3\theta$$

$$= \frac{2}{3} 4\pi \left(\frac{d\sigma_c}{d\zeta}\right)_{\max} \tag{7.92}$$

where $(d\sigma_c/d\zeta)_{\max}$ is the maximum value of $(d\sigma_c/d\zeta)$ which occurs at $\psi = 0$.

The quantity $(d\sigma_c/d\zeta)_{\max}$ is found from (7.90) after substitution for σ_1 from (7.88), where σ_1 was calculated for forward scattering where the cross-section has its maximum value. If the lineshape is Lorentzian, from (3.22) we have that

$$\chi_R'' = [\chi_R''(\omega_v = \Omega)]\pi(\Delta\omega_L)\frac{g_L(\omega, \Omega)}{2} \quad \text{where} \quad \int_{-\infty}^{\infty} g_L(\omega, \Omega)\, d\omega = 1$$

The total cross-section as determined from (7.88), (7.90), and (7.92) is

$$\sigma_c = \frac{16\pi^3 \hbar(\Delta\omega_L)\eta_s[\chi_R''(\omega_v = \Omega)]}{3\epsilon_0 \eta_i \lambda_i \lambda_s^3 N_V} \tag{7.93}$$

With a helium-neon laser as a source ($\lambda_i = 6328$ Å), a measured value of $\sigma_c = 5.6 \times 10^{-29}$ cm² is obtained for the 992 cm⁻¹ line in benzene.[32] From this the value, $\chi_R''(\omega_v = \Omega)$ is found to be 8×10^{-21} MKS, calculated with $N_V = 6.78 \times 10^{21}$ cm⁻³, $\Delta\omega_L$ = line width = 2.5 cm⁻¹, and $\lambda_s = 6750$ Å.

7.7 BRILLOUIN EFFECT

The Brillouin effect is a nonlinear light-scattering process in which the light is scattered from an acoustic vibration with a shift in the frequency of the light corresponding to that of the acoustic mode. The effect results from the modulation of the refractive index of a medium by the propagation of a sound wave through it. The conservation of energy (frequency) and momentum (wave vector) is shown in Figure 7.29.

The amount of light scattered into the shifted frequency ω_s is generally many orders of magnitude below that of the incident beam. However, if the intensity of the incident light is high enough, sufficient Brillouin scattering may take place for stimulated Brillouin scattering to occur, and thus the strength of the scattered beam may approach that of the incident beam. The remarks in Section 7.6 contrasting spontaneous and stimulated Raman scattering apply also to the Brillouin effect. The stimulated Brillouin effect has been observed in solids, liquids, and gases. In the analysis that follows we shall consider the monatomic crystal. The approach is easily extended to cover a continuum.

The Brillouin effect is similar to the Raman effect except that acoustic vibrations play the role of the optical vibrations. As a result of this difference, the frequency shifts are much smaller, on the order of a few GHz rather than a few hundred Angstroms. In addition, the frequency shift, rather than being a fixed amount as in Raman scattering, depends on the angle between the incident and scattered beams, the maximum shift occurring for back-scattered light. The reason for this is that the frequency of the acoustic wave involved in the interaction depends on the k value, as is apparent from the dispersion characteristics shown in

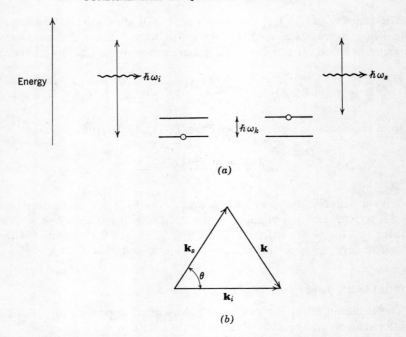

figure 7.29 Conservation of energy (frequency) and momentum (wave vector) in the Brillouin scattering process: (a) Photons in the incident beam are converted into frequency-shifted scattered photons with an accompanying change of state of the medium in which a phonon of energy $\hbar\omega_k$ is created; the lengths of the vertical arrows are proportional to the photon and phonon energies. (b) Wave vector conservation, $\mathbf{k}_i = \mathbf{k}_s + \mathbf{k}$.

Figures 7.10 and 7.15; and the k value, in turn, depends on the angles determined by wave vector conservation.

From the expression for the dispersion characteristic in a monatomic crystal, (7.3), for a longitudinal acoustic wave with $ka \ll \pi$, we have

$$\omega_k = 2\left(\frac{C_L}{M}\right)^{1/2} \left|\sin\frac{ka}{2}\right| \approx \left(\frac{C_L}{M}\right)^{1/2} ka \equiv v_a k \tag{7.94}$$

where k is the wave vector of the vibrational wave and v_a is the acoustic velocity in the crystal. The condition $ka \ll \pi$ is assured by wave vector conservation as indicated in Figure 7.29b, since $|k_s a|, |k_i a| \ll \pi$ for optical waves. With $\omega_k = \omega_i - \omega_s$, the angular dependence of the frequency shift is found by substitution in (7.94) of the value for k determined by the geometrical construction of Figure

7 INTERACTIONS BETWEEN RADIATION AND PHONONS

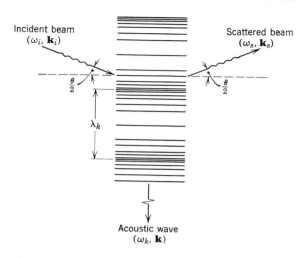

figure 7.30 Brillouin scattering seen as diffraction from a moving grating formed by a sound wave.

7.29. Since $k_i = \omega_i \eta_i / c$ and $k_s \approx k_i$, the result is

$$\omega_i - \omega_s = \frac{2\omega_i \eta_i v_a}{c} \sin \frac{\theta}{2} \qquad (7.95)$$

where θ is the angle between the incident and scattered-light beams. As mentioned above, the maximum shift occurs for back-scattering or $\theta = \pi$. The angular dependence of the frequency shift just derived is displayed graphically in the dispersion vector diagrams of Figure 7.15.

For yet another approach to the derivation of (7.95) that is especially meaningful from the physical standpoint, we can consider the scattering of light from a moving diffraction grating formed by a sound wave as shown in Figure 7.30. The first-order Bragg reflection condition† is given by

$$2\lambda_k \sin \frac{\theta}{2} = \lambda_i$$

where λ_i is the wavelength of the incident beam in the medium and λ_k is both the acoustic wavelength and grating spacing. Since the light is reflected from a moving

---

† The Bragg condition for reflection from a series of planes is that condition whereby reflections of an incident wave from adjacent planes add constructively. The nth order Bragg condition corresponds to a path difference between interfering waves of n wavelengths.

object, the frequency is shifted by the Doppler effect in accordance with the expression

$$\frac{\omega_s}{\omega_i} = 1 - \frac{2v_a}{c/\eta_i}\sin\frac{\theta}{2}$$

which agrees with (7.95).

7.7.1 Hamiltonian

The analysis of the Brillouin effect proceeds in a manner similar to that for the Raman effect presented in the previous section, except that acoustic rather than optical vibrations are responsible for the fluctuations in the polarizability or refractive index. The acoustic waves may be either transverse or longitudinal. We shall consider the case where the electric fields of the incident and scattered waves are transverse and polarized in the same direction, and the acoustic wave is longitudinal. The medium is the monatomic crystal discussed in Sections 7.2.1 and 7.3.1.

As mentioned above, much of the procedure used and many of the expressions derived in this section for the Brillouin effect closely parallel the development of the material for the Raman effect. Therefore, let us highlight at the outset the major differences. In the Raman effect the vibrational mode from which scattering takes place is an optical mode, which is a vibration of a single molecule or unit cell that is essentially independent of those of its neighbors. This is borne out by the fact that the ωk dispersion characteristic for an optical mode is flat in the region of interest, in which case the group velocity $v_g = d\omega/dk$ is zero. This implies that energy does not propagate, which is due to the lack of coupling between neighbors. As a result, the analysis for the Raman effect can proceed in terms of variables Q, $a^\dagger$, and a associated with the vibration of a single molecule or unit cell because each molecular or unit cell vibration is independent of its neighbors.

In the Brillouin effect where acoustic rather than optical modes are involved, on the other hand, we must consider the interaction between neighbors. An acoustic mode is inherently a collective phenomenon involving tightly coupled neighbors. The group velocity $v_g = d\omega/dk$ is finite for an acoustic mode and indicates the ability of an acoustic wave to propagate energy as a result of the intermolecular coupling. In the analysis of the Brillouin effect in a crystal, therefore, sums over lattice sites appear and the acoustic mode variables q_l, $a_k^\dagger$, and a_k are defined in terms of a range of lattice points.

The electronic polarizability α_m at lattice site m due to the propagation of an acoustic wave,

$$q_m = \tfrac{1}{2}\tilde{q}e^{i(\omega_k t - kma)} + \text{c.c.}$$

fluctuates to first order as

$$\alpha_m = \alpha_m^0 + \sum_l \left(\frac{\partial \alpha_m}{\partial(q_{l+1} - q_l)}\right)_0 (q_{l+1} - q_l) \qquad (7.96)$$

where q_m is the displacement from equilibrium of an atom at lattice site m. The subscript '0' on the derivative indicates that the derivative is to be evaluated at the equilibrium position of the lattice structure. The change in polarizability has been expressed as a function of the relative displacements of neighbor atoms, and the polarizability at lattice site m is assumed to be affected to some degree by the relative displacements of all of the atoms in the lattice. Since the induced electric dipole moment is given by the product of the polarizability and the electric field, that part of the induced dipole moment at the mth site associated with the fluctuating part of the polarizability is given by

$$\mu_m = \alpha'_m E_m$$

where

$$\alpha'_m = \sum_l \left(\frac{\partial \alpha_m}{\partial(q_{l+1} - q_l)}\right)_0 (q_{l+1} - q_l) \tag{7.97}$$

The polarizability at site m given by (7.97) is expressed in terms of a summation of contributions caused by relative displacements of atoms at sites labeled by l. Appendix 9 shows that (7.97) can be rewritten in the form

$$\alpha'_m = C_b(q_{m+1} - q_m) \tag{7.98}$$

where

$$C_b = \sum_l \left(\frac{\partial \alpha_0}{\partial(q_{l+1} - q_l)}\right)_0 e^{-ikla}$$

That is, the polarizability at lattice site m can be expressed in terms of the relative atomic displacement at that site alone. This result simplifies the analysis considerably.

The interaction energy term at site m in the presence of an electric field takes the form (7.59) characteristic of the potential energy stored in a polarizable medium

$$\mathcal{H}'_m = -\tfrac{1}{2}\alpha'_m E_m^2 = -\tfrac{1}{2}C_b(q_{m+1} - q_m)E_m^2 \tag{7.99}$$

7.7.2 Quantum Mechanical Treatment

In this section we derive the equations of motion that describe the Brillouin effect. As in the analysis of the Raman effect, the problem is treated semiclassically, that is, the fields are treated classically whereas the medium, in this case the acoustic mode system, is treated quantum mechanically. As we showed in some detail in Section 7.6.2 in connection with the Raman effect, semiclassical treatment of the fields provides for a description of the *stimulated* Brillouin effect only; the spontaneous effect is accounted for by means of the "extra photon" approach, discussed in 7.6.2.

For the propagation of an acoustic mode of wavenumber k in a monatomic crystal, the unperturbed part of the Hamiltonian is given by (7.25):

$$\mathcal{H}_0 = \hbar\omega_k(a_k^\dagger a_k + \tfrac{1}{2}) \tag{7.100}$$

where ω_k is the frequency of the acoustic mode, which depends on the wavenumber k.

On the basis of (7.99), the interaction term for the problem is taken to be of the form

$$\mathscr{H}' = -\tfrac{1}{2}C_b \sum_{l=1}^{\mathscr{N}} (q_{l+1} - q_l)E_l^2 \tag{7.101}$$

where the operator q_l corresponding to an acoustic mode is given by (7.18):

$$q_l = \left(\frac{\hbar}{2\mathscr{N}M\omega_k}\right)^{1/2} [a_k e^{ikla} + a_k^\dagger e^{-ikla}] \tag{7.102}$$

The sum over $\mathscr{N}$ values of l in (7.101) covers the range of lattice points inside the volume over which periodic boundary conditions have been defined, as discussed in Section 7.3.1.

The equation of motion for the expectation value of the acoustic vibrational amplitude at a given lattice site m, $\langle q_m \rangle$, is found by the use of (1.51):

$$\langle \ddot{q}_m \rangle + \frac{2}{T_2} \langle \dot{q}_m \rangle + \frac{1}{T_2^2} \langle q_m \rangle = -\frac{1}{\hbar^2} \langle [[q_m, \mathscr{H}], \mathscr{H}] \rangle \tag{7.103}$$

Evaluation of the inside commutator in (7.103) with the aid of (7.100) and (7.102) yields

$$[q_m, \mathscr{H}] = [q_m, \mathscr{H}_0 + \mathscr{H}'] = [q_m, \mathscr{H}_0] + [q_m, \mathscr{H}'] = [q_m, \mathscr{H}_0]$$

$$= \hbar\omega_k \left(\frac{\hbar}{2\mathscr{N}M\omega_k}\right)^{1/2} (a_k e^{ikma} - a_k^\dagger e^{-ikma})$$

since $\mathscr{H}'$ commutes with q_m. Evaluation of the outside commutator proceeds similarly, yielding

$$[[q_m, \mathscr{H}], \mathscr{H}] = \hbar^2 \omega_k^2 q_m - \frac{C_b \hbar^2}{4\mathscr{N}M} \sum_{l=1}^{\mathscr{N}} \{E_l^2[(e^{-ika} - 1)e^{ik(m-l)a}$$

$$+ (e^{ika} - 1)e^{-ik(m-l)a}]\}(a_k a_k^\dagger - a_k^\dagger a_k)$$

$$= \hbar^2 \omega_k^2 q_m - \frac{C_b \hbar^2}{4\mathscr{N}M} \sum_{l=1}^{\mathscr{N}} \{E_l^2[(e^{-ika} - 1)e^{ik(m-l)a} + (e^{ika} - 1)e^{-ik(m-l)a}]\}$$

For the purpose of physical description to be considered later in this section, we shall sometimes prefer the form containing the operator $(a_k a_k^\dagger - a_k^\dagger a_k)$, which can be eliminated in the above equation by use of the commutator relation $[a_k, a_k^\dagger] = 1$.

7 INTERACTIONS BETWEEN RADIATION AND PHONONS 249

With substitution of the above, the equation of motion (7.103) becomes

$$\langle \ddot{q}_m \rangle + \Delta\omega_L \langle \dot{q}_m \rangle + \omega_k^2 \langle q_m \rangle$$

$$= \frac{C_b}{4\mathcal{N}M} \sum_{l=1}^{\mathcal{N}} \{E_l^2[(e^{-ika} - 1)e^{ik(m-l)a} + (e^{ika} - 1)e^{-ik(m-l)a}]\}(\langle n_k \rangle + 1)$$

$$- \frac{C_b}{4\mathcal{N}M} \sum_{l=1}^{\mathcal{N}} \{E_l^2[(e^{-ika} - 1)e^{ik(m-l)a} + (e^{ika} - 1)e^{-ik(m-l)a}]\}\langle n_k \rangle \quad (7.104a)$$

$$= \frac{C_b}{4\mathcal{N}M} \sum_{l=1}^{\mathcal{N}} \{E_l^2[(e^{-ika} - 1)e^{ik(m-l)a} + (e^{ika} - 1)e^{-ik(m-l)a}]\} \quad (7.104b)$$

where $\Delta\omega_L = 2/T_2$ and we take $\omega_k^2 \gg 1/T_2^2$. The coefficients that appear in the long form of (7.104a) are functions of $\langle n_k \rangle = \langle a_k^\dagger a_k \rangle$, the expectation value of the number of phonons in the k mode. We shall find this longer form more suitable for the description of the physical processes involved, while the shorter form of (7.104b) is more convenient for mathematical manipulation.

If there are N_V unit cells per unit volume, the nonlinear polarization at the mth lattice site can be written, from (7.98) as

$$P_m = \frac{\langle \mu_m \rangle}{(\text{volume of unit cell})} = N_V \langle \mu_m \rangle = N_V \langle \alpha'_m \rangle E_m = N_V C_b (\langle q_{m+1} \rangle - \langle q_m \rangle) E_m$$

(7.105)

Equations 7.104 and 7.105 describe the behavior of the medium in the presence of a field. To this we add the wave equation (2.50) for the propagation of the field, neglecting loss,

$$\nabla \times (\nabla \times E) + \frac{\eta^2}{c^2} \frac{\partial^2 E}{\partial t^2} = -\mu_0 \frac{\partial^2 P}{\partial t^2} \quad (7.106)$$

As discussed in connection with the Raman effect in the previous section, local field correction factors, which are introduced into (7.104) and (7.105) to make them compatible with the macroscopic field equation (7.106), can be absorbed into C_b. Equations 7.104–7.106 can therefore be taken as they stand to provide a self-consistent approach to the Brillouin scattering problem. The acoustic mode is described by (7.104) as a harmonic oscillator of frequency ω_k that can be driven near resonance by a product of an incident field and a scattered Stokes field whose frequencies differ by the vibrational frequency. The Stokes polarization needed to generate the Stokes field in (7.106) is, in turn, provided by the mixing of the vibrational mode and incident field as indicated in (7.105).

Equations 7.104–7.106 for the Brillouin effect are essentially identical in form to 7.65–7.67 for the Raman effect. The primary difference is the appearance of a sum over lattice points in the driving term of the vibrational equation (7.104) for the Brillouin effect, whereas a single term appears in the vibrational equation

(7.65) for the Raman effect. However, we shall show that in the solution the sum reduces to a single term with the result that the solutions to the basic equations of the Raman and Brillouin effects are identical.

The solution to the equations of motion (7.104)–(7.106) is found by assuming solutions of the form

$$\langle q_m \rangle = \frac{\tilde{q}}{2} e^{i(\omega_v t - kma)} + \text{c.c.}$$

$$E_m = \frac{\tilde{E}_i}{2} e^{i(\omega_i t - \mathbf{k}_i \cdot \mathbf{r}_m)} + \frac{\tilde{E}_s}{2} e^{i(\omega_s t - \mathbf{k}_s \cdot \mathbf{r}_m)} + \text{c.c.} \quad (7.107)$$

$$P_m = \frac{\tilde{P}_s}{2} e^{i(\omega_s t - \mathbf{k}_s \cdot \mathbf{r}_m)} + \text{c.c.}$$

where the subscripts v, i, and s denote vibrational, incident, and scattered Stokes waves, respectively, m refers to the lattice site, and $\mathbf{r}_m$ is the distance from a fixed origin to the mth lattice site. The wave vectors are allowed to be complex to allow for growth except that we take $\mathbf{k}_i$ to be real, that is, we assume negligible depletion of the incident beam. We also assume the frequency condition $\omega_s = \omega_i - \omega_v$ determined by energy conservation as discussed in Section 7.4.2.

In the solution of (7.104) it is necessary, following substitution of (7.107), to evaluate a summation on the right-hand side, which takes the form

$$S = \sum_{l=1}^{\mathcal{N}} \{[(e^{-ika} - 1)e^{ik(m-l)a} + (e^{-ika} - 1)e^{-ik(m-l)a}]e^{-i(\mathbf{k}_i - \mathbf{k}_s^*) \cdot \mathbf{r}_l}\}$$

$$= (e^{-ika} - 1)e^{ikma} \sum_{l=1}^{\mathcal{N}} e^{-ikla} e^{-i(\mathbf{k}_i - \mathbf{k}_s^*) \cdot \mathbf{r}_l} + (e^{ika} - 1)e^{-ikma} \sum_{l=1}^{\mathcal{N}} e^{ikla} e^{-i(\mathbf{k}_i - \mathbf{k}_s^*) \cdot \mathbf{r}_l}$$

From the lattice sum rule given by (7.14),

$$\sum_{l=1}^{\mathcal{N}} e^{i(k-k')la} = \mathcal{N} \delta_{kk'}$$

the above summation reduces to

$$S = \mathcal{N}(e^{ika} - 1)e^{-ikma}; \quad \text{for} \quad (\mathbf{k}_i - \mathbf{k}_s^*) \cdot \mathbf{r}_l = kla$$
$$= 0; \quad \text{otherwise}$$

Since $\mathbf{k}$ is the wave vector for the acoustic wave, we observe that S is nonzero only for wave vector conservation.

From the solution of (7.104)–(7.106), obtained by equating factors of like exponentials in frequency, we find that the polarization and field at the Stokes frequency are given by expressions identical in form to those applicable to the Raman effect,

$$\tilde{P}_s = \epsilon_0 \chi_B(\omega_v) |\tilde{E}_i|^2 \tilde{E}_s$$

$$\left(k_s^2 - \frac{\eta_s^2 \omega_s^2}{c^2}\right) \tilde{E}_s = \mu_0 \omega_s^2 \epsilon_0 \chi_B(\omega_v) |\tilde{E}_i|^2 \tilde{E}_s$$

where χ_B is the Brillouin susceptibility defined by

$$\chi_B(\omega_v) = \frac{N_V C_b^2}{\epsilon_0 8M} \frac{k^2 a^2}{\omega_k^2 - \omega_v^2 - i(\Delta\omega_L)\omega_v} \tag{7.108}$$

In the derivation of the above we have taken $\sin^2(ka/2) \approx k^2a^2/4$, which holds for wave vectors involved in Brillouin scattering since ka is small. Since $\chi_B(\omega_v)$ relates the polarization to a cubic product of fields, the Brillouin susceptibility for the general three-dimensional case is a fourth-rank tensor property of the scattering medium.

The Brillouin susceptibility can be expressed in terms of measurable parameters associated with the photoelastic effect. In the photoelastic effect a change in refractive index caused by strain is expressed in the form[33]

$$\Delta B_{ij} = p_{ijrs} \times (\text{strain})_{rs} \tag{7.109}$$

where the p_{ijrs} are the elasto-optical coefficients and B_{ij} is the relative dielectric impermeability tensor at optical frequencies defined by $B_{ij} = \epsilon_0\, \partial E_i/\partial D_j$. For the case considered in this section the electric fields of the incident and scattered waves are transverse to the direction of propagation and polarized in the same direction, and the acoustic wave is longitudinal. Later in the section we shall consider the specific example of the excitation of a longitudinal acoustic wave along the x axis of quartz. The incident and scattered waves are assumed to propagate along the x axis and are polarized along the y axis. Therefore, in the following development we shall consider the relevant coefficient to be p_{yyxx}. Therefore we have

$$p_{yyxx} \times (\text{strain})_{xx} = \Delta B_{yy} = \Delta\left(\frac{\epsilon_0}{\epsilon_{yy}}\right) = \Delta\left(\frac{1}{\kappa_{yy}}\right) = \frac{-\Delta\kappa_{yy}}{\kappa_{yy}^2} \tag{7.110}$$

where we have used the fact that for a field along a principal axis, $D_i = \epsilon_{ii}E_i$, so that $B_{ii} = \epsilon_0/\epsilon_{ii}$. We have also the general relationship

$$D_y = \epsilon_{yy}E_y + P_y = (\epsilon_{yy} + \Delta\epsilon_{yy})E_y$$

from which we obtain

$$P_y = \Delta\epsilon_{yy}E_y = \epsilon_0 \Delta\kappa_{yy}E_y$$

With P_y given by (7.105), we find that $\Delta\kappa_{yy}$ in the above expression is given by

$$\Delta\kappa_{yy} = \frac{N_V C_b a}{\epsilon_0} \frac{\langle q_{m+1}\rangle - \langle q_m\rangle}{a} = \frac{N_V C_b a}{\epsilon_0} \times (\text{strain})_{xx}$$

From (7.110) and the above we therefore have the relationship

$$C_b = \frac{-\kappa_{yy}^2 p_{yyxx}\epsilon_0}{N_V a}$$

Substitution of the above into (7.108) yields

$$\chi_B = \frac{\kappa_{yy}^4 \epsilon_0}{8\rho_m} \frac{k^2 p_{yyxx}^2}{\omega_k^2 - \omega_v^2 - i(\Delta\omega_L)\omega_v} \qquad (7.111)$$

where $\rho_m = N_V M$ is the mass density, κ_{yy} is the dielectric constant, k is the magnitude of the acoustic wave vector, and p_{yyxx} is the relevant elasto-optical coefficient.

Following the development used in the analysis of the Raman effect in Section 7.6.4, we find that the Stokes intensity satisfies the growth equation

$$\frac{\partial I_s}{\partial z} = g_s I_s \qquad (7.112)$$

where the power gain g_s is given by

$$g_s = \frac{2\pi \chi_B'' |\tilde{E}_i|^2}{\lambda_s \eta_s} = \frac{4\pi \chi_B'' I_i}{\lambda_s \eta_s \eta_i \epsilon_0 c} \quad m^{-1} \qquad (7.113)$$

In the above expression χ_B'' is the imaginary part of the Brillouin susceptibility, $\chi_B = \chi_B' + i\chi_B''$, λ_s is the free-space wavelength of the Stokes radiation, and $I_i = \eta_i \epsilon_0 c |\tilde{E}_i|^2/2$ is the intensity of the incident wave.

Since growth in the Stokes intensity is driven by the Stokes intensity itself, the effect being described is an induced effect, termed the stimulated Brillouin effect.

One of the first materials in which stimulated Brillouin scattering was observed[34] was α-quartz, a uniaxial crystal, trigonal in structure, belonging to the point group 32. The experimental arrangement and results are shown in Figure 7.31. The incident beam consisted of a 50-megawatt 30-nanosecond pulse from a Q-switched ruby laser. One of the orientations for the experiment consists of an incident wave propagating along the x axis of the crystal as an ordinary ray (that is, polarized perpendicular to the c axis) which results in the excitation of a longitudinal acoustic wave (also along the x axis) and a back-scattered Stokes wave of the same polarization as that of the incident wave. For this case[35] the following constants apply:

$\kappa_{yy} = \eta_{yy}^2 = (1.54)^2 = 2.37$ (dielectric constant)

$p_{yyxx} = 0.098$ (elasto-optical coeff.)

$\rho_m = 2.65$ gm/cm³ (mass density)

$\Gamma_a = 325$ cm⁻¹ (acoustic attenuation constant for the elastic displacement)[36]

$C_{11} = 8.5 \times 10^{11}$ ergs/cm³ (elastic stiffness constant)

$v_a = (C_{11}/\rho_m)^{1/2} = 5.66 \times 10^5$ cm/sec (acoustic velocity)

$\Delta\omega_L = \Gamma_a v_a = 1.84 \times 10^8$ rad/sec (acoustic linewidth).

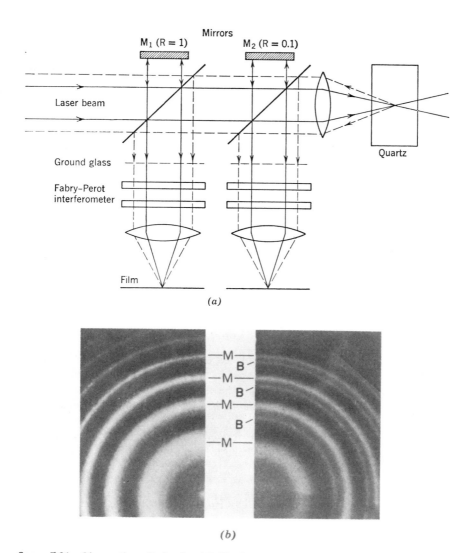

figure 7.31 Observation of stimulated Brillouin scattering in the backward direction.[34] (a) Experimental arrangement. A small percentage of the laser beam is deflected by the beam splitters to the mirrors M_1 and M_2 where it is reflected downward. The stimulated Brillouin emission in the back direction is deflected downward by the same beam splitters, from the other side. The Fabry-Perot interferometer converts frequency differences to angular differences, thus permitting discrimination spatially between laser and Brillouin waves on film. (b) Fabry-Perot interferograms of the laser radiation (rings labeled L) and of the Brillouin scattered radiation (rings labeled B). The interferogram on the left was obtained from M_1 and the one on the right was from M_2. With the lower reflectivity mirror, M_2, the Brillouin and laser intensities are more nearly equal.

For scattering in the back direction (7.95) yields a Brillouin frequency shift of 25.1 GHz or 0.84 cm^{-1}, which compares with a measured value of 0.85 cm^{-1}.[34] From (7.94) we have that $k = \omega_k/v_a$ which, with the aid of the above list of constants, permits us to determine the Brillouin susceptibility given by (7.111). On resonance, where $\omega_k = \omega_v$, Equation 7.111 gives $\chi_B = i\chi_B'' = i\,3.3 \times 10^{-21}$ (MKS units). The Brillouin gain for this case is found from (7.113) to be $g_s = 10^{-9}I_i$, where g_s is in cm^{-1} and I_i is the incident intensity in W/cm^2. Therefore, an incident power of 1 gigawatt/cm^2 yields a gain of 1 cm^{-1} which is sufficient to overcome typical optical losses (<0.1 cm^{-1}) and thereby permit the buildup of stimulated Brillouin emission.

The equations appropriate for Brillouin scattering are identical with those for the Raman effect described in the previous section. We have only to substitute χ_B for χ_R in order that the Raman equations beginning with (7.78) apply to the Brillouin effect. A description of the Brillouin effect in terms of photons and phonons is obtained from the substitution of (7.78) into (7.112). In terms of the photon variables $\langle n_i \rangle$ and $\langle n_s \rangle$, the expectation values of the number of photons in incident and scattered traveling-wave modes, respectively, we have

$$\frac{\partial \langle n_s \rangle}{\partial z} = K_s \langle n_s \rangle \langle n_i \rangle \qquad (7.114)$$

where K_s is given by (7.80) with χ_R'' replaced by χ_B''. As discussed following (7.80), the above expression, although expressed in terms of photon variables, is a semiclassically derived expression in which the fields have not been quantized. The substitution for instance, of $\langle n_s \rangle$ for I_s is therefore simply a substitution of variables.

Equation 7.114, which has been derived on the basis of the short form of (7.104b), may be rewritten in terms of the longer form (7.104a) in which $\langle n_k \rangle$, the number of phonons in the vibrational mode, appears:

$$\frac{\partial \langle n_s \rangle}{\partial z} = K_s(\langle n_k \rangle + 1)\langle n_s \rangle \langle n_i \rangle - K_s \langle n_k \rangle \langle n_s \rangle \langle n_i \rangle \qquad (7.115)$$

Equations 7.114 and 7.115 describe the induced or stimulated Stokes emission process, since the participation of Stokes photons is required. A description of the spontaneous Brillouin effect can be obtained as in the analysis of the Raman effect by following the $\langle n \rangle \rightarrow \langle n \rangle + 1$ rule for emission processes, which converts expressions derived on the basis of classical fields to the form required by field quantization. The expanded equation then takes the form

$$\frac{\partial \langle n_s \rangle}{\partial z} = K_s(\langle n_k \rangle + 1)(\langle n_s \rangle + 1)\langle n_i \rangle - K_s \langle n_k \rangle \langle n_s \rangle (\langle n_i \rangle + 1) \qquad (7.116)$$

Equation 7.116 for the Brillouin effect is identical in form to (7.81) for the Raman effect. As we saw in connection with the Raman effect, the first term on the right-hand side of the above equation corresponds to the process shown in Figure 7.27a in which an incident photon is converted into a scattered Stokes photon accompanied by emission of a phonon; the second term corresponds to the simultaneous annihilation of a Stokes photon and a phonon to produce a photon at the frequency of the incident wave, ω_i, as shown in Figure 7.27b. Thus we have a description of the Brillouin effect in terms of the variables of a fully-quantized approach, that is, photons and phonons.

For spontaneous Brillouin scattering $n_s, n_k \ll 1$ and (7.116) becomes

$$\frac{\partial \langle n_s \rangle}{\partial z} = K_s \langle n_i \rangle$$

Therefore, as in the Raman case, the spontaneous emission is proportional to the incident intensity and the interaction length.

The Brillouin effect is of research interest because of the information it yields about the acoustic and photoelastic properties of matter, and it has found practical application in such devices as frequency translators for optical heterodyne systems and light beam deflection systems for optical scanning.

7.8 SELF-FOCUSING

An important variable that appears in gain expressions for the Raman and Brillouin effects is the power per unit area of the optical beams. In experimental studies it has been found that the effective gain constants are often one or two orders of magnitude greater than expected. Such results indicate that the effective intensities may be greater than expected.[37-56]

The anomalous gain effects result from a phenomenon known as self-focusing or self-trapping. This phenomenon is a result of the fact that the index of refraction η is higher in regions of high optical intensity than in regions of low intensity because of nonlinear effects.[41,42] The increase in the index of refraction in regions of high intensity produces lens effects that result in the formation of intense filaments 2–100 μ in diameter. As a result, the effective intensity is much greater than one would expect in the absence of self-focusing.

The increase in the effective intensity results in lowered threshold requirements for a given incident power density in, for example, stimulated Raman emission. In fact, in self-focusing liquids the threshold depends not so much on the strength of the Raman scattering cross-section as on the self-focusing properties of the liquid.[48]

The self-focusing phenomenon in an isotropic medium can be described in terms of an intensity-dependent index of refraction of the form

$$\eta = \eta_0 + \eta_2 E^2 \qquad (7.116)$$

The nonlinear coefficient η_2 may be caused by the ac Kerr effect, which involves the reorientation of molecules with anisotropic polarizabilities; by the electrostrictive effect, which involves field-induced macroscopic density changes in the medium; and by nonlinearities in the electronic polarizability.[54] For liquids the first two effects may be of comparable size while the third may be much smaller. For solids, in which molecular reorientation is frozen, electrostrictive effects predominate.

For the Kerr and electrostrictive effects the nonlinear dielectric response time is much slower than the period of the optical oscillations. Therefore, only the dc component of $\eta_2 E^2$ in (7.116) need be retained. With

$$E = \tfrac{1}{2}\tilde{E}e^{i(\omega t - kz)} + \text{c.c.}$$

(7.116) becomes

$$\eta = \eta_0 + \frac{\eta_2}{2}|\tilde{E}|^2 = \eta_0 + \Delta\eta \qquad (7.117)$$

In what follows we shall always assume that the intensity-dependent correction term $\Delta\eta$ is very small with respect to the unperturbed refractive index η_0.

The basic features of self-focusing can be understood in terms of the picture shown in Figure 7.32. The maximum-angle ray in a diffraction-limited beam spreads with a half-angle of roughly

$$\theta \approx \frac{1.22\lambda}{2\eta_0 D} \qquad (7.118)$$

where θ is measured between peak and half-power points of the cross-sectional intensity distribution, and D is the initial beam diameter. A propagating ray of this angle will just be self-trapped by total internal reflection at the beam boundary if the index of refraction in the beam, η, exceeds that outside the beam, η_0, by a

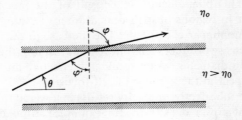

figure 7.32 Self-trapping in a region of high refraction index. From Snell's law, $\eta_0 \sin \varphi = \eta \sin \varphi'$, trapping because of total internal reflection from the beam boundary occurs when $\varphi = \pi/2$, or $\eta_0 = \eta \sin(\pi/2 - \theta) = \eta \cos \theta$.

critical amount. From Snell's law, the ray will be trapped provided that

$$\eta_0 = \eta \cos\theta \approx \eta\left(1 - \frac{\theta^2}{2}\right)$$

or

$$\theta \approx \left(\frac{2\Delta\eta}{\eta_0}\right)^{1/2} \tag{7.119}$$

where $\Delta\eta = \eta - \eta_0$. Equating (7.118) and (7.119), for the required differential in refractive indices we have

$$\Delta\eta = \left(\frac{1.22\lambda}{D}\right)^2 \frac{1}{8\eta_0} \tag{7.120}$$

According to (7.117), the differential refractive index depends on the field as $\Delta\eta = \eta_2 |\tilde{E}|^2/2$. Therefore the critical field for trapping can be found with the aid of (7.120), from which we can determine that trapping occurs when the power carried by the beam, $P = (\eta_0 \epsilon_0 c |\tilde{E}|^2/2)(\pi D^2/4)$, reaches the critical value

$$P_{cr} = \frac{\pi(1.22\lambda)^2 \epsilon_0 c}{32\eta_2} \tag{7.121}$$

A detailed computer calculation based on the solution of the wave equation yields a nearly identical result,[49]

$$P_{cr} = \frac{5.763\lambda^2 \epsilon_0 c}{4\pi^2 \eta_2} \tag{7.122}$$

Of special interest in the above expressions is the fact that a diffraction-limited beam is self-trapped regardless of its size or intensity, provided only that the total power carried by the beam exceeds the critical amount. Therefore a beam may be self-trapped at any arbitrary diameter, and the self-trapping occurs at a critical power level independent of the beam diameter. If the area of a trapped beam carrying the critical power were to be doubled, for example, which halves the intensity and therefore halves the nonlinear index effect, the beam would remain trapped because the change in refractive index required for trapping also halves.

If the power exceeds the critical value, beam rays will be reflected back toward the axis. This leads to the concept of a focusing distance, z_f, which is the distance the beam travels before it comes to a focus because of self-focusing effects. To study this concept, it is useful to consider a given intensity distribution across the beam. To be specific, we shall consider a beam with a Gaussian intensity distribution. For a Gaussian beam the square of the electric field has a cross-sectional distribution

$$|\tilde{E}|^2(r) = |\tilde{E}_m|^2 e^{-2r^2/a^2} \tag{7.123}$$

where $\tilde{E}_m$ is the magnitude of the electric field at beam center, and a is the beam radius at which the intensity drops to $e^{-2} = 0.135$ of its on-axis value. The total power P carried by the beam is given by

$$P = \frac{\eta_0 \epsilon_0 c}{2} 2\pi \int |\tilde{E}|^2 r \, dr$$
$$= \left(\frac{\pi a^2}{2}\right)\left(\frac{\eta_0 \epsilon_0 c}{2}\right) |\tilde{E}_m|^2 \tag{7.124}$$

The radius of curvature, R, of a ray in a medium of variable refractive index is given by[57] (see Problem 7.9)

$$\frac{1}{R} = \frac{1}{\eta(r)} \frac{\partial \eta(r)}{\partial r} \approx \frac{1}{\eta_0} \frac{\partial \eta(r)}{\partial r} \tag{7.125}$$

Since the refractive index is related to the field by (7.117), we have

$$\eta = \eta_0 + \frac{\eta_2}{2} |\tilde{E}_m|^2 e^{-2r^2/a^2}$$
$$\approx \eta_0 + \frac{\eta_2}{2} |\tilde{E}_m|^2 (1 - 4r^2/a^2) \tag{7.126}$$

where the second expression is a two-term Taylor's series expansion of $|\tilde{E}|^2$ in the variable r^2, which is valid in the high-intensity region of interest in the central portion of the beam. Substitution of (7.126) into (7.125) yields an expression for the radius of curvature of a ray at radius r:

$$R = \frac{a^2 \eta_0}{4\eta_2 E_m^2 r} \tag{7.127}$$

Reference to Figure 7.33 indicates that if a parallel ray at radius r were to follow a circle with the initial radius of curvature, R, it would intercept the axis at a distance

$$z_f \simeq \sqrt{2rR} \tag{7.128}$$

We shall, therefore, take this distance as defining the focusing distance z_f. Combination of (7.127) and (7.128) then leads to the expression

$$z_f = \frac{a}{2} \left(\frac{2\eta_0}{\eta_2}\right)^{1/2} \frac{1}{|\tilde{E}_m|} \tag{7.129}$$

The above expression was derived on the basis of initially parallel rays. However, a Gaussian beam, rather than being composed of parallel rays, tends to spread because of diffraction. The tendency to spread counteracts the self-focusing phenomenon. When the two effects just counterbalance each other,

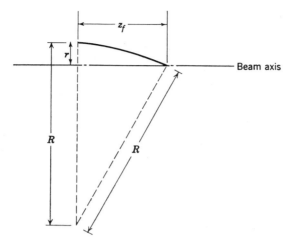

figure 7.33 Focusing of a ray in a medium of variable refractive index. The distance z_f is given by $z_f = \sqrt{R^2 - (R-r)^2} \approx \sqrt{2rR}$ for $r \ll R$.

self-trapping without beam spread occurs. In this case, the beam does not come to a focus. Therefore we are led to correct (7.129) so that for powers above the critical power for self-trapping, the net focusing distance after correction for diffraction spreading is of the form

$$z_{\text{net}} = \frac{a}{2}\left(\frac{2\eta_0}{\eta_2}\right)^{1/2} \frac{1}{|\tilde{E}_m| - |\tilde{E}_m|_{\text{cr}}} \tag{7.130}$$

where $|\tilde{E}_m|_{\text{cr}}$ is determined from (7.122) and (7.124). When $|\tilde{E}_m| = |\tilde{E}_m|_{\text{cr}}$, $z_{\text{net}} = \infty$, which is just the case of self-trapping without beam spread.

As a numerical example let us consider the self-focusing liquid carbon disulfide (CS_2), for which $(\eta_2)_{\text{esu}} \sim 10^{-11}$ [$(\eta_2)_{\text{MKS}} = (0.11 \times 10^{-8})(\eta_2)_{\text{esu}} = 1.1 \times 10^{-20}$]. The critical power for self-trapping of a ruby laser beam ($\lambda = 6943$ Å) is, from (7.122), $P_{\text{cr}} \sim 17$ kW. For a 1 MW beam with a 2-mm diameter, we have a net self-focusing length, from (7.130), $z_{\text{net}} \sim 60$ cm.

The self-focusing phenomenon has been observed experimentally, both by direct observation and by inference based on threshold data for the stimulated Raman effect. For the latter, the threshold for stimulated Raman emission in a strongly self-focusing liquid occurs when the pump laser power reaches the value required to bring the self-focusing length just inside the Raman cell. High-intensity filaments can then be formed, within which stimulated Raman emission

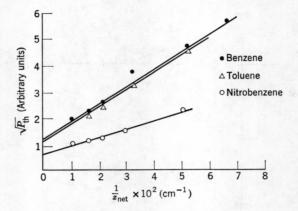

figure 7.34 From self-focusing theory, the threshold power for the formation of filaments in a liquid column of length z_{net} is given by (7.131) where z_{net} is the distance from the entrance plane at which the optical beam comes to a focus because of self-focusing effects. Thus, if the threshold for stimulated Raman emission corresponds to the threshold for filament formation, a plot of the square root of the threshold laser power for stimulated emission versus z_{net} should yield straight lines, and this is seen to be the case.[48]

takes place. To see this, we write (7.130) in the form

$$P^{1/2} = P_{cr}^{1/2} + \frac{A}{z_{net}} \tag{7.131}$$

where, from (7.124) and (7.130),

$$A = a \left(\frac{\pi \epsilon_0 c \eta_0^2}{8 \eta_2} \right)^{1/2}$$

According to the above equation, plots of $P^{1/2}$ versus $1/z_{net}$ for various materials should yield straight lines, and the critical power for self-trapping should be obtainable from the vertical intercepts of these straight lines. Striking confirmation of these considerations is presented in Figure 7.34. In the experimental arrangement z_{net} is set arbitrarily by the choice of cell length, and $P^{1/2}$ is that power required to initiate filament formation and, therefore, the onset of stimulated Raman emission.

REFERENCES

1. M. Born and K. Huang, *Dynamical Theory of Crystal Lattices*, Oxford University Press, London, 1954, pp. 79ff.
2. R. A. Smith, *Wave Mechanics of Crystalline Solids*, Chapman and Hall, London, 1961, pp. 204ff.
3. C. Kittel, *Quantum Theory of Solids*, Wiley, New York, 1963, pp. 21ff.
4. C. B. Walker, "X-Ray Study of Lattice Vibrations in Aluminum," *Phys. Rev.*, **103**, 547, 1956.
5. A. D. B. Woods, B. N. Brockhouse, R. A. Cowley, and W. Cochran, "Lattice Dynamics of Alkali Halide Crystals—II. Experimental Studies of KB_2 and NaI," *Phys. Rev.*, **131**, 1025, 1963.
6. A. Messiah, *Quantum Mechanics*, North-Holland Publishing Company, Amsterdam, 1961, Vol. 1, pp. 50ff.
7. C. Kittel, *Introduction to Solid-State Physics*, 3rd ed., Wiley, New York, 1966, pp. 187ff.
8. Reference 7, p. 134.
9. F. A. Jenkins and H. E. White, *Fundamentals of Optics*, 3rd ed., McGraw-Hill, New York, 1957, p. 511.
10. Reference 1, pp. 91ff.
11. Reference 7, p. 156.
12. A. S. Barker, Jr., "Transverse and Longitudinal Optic Mode Study in MgF_2 and ZnF_2," *Phys. Rev.*, **136**, A1290, 1964.
13. C. V. Raman and R. S. Krishnan, "A New Type of Secondary Radiation," *Nature*, **121**, 501, 1928.
14. G. Landsberg and L. Mandelstam, "Zuschriften—Eine neue Erscheinung bei der Lichtzerstreuung in Krystallen" *Naturwissenshaften*, **16**, 557, 1928.
15. E. J. Woodbury and W. K. Ng, "Ruby Laser Operation in the Near IR," *Proc. IRE* (Corr.), **50**, 2367, 1962.
16. G. Eckhardt, R. W. Hellwarth, F. J. McClung, S. E. Schwarz, and D. Weiner, "Stimulated Raman Scattering from Organic Liquids," *Phys. Rev. Letters*, **9**, 455 (1962).
17. E. J. Woodbury, "Raman Laser Action in Organic Liquids," in *Quantum Electronics Proceedings of the Third International Congress*, ed. by P. Grivet and N. Bloembergen, Columbia University Press, New York, **2**, 1576, 1964.
18. M. Geller, D. P. Bortfeld, and W. R. Sooy, "New Woodbury-Raman Laser Materials," *Appl. Phys. Letters*, **3**, 36, 1963.
19. B. P. Stoicheff, "Characteristics of Stimulated Raman Radiation Generated by Coherent Light," *Phys. Letters*, **7**, 186, 1963.
20. G. Eckhardt, D. P. Bortfeld, and M. Geller, "Stimulated Emission of Stokes and Anti-Stokes Raman Lines from Diamond, Calcite, and α-Sulfur Single Crystals," *Appl. Phys. Letters*, **3**, 137, 1963.
21. R. W. Minck, R. W. Terhune, and W. G. Rado, "Laser-Stimulated Raman Effect and Resonant Four-Photon Interactions in Gases H_2, D_2, and CH_4," *Appl. Phys. Letters*, **3**, 181, 1963.
22. R. W. Hellwarth, "Theory of Stimulated Raman Scattering," *Phys. Rev.*, **130**, 1850, 1963.
23. R. W. Hellwarth, "Analysis of Stimulated Raman Scattering of a Giant Laser Pulse," *Appl. Optics*, **2**, 847, 1963.
24. H. J. Zeiger and P. E. Tannenwald, "The Raman Maser," *Quantum Electronics Proceedings of the Third International Congress*, ed. by P. Grivet and N. Bloembergen, Columbia University Press, New York, **2**, 1588, 1964.

25. E. Garmire, F. Pandarese, and C. H. Townes, "Coherently Driven Molecular Vibrations and Light Modulation," *Phys. Rev. Letters*, **11,** 160, 1963.
26. G. Eckhardt, "Selection of Raman Laser Materials," *IEEE J. Quant. Elect.*, **QE-2,** 1, 1966.
27. M. D. Martin and E. L. Thomas, "Infrared Difference Frequency Generation," *IEEE J. Quant. Electr.*, **QE-2,** 196, 1966.
28. G. Placzek, *Marx Handbuch Der Radiologie*, E. Marx, ed., 2nd ed., Akademische Verlagsgesellschaft, Leipzig, 1934, Vol. 6, Part II, p. 205.
29. G. Herzberg, *Infrared and Raman Spectra at Polyatomic Molecules*, vol. 2, Van Nostrand, New York, 1945, p. 109.
30. J. A. Stratton, *Electromagnetic Theory*, p. 206, McGraw-Hill, New York, 1941, pp. 112ff.
31. Reference 1, p. 310.
32. T. C. Damen, R. C. C. Leite, and S. P. S. Porto, "Angular Dependence of the Raman Scattering from Benzene Excited by the He-Ne CW Laser," *Phys. Rev. Letters*, **14,** 9, 1965.
33. J. F. Nye, *Physical Properties of Crystals*, Oxford, 1957, pp. 243ff.
34. R. Y. Chiao, C. H. Townes, and B. P. Stoicheff, "Stimulated Brillouin Scattering and Generation of Intense Hypersonic Waves," *Phys. Rev. Letters*, **12,** 592, 1964.
35. *American Institute of Physics Handbook*, ed. by D. E. Gray, McGraw-Hill, New York, 1965.
36. N. M. Kroll, "Excitation of Hypersonic Vibrations by Means of Photoelastic Coupling of High-Intensity Light Waves to Elastic Waves," *J. Appl. Phys.*, **36,** 34, 1965.
37. F. J. McClung, W. G. Wagner, and D. Weiner, "Mode-Structure Independence of Stimulated Raman-Scattering Conversion Efficiencies," *Phys. Rev. Letters*, **15,** 96, 1965.
38. D. Weiner, S. E. Schwarz, and F. J. McClung, "Comparison of Observed and Predicted Stimulated Raman Scattering Conversion Efficiencies," *J. Appl. Phys.*, **36,** 2395, 1965.
39. G. Bret and G. Mayer, "Forward Emission of Raman Radiation in Various Liquids," *Proc. of the Int'l Conf. on the Phys. of Quant. Elect.*, ed. by P. L. Kelley, B. Lax, and P. E. Tannenwald, McGraw-Hill, New York, 1966.
40. G. Hauchecorne and G. Mayer, "Effets de l'anisotropie moléculaire sur la propagation d'une lumière intense," *Compt. Rend.*, **261,** 4014, 1965.
41. R. Y. Chiao, E. Garmire, and C. H. Townes, "Self-Trapping of Optical Beams," *Phys. Rev. Letters*, **13,** 479, 1964.
42. P. L. Kelley, "Self-Focusing of Optical Beams," *Phys. Rev. Letters*, **15,** 1005, 1965.
43. G. A. Askar'yan, "Effects of the Gradient of a Strong Electromagnetic Beam on Electrons and Atoms," *Soviet Phys. JETP*, **15,** 1088, 1962.
44. V. I. Talanov, "Propagation of Short Electromagnetic Impulses in an Active Medium," *Radiophysics*, **7,** 254, 1964.
45. N. F. Pilipetskii and A. R. Rustamov, "Observation of Self-Focusing of Light in Liquids," *JETP Letters*, **2,** 55, 1965.
46. Y. R. Shen and Y. J. Shaham, "Beam Deterioration and Stimulated Raman Effect," *Phys. Rev. Letters*, **15,** 1008, 1965.
47. P. Lallemand and N. Bloembergen, "Self-Focusing of Laser Beams and Stimulated Raman Gain in Liquids," *Phys. Rev. Letters*, **15,** 1010, 1965.
48. C. C. Wang, "Length-Dependent Threshold for Stimulated Raman Effect and Self-Focusing of Laser Beams in Liquids," *Phys. Rev. Letters*, **16,** 344, 1966.
49. E. Garmire, R. Y. Chiao, and C. H. Townes, "Dynamics and Characteristics of the Self-Trapping of Intense Light Beams," *Phys. Rev. Letters*, **16,** 347, 1966.
50. N. Bloembergen and P. Lallemand, "The Influence of Self-Focusing on the Stimulated Brillouin, Raman, and Rayleigh Effects," *IEEE Jour. Quant. Elect.*, **QE-2,** 246, 1966.
51. D. H. Close, C. R. Giuliano, R. W. Hellwarth, L. D. Hess, F. J. McClung, and W. G. Wagner, "The Self-Focusing of Light of Different Polarizations," *IEEE Jour. Quant. Elect.*, **QE-2,** 553, 1966.

52. A. Javan and P. L. Kelley, "Possibility of Self-Focusing due to Intensity Dependent Anomalous Dispersion," *IEEE Jour. Quant. Elect.*, **QE-2**, 470, 1966.
53. R. Y. Chiao, M. A. Johnson, S. Krinsky, H. A. Smith, C. H. Townes, and E. Garmire, "A New Class of Trapped Light Filaments," *IEEE Jour. Quant. Elect.*, **QE-2**, 467, 1966.
54. C. C. Wang, "Nonlinear Susceptibility Constants and Focusing of Optical Beams in Liquids," *Phys. Rev.*, **152**, 149, 1966.
55. R. W. Hellwarth, "Effect of Molecular Redistribution on the Nonlinear Refractive Index of Liquids," *Phys. Rev.*, **152**, 156, 1966.
56. R. G. Brewer and C. H. Townes, "Standing Waves in Self-Trapped Light Filaments," *Phys. Rev. Letters*, **18**, 196, 1967.
57. M. Born and E. Wolf, *Principles of Optics*, 3rd ed., Pergamon Press, New York, 1965, p. 122.

PROBLEMS

7.1 Show that in the limit of long-wavelength vibrations ($k \to 0$) the difference equation (7.1) can be replaced by a differential equation of the form

$$\rho_l \frac{\partial^2 q}{\partial t^2} = C \frac{\partial^2 q}{\partial x^2}$$

which is the propagation equation for acoustic waves on a homogeneous line of linear mass density ρ_l and elastic stiffness C.

7.2 Consider the case of an acoustic mode that satisfies periodic boundary conditions as discussed in Section 7.3.1. Show by summing over the contributions from all atoms that acoustic modes other than that for which $\mathbf{k} = 0$ do not carry physical momentum. What does the $\mathbf{k} = 0$ mode correspond to physically?

7.3 Show that the Hamiltonian (7.23) leads to the equations of motion (7.17) directly from Hamilton's equations.

7.4 The energy of an electron is given by the relativistic expression

$$\mathscr{W} = p^2 c^2 + m_0^2 c^4$$

where p is the momentum of the electron, m_0 is the rest mass, and c is the velocity of light.

(a) Draw the dispersion ($\mathscr{W}p$) diagram for the electron indicating the asymptotes and intercepts.

(b) In Compton scattering an incident photon is scattered from an electron with a shift in frequency. The differences in energy and momentum of the incident and scattered photons are taken up by electron recoil. Recalling that energy and momentum of a photon are related by $\mathscr{W} = pc$, draw the dispersion vector conservation diagrams for (i) back-scattering from a stationary electron, (ii) back-scattering from an electron moving opposite to the incident photon.

7.5 The crystal LiNbO$_3$ has an infrared-absorbing optical mode at $(1/\lambda) = 628$ cm^{-1} of strength $\Delta\kappa = 2.55$ and a high-frequency dielectric constant $\kappa_\infty = 4.6$.

(a) Plot the dispersion characteristic (ωk diagram) for the region

$$500 \text{ cm}^{-1} < 1/\lambda < 610 \text{ cm}^{-1}.$$

(b) If one performed a Raman scattering experiment in which a He-Ne laser beam ($\lambda = 6328$ Å) was scattered from this mode, wave vector conservation would result in a Stokes output whose frequency was a function of the angle between the Stokes and laser beams. Obtain the angular dependence of the Stokes output frequency.

7.6 In order to reach threshold for stimulated Stokes emission in a cavity, it was shown that the single-pass gain must exceed the single-pass loss δ.

(a) Discuss whether the threshold is lowered by increasing the length L of the Raman medium when the loss δ is due to (i) fixed mirror losses, (ii) absorption loss in the Raman medium.

(b) Calculate the threshold for stimulated Stokes scattering of a ruby laser beam ($\lambda = 6943$ Å) from the $(1/\lambda) = 628$ cm^{-1} optical vibrational mode of the crystal LiNbO$_3$ (lithium niobate). Assume a 1.1% loss per pass ($\delta = 0.011$) and a 0.55-cm crystal. Take $\eta_s = \eta_i = 2.2$ and $\chi_R''(\omega_v = \Omega) = 3.6 \times 10^{-20}$ MKS units.

7.7 Raman scattering is possible from electronic states as well as from vibrational modes. In this case an incident photon is annihilated and a Stokes photon gener-

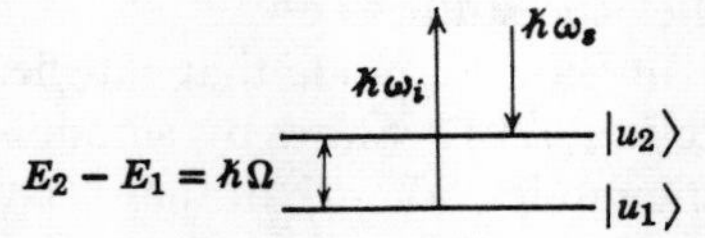

figure P7.7 Energy-level diagram for Raman scattering from a two-level electronic system.

ated with the difference in energy absorbed by the electron, as shown in Figure P7.7. The Hamiltonian for such a system is given by $\mathcal{H} = \mathcal{H}_0 + \mathcal{H}'$, where

$$\mathcal{H}_0 = \begin{pmatrix} E_1 & 0 \\ 0 & E_2 \end{pmatrix}$$

and

$$\mathcal{H}' = -(\tfrac{1}{2})\alpha' E^2 = -(\tfrac{1}{2}) \begin{pmatrix} 0 & \alpha'_{12} \\ (\alpha'_{12})^* & 0 \end{pmatrix} E^2$$

The nonlinear dipole moment associated with the Raman effect is still $\mu = \alpha' E$.

(a) Show that the equation of motion for the expectation value of the polarizability, $\langle \alpha' \rangle$, is given by

$$\langle \ddot{\alpha}' \rangle + \Delta\omega_L \langle \dot{\alpha}' \rangle + \Omega^2 \langle \alpha' \rangle = \frac{\Omega}{\hbar} |\alpha'_{12}|^2 (\rho_{11} - \rho_{22}) E^2$$

(b) Show that the Raman susceptibility, assuming N_V atoms per unit volume, is given by

$$\chi_R(\omega_v) = \frac{\Omega \overline{(N_1 - N_2)|\alpha'_{12}|^2}}{2\hbar\epsilon_0} \frac{1}{\Omega^2 - \omega_v^2 - i\omega_v(\Delta\omega_L)}$$

where $N_1 - N_2 = \overline{N_V(\rho_{44} - \rho_{22})}$ is the population difference per unit volume, and the overbar indicates a spatial average.

7.8 The dispersion relation for a long-wavelength acoustic wave is given by $\omega_k = v_a k$, where v_a is the velocity of sound in the medium of interest. With the aid of the geometrical construction of Figure 7.29 for wave vector conservation, derive the Brillouin frequency shift expression (7.95).

7.9 Show that the radius of curvature, R, of a ray in a medium of variable refractive index $\eta(r)$ is given by

$$\frac{1}{R} = \frac{1}{\eta(r)} \frac{\partial \eta(r)}{\partial r}$$

(Hint: Consider a limiting case of Snell's Law.)

ELECTRONS IN CRYSTALS
8

8.1 INTRODUCTION:

In this chapter we shall consider interactions between electrons and electromagnetic fields in crystalline media. When atoms are brought together to form a periodic array as in a crystal, the discrete energy states of the isolated atom are broadened into energy bands. The existence of a continuum of energy states means that the evaluation of transition probability or conductivity requires an integration over energy intervals. This is in contrast with the previously performed analyses, in which transition probability was determined from a summation of contributions involving distinct energy levels.

Section 8.2 presents a discussion of the behavior of an electron in a crystal in the absence of an applied field. The first part of this section will be concerned with an ideal crystal in which there are no impurities or lattice vibrations, and the second portion will include reflection and trapping of electrons by impurities. With the background material provided in Section 8.2, we shall then be in a position to study the interaction between radiation and electrons in crystals. The influence of a dc or low-frequency field is discussed in Section 8.3, and band-to-band transitions resulting from infrared or optical signals are considered in Section 8.4. Photoconductivity, that portion of the conductivity that results from the absorption of light, is studied in Section 8.5. The semiconductor injection laser is analyzed in Section 8.6.

8.2 ELECTRONS IN CRYSTALS WITHOUT AN APPLIED FIELD

In this section we shall study the behavior of electrons in crystalline media in the absence of an applied field. First an ideal crystal will be considered in which no impurities exist. The subject matter will include energy band theory, Brillouin zones, Bloch functions, momentum, and Bragg reflection. Imperfect crystals will then be discussed from the standpoint of reflection and trapping by impurities.

8 ELECTRONS IN CRYSTALS

figure 8.1 An infinite, one-dimensional array of identical atoms separated by a distance a.

This material provides a background for subsequent sections, in which fields are applied to the crystal.

8.2.1 Energy Bands

If we consider an infinite, one-dimensional array of identical atoms as shown in Figure 8.1, for a large interatomic spacing the wave function characterizing an electron located near a given atom will approximate closely the wave function for an electron near an isolated atom. In this case the energy eigenvalues are discrete, with a small amount of line broadening resulting from relaxation mechanisms. However, as the atoms are brought closer together, the energy levels for an electron broaden into energy bands, even in the absence of relaxation processes. This conversion from an energy level to an energy band is illustrated in Figure 8.2, in which energy is plotted as a function of interatomic spacing.

Let us determine the possible energies for an electron near an atom, when the influence of neighboring atoms is just beginning to be felt. The Hamiltonian is

$$\mathcal{H} = \mathcal{H}_0 + \mathcal{H}' \tag{8.1}$$

where $\mathcal{H}_0$ is the Hamiltonian for a given isolated atom and $\mathcal{H}'$ is a perturbation term resulting from all other atoms.

We can obtain approximate values for energy eigenvalues by making a reasonable choice for the eigenfunction $\psi(x)$ and evaluating energy from the wave equation. A reasonable choice for $\psi(x)$ is one for which

$$\psi(x + la) \simeq \varphi(x) \tag{8.2}$$

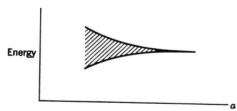

figure 8.2 Electron energy for an electron near an atom as a function of interatomic spacing. The electron may assume any value of energy included in the shaded area.

where $\varphi(x)$ is the eigenfunction for an isolated atom, l is an integer, a is the interatomic spacing, and $x = 0$ gives the location of one of the atoms. That is, $\psi(x)$ is chosen so that at each atomic site the eigenfunction is approximately that of an isolated atom.

With the atoms far enough apart so that there is little interaction, the atomic wave function $\varphi(x)$ is barely discernible by its neighbor, and thus the condition $|\varphi(0)| \gg |\varphi(\pm a)|$ is satisfied. Therefore a function that satisfies (8.2) is

$$\psi(x) = \sum_l c(x_l)\varphi(x - x_l) \tag{8.3}$$

where $x_l \equiv la$. At each atomic site, $\psi(x)$ in (8.3) is approximately equal to the atomic wave function because of the small overlap between the atomic functions.

Using Dirac notation, we may rewrite (8.3) as

$$|\psi\rangle = \sum_l c(l) |l\rangle \tag{8.4}$$

with an assumed normalization condition

$$\langle l | l \rangle = 1 \tag{8.5}$$

The ket vector $|l\rangle$ represents an atomic wave function at lattice site l. Because of the small amount of overlap between the wave functions associated with adjacent atoms, we have

$$\langle m | l \rangle \ll 1 \quad \text{for} \quad m \neq l \tag{8.6}$$

The wave equation is given by

$$(\mathcal{H}_0 + \mathcal{H}') |\psi\rangle = E |\psi\rangle \tag{8.7}$$

where E is the eigenvalue that is to be determined. The summation in (8.4) may be substituted for $|\psi\rangle$ in (8.7), and with the relationship

$$\mathcal{H}_0 |l\rangle = E_0 |l\rangle \tag{8.8}$$

where E_0 is the energy eigenvalue for an isolated atom, we find that

$$\sum_l (E_0 + \mathcal{H}')c(l) |l\rangle = E \sum_l c(l) |l\rangle \tag{8.9}$$

If we form the product of the bra vector $\langle m|$ with both sides of (8.9), the result is found to be

$$c(m)E_0 + \sum_l c(l)\mathcal{H}'_{ml} = c(m)E \tag{8.10}$$

where we have used the condition (8.6) that $\langle m | l \rangle \simeq \delta_{ml}$.

For equidistant atoms there is no distinction between $\mathcal{H}'_{m,m+1}$ and $\mathcal{H}'_{m,m-1}$, so that $\mathcal{H}'_{m,m+1} = \mathcal{H}'_{m,m-1}$. Nearest neighbors will have the greatest influence,

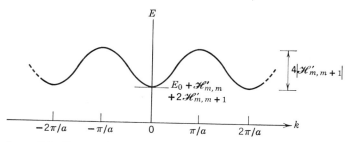

figure. 8.3 Energy eigenvalue E as a function of k for an electron in a one-dimensional array of atoms.

and therefore a reasonable approximation is to include only the terms $l = m$, $m + 1$, and $m - 1$ in the summation in (8.10). Then (8.10) becomes

$$c(m)[E_0 + \mathcal{H}'_{mm}] + [c(m-1) + c(m+1)]\mathcal{H}'_{m,m+1} = c(m)E \quad (8.11)$$

Equation 8.11 is a linear, homogeneous difference equation with constant coefficients that may be solved for $c(m)$ by assuming a solution of the form

$$c(m) = e^{ikma} \quad (8.12)$$

where k is a constant. The substitution of (8.12) into (8.11) yields the relationship between E and k when there is weak interaction between nearest neighbors:

$$E = E_0 + \mathcal{H}'_{mm} + 2\mathcal{H}'_{m,m+1} \cos ka \quad (8.13)$$

If E were known, then k could be determined, or vice versa. If $\mathcal{H}'_{m,m+1}$ is negative, a plot of E as a function of k is as shown in Figure 8.3. If $\mathcal{H}'_{m,m+1} > 0$, then the top of the energy band occurs at $k = 0$ and the minimum occurs at $k = \pi/a$. Note that as the atoms are moved farther apart, both $\mathcal{H}'_{mm}$ and $\mathcal{H}'_{m,m+1}$ approach zero, so that $E \to E_0$.

The upper and lower limits of the energy band occur at values of k for which

$$k = \frac{n\pi}{a} \quad (8.14)$$

where n is an integer. In Section 8.2.3 we shall see that (8.14) corresponds to Bragg reflection, a condition that was previously considered on page 245 with reference to Brillouin scattering. For a three-dimensional array of atoms distributed along orthogonal axes, the values of $\mathbf{k}$ that give the limits on the energy band are

$$\mathbf{k} \cdot \mathbf{d} = \pi d^2 \quad (8.15)$$

where

$$\mathbf{k} = \mathbf{1}_x k_x + \mathbf{1}_y k_y + \mathbf{1}_z k_z$$
$$d^2 = \mathbf{d} \cdot \mathbf{d}$$

and

$$\mathbf{d} = \mathbf{1}_x \frac{n_x}{a} + \mathbf{1}_y \frac{n_y}{b} + \mathbf{1}_z \frac{n_z}{c}$$

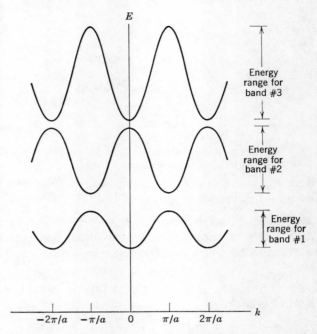

figure 8.4 Energy band diagram. For each value of k there are many values for E, corresponding to different energy bands.

The constants a, b, and c are the interatomic distances along the x, y, and z directions, respectively, and n_x, n_y, and n_z are integers. The unit vectors along the x, y, and z directions are $\mathbf{1}_x$, $\mathbf{1}_y$, and $\mathbf{1}_z$, respectively.

Figure 8.3 shows a single energy band resulting from the broadening of a single atomic energy eigenvalue E_0. If additional atomic eigenvalues were to be considered, this would yield additional energy bands, which broaden as E increases. The reason for the broadening is that the atomic wave functions for higher energies have greater spatial extent. This causes more overlap between adjacent atomic wave functions for a fixed separation, and more overlap results in a larger value for $|\mathcal{H}'_{m,m+1}|$. The energy diagram for several energy bands is illustrated in Figure 8.4.

The assumptions that have been made (such as only considering symmetric, weak interaction between nearest neighbors) have resulted in an energy-versus-k curve that is symmetric about $k = 0$, but in general energy bands need not be symmetric about $k = 0$, when these assumptions are removed.

Equation 8.15 defines surfaces in $\mathbf{k}$ space, and these surfaces correspond to the limits of an energy band. The volumes enclosed by these surfaces are known as

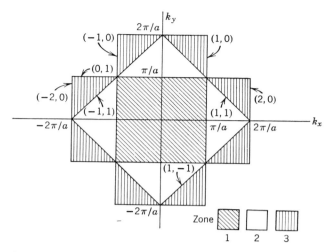

figure 8.5 Two-dimensional Brillouin zones. The bracketed integers adjacent to the lines indicate the values for n_x and n_y for each line; that is, the $n_x = 1$, $n_y = 0$ line is designated by (1, 0). This diagram indicates the lines in **k** space for which Bragg reflection occurs.

Brillouin zones. For a two-dimensional problem in which $n_z = 0$ and $b = a$, (8.15) reduces to

$$k_x n_x + k_y n_y = \frac{\pi}{a}(n_x^2 + n_y^2) \tag{8.16}$$

Each set of values for n_x and n_y defines a line in the $k_x - k_y$ plane that corresponds to a band edge, and the areas enclosed by the lines are the two-dimensional Brillouin zones. The first three Brillouin zones for a two-dimensional array are shown in Figure 8.5.

8.2.2 Bloch Functions

The substitution of $c(l)$ as given by (8.12) into (8.3) yields an expression for the eigenfunction:

$$\psi(x) = \sum_l e^{ilka} \varphi(x - la) \tag{8.17}$$

This $\psi(x)$ satisfies the difference equation

$$\psi(x + a) = e^{ika}\psi(x) \tag{8.18}$$

as may be shown by writing out $\psi(x + a)$ from (8.17). Equation 8.18 might have been written directly from symmetry considerations, since the probability $\psi\psi^*$ of finding an electron at x must be the same as the probability of finding it at $x + a$

for an infinite array of identical atoms. Therefore the difference equation of (8.18) applies to any periodic structure with periodicity a. The most general form of the solution to (8.18) is

$$\psi(k, x) = e^{ikx}u(k, x) \tag{8.19}$$

where

$$u(k, x + na) = u(k, x) \tag{8.20}$$

and n is an integer. The function $u(k, x)$ has the same symmetry as the crystal, in that translation by an integral number of atomic spacings does not alter the value of the function. Functions of the form given by (8.19) are known as Bloch functions. The three-dimensional representation for the Bloch function is

$$\psi(\mathbf{k}, \mathbf{r}) = e^{i\mathbf{k}\cdot\mathbf{r}}u(\mathbf{k}, \mathbf{r}) \tag{8.21}$$

where $\mathbf{r}$ is a vector in coordinate space and $u(\mathbf{k}, \mathbf{r})$ possesses the translational symmetry of the crystal.

For each value of $\mathbf{k}$, (8.21) does not specify a unique eigenfunction because of a multiplicity of eigenvalues resulting from the various energy bands, as shown in Figure 8.4. Therefore it is necessary to indicate the band with which the eigenfunction is associated:

$$\psi_B(\mathbf{k}, \mathbf{r}) = e^{i\mathbf{k}\cdot\mathbf{r}}u_B(\mathbf{k}, \mathbf{r}) \tag{8.22}$$

where the subscript B specifies the energy band.

It is desirable to have an orthonormal set of eigenfunctions. This may be accomplished by imposing periodic boundary conditions so that, for the one-dimensional case, we require that

$$\psi_B(k, x + Ga) = \psi_B(k, x) \tag{8.23}$$

where G is an integer that is much larger than 1. The imposed condition is that the eigenfunction repeat itself when translated by a large integral number of atomic spacings. (The application of periodic boundary conditions is discussed in more detail in Section 7.3.1.) From (8.22) and (8.23) we have

$$\psi_B(k, x + Ga) = e^{ik(x+Ga)}u_B(k, x + Ga) = e^{ikx}u_B(k, x)$$

Since $u_B(k, x + Ga) = u_B(k, x)$, it is necessary that

$$kGa = 2n\pi \tag{8.24}$$

for integer values of n.

With periodic boundary conditions we may prove, by means of the wave equation, that eigenfunctions corresponding to different eigenvalues are orthogonal:

$$\int_0^{Ga} \psi_{B'}^*(k', x)\psi_B(k, x)\,dx = 0 \quad \text{for} \quad B' \neq B \quad \text{or} \quad k' \neq k$$

8 ELECTRONS IN CRYSTALS

An appropriate constant may be included in $u_B(k, x)$ so that when $B' = B$ and $k' = k$ the integral is normalized to unity.

In three dimensions the periodic boundary condition is that the eigenfunction repeat when translated by a large integral number of unit cells. The unit cell is a volume that reproduces the crystal by translations in various directions. The orthogonality condition is then given by

$$\int_F \psi_{B'}^*(\mathbf{k}', \mathbf{r}) \psi_B(\mathbf{k}, \mathbf{r}) \, dV = \delta_{B'B} \delta_{\mathbf{k}'\mathbf{k}} \tag{8.25}$$

where the integral over the volume F, termed the fundamental domain, includes many unit cells. In Dirac notation we may write $\psi_B(\mathbf{k}, \mathbf{r})$ as $|\mathbf{k}, B\rangle$, so that the orthogonality condition is

$$\langle \mathbf{k}', B' | \mathbf{k}, B \rangle = \delta_{B'B} \delta_{\mathbf{k}'\mathbf{k}}$$

Since the dimensions of the volume F are chosen arbitrarily, they have no physical significance, and so any expression for a measurable quantity will not involve the dimensions of F.

Periodic boundary conditions result in discrete values for k, but with G chosen sufficiently large the increment between different k values becomes small and so k space may be considered a continuum. Equation 8.25 then becomes

$$\int \psi_{B'}^*(\mathbf{k}', \mathbf{r}) \psi_B(\mathbf{k}, \mathbf{r}) \, dV = \delta_{B'B} \, \delta(\mathbf{k}' - \mathbf{k}) \tag{8.26}$$

where $\delta(\mathbf{k}' - \mathbf{k})$ is the Dirac delta function.

8.2.3 Momentum

For the electron in free space where

$$\mathcal{H} = \frac{\mathbf{p}^2}{2m} \tag{8.27}$$

the eigenfunctions of $\mathcal{H}$ are

$$\psi(\mathbf{k}, \mathbf{r}) = e^{i\mathbf{k}\cdot\mathbf{r}} \tag{8.28}$$

The $\langle \mathbf{p} \rangle$ for an electron in a given energy eigenstate is

$$\langle \mathbf{p} \rangle = \langle \mathbf{k} | -i\hbar \nabla | \mathbf{k} \rangle = \hbar \mathbf{k} \tag{8.29}$$

Thus the momentum of a free electron is $\hbar$ times the wave vector $\mathbf{k}$.

If a free electron impinges on a crystal with a $\mathbf{k}$ value corresponding to a band edge, Bragg reflection occurs. As illustrated for normal incidence in Figure 8.6, a wave incident on a crystal is totally reflected when the distance traveled by a wave reflected from an inner atomic layer is in phase with the wave reflected from the surface. From Figure 8.6 we see that the Bragg reflection condition is

$$2a = n\lambda \tag{8.30}$$

for integer values of n. Since $\lambda = h/p$ for an electron, from (8.29) we have, for the one-dimensional case,

$$2a = n\left(\frac{2\pi}{k}\right) \tag{8.31}$$

The value for k given by (8.31) is the same as that given by (8.14), so that an incident electron with a k value corresponding to a band edge is reflected from the crystal.

Let us now determine the momentum of an electron in a given energy eigenstate inside the crystal. If we operate on the wave equation with $\nabla_\mathbf{k}$, where

$$\nabla_\mathbf{k} \equiv \mathbf{1}_x \frac{\partial}{\partial k_x} + \mathbf{1}_y \frac{\partial}{\partial k_y} + \mathbf{1}_z \frac{\partial}{\partial k_z}$$

and $\mathbf{1}_i$ is the unit vector in the i direction, since $\mathcal{H}$ is not a function of $\mathbf{k}$ we have

$$\mathcal{H} \nabla_\mathbf{k} |\mathbf{k}, B\rangle = E(\mathbf{k}, B) \nabla_\mathbf{k} |\mathbf{k}, B\rangle$$
$$+ |\mathbf{k}, B\rangle \nabla_\mathbf{k} E(\mathbf{k}, B) \tag{8.32}$$

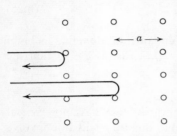

figure 8.6 Bragg reflection for normal incidence. The wave reflected from an inner layer is in phase with the wave reflected from the surface.

The scalar product of (8.32) with $\langle \mathbf{k}, B|$ gives

$$\langle \mathbf{k}, B| (\mathcal{H} - E) \nabla_\mathbf{k} |\mathbf{k}, B\rangle = \nabla_\mathbf{k} E \tag{8.33}$$

where we have written $E(\mathbf{k}, B)$ as E, and the $E \nabla_\mathbf{k} |\mathbf{k}, B\rangle$ term on the right-hand side of (8.32) has been transposed to the left-hand side of (8.33).

To evaluate the left-hand side of (8.33), let us return to the representation of the eigenfunction as a function of the spatial coordinates. We find that, for the Bloch function,

$$\nabla_\mathbf{k} \psi_B(\mathbf{k}, \mathbf{r}) = i\mathbf{r} \psi_B(\mathbf{k}, \mathbf{r}) + e^{i\mathbf{k} \cdot \mathbf{r}} \nabla_\mathbf{k} u_B(\mathbf{k}, \mathbf{r}) \tag{8.34}$$

Thus we may write the left-hand side of (8.33) as

$$\langle \mathbf{k}, B| (\mathcal{H} - E) \nabla_\mathbf{k} |\mathbf{k}, B\rangle = \int_F dV \psi_B^* (\mathcal{H} - E) i\mathbf{r} \psi_B$$
$$+ \int_F dV \psi_B^* (\mathcal{H} - E) e^{i\mathbf{k} \cdot \mathbf{r}} \nabla_\mathbf{k} u_B \tag{8.35}$$

Since $(\mathcal{H} - E)$ is an Hermitian operator and the functions in the integrand for the second term on the right-hand side of (8.35) obey the periodic boundary conditions, we may write

$$\int_F dV \psi_B^* (\mathcal{H} - E) e^{i\mathbf{k} \cdot \mathbf{r}} \nabla_\mathbf{k} u_B = \int_F dV ((\mathcal{H} - E) \psi_B)^* e^{i\mathbf{k} \cdot \mathbf{r}} \nabla_\mathbf{k} u_B = 0 \tag{8.36}$$

Although $(\mathcal{H} - E)$ is Hermitian, it is not correct to move the operator to the left if the functions do not obey the periodic boundary conditions (see Problem 1.9)

Since $i\mathbf{r}\psi_B$ is not periodic, the procedure used in (8.36) may not be used to evaluate the first term on the right-hand side of (8.35). To obtain an equivalent expression for this term, note that

$$\mathscr{H} i\mathbf{r}\psi_B = \left(-\frac{\hbar^2}{2m}\nabla^2 + \mathscr{V}\right) i\mathbf{r}\psi_B \qquad (8.37)$$

With the vector identity

$$\nabla^2(\mathbf{r}\psi_B) = \mathbf{r}\nabla^2\psi_B + 2\nabla\psi_B$$

(8.37) becomes

$$\mathscr{H} i\mathbf{r}\psi_B = i\mathbf{r}\mathscr{H}\psi_B + \frac{\hbar}{m}(-i\hbar\nabla)\psi_B \qquad (8.38)$$

Therefore, from (8.38),

$$\int_F dV\,\psi_B^*(\mathscr{H} - E)i\mathbf{r}\psi_B = \int_F dV\,\psi_B^* i\mathbf{r}(\mathscr{H} - E)\psi_B + \int_F dV\,\psi_B^* \frac{\hbar}{m}(-i\hbar\nabla)\psi_B \qquad (8.39)$$

The first term on the right-hand side of (8.39) is zero because the wave equation is $(\mathscr{H} - E)\psi_B = 0$. Substitution of (8.39) into (8.35) yields

$$\langle \mathbf{k}, B|\,(\mathscr{H} - E)\nabla_\mathbf{k}\,|\mathbf{k}, B\rangle = \frac{\hbar}{m}\langle \mathbf{k}, B|\,\mathbf{p}\,|\mathbf{k}, B\rangle \qquad (8.40)$$

where we have used the identity

$$\int_F dV\,\psi_B^* \frac{\hbar}{m}(-i\hbar\nabla)\psi_B = \frac{\hbar}{m}\langle \mathbf{k}, B|\,\mathbf{p}\,|\mathbf{k}, B\rangle$$

From (8.33) we now have

$$\langle \mathbf{k}, B|\,\mathbf{p}\,|\mathbf{k}, B\rangle = \frac{m}{\hbar}\nabla_\mathbf{k} E \qquad (8.41)$$

Equation 8.41 gives the momentum of an electron in a given energy eigenstate in terms of the **k**-space derivative of the eigenvalue. Figure 8.7 is a one-dimensional illustration of $\langle \mathbf{p}\rangle$ as a function of **k**.

The electron momentum may be either positive or negative depending upon the sign of the slope of the energy curves. At the edges of the energy band the momentum is zero.

8.2.4 Reflection by Impurities†

Thus far we have considered the behavior of an electron in an ideal crystal, where there is equal spacing between identical atoms. However, it is impossible to fabricate crystals without impurities, and a lower limit on impurity concentration

† The material presented in Sections 8.2.4 and 8.2.5 is based upon the presentation in Chapter 13 of R. P. Feynman, R. B. Leighton, and M. Sands, *The Feynman Lectures on Physics*, Vol. III, Addison-Wesley, Reading, Mass., 1965.

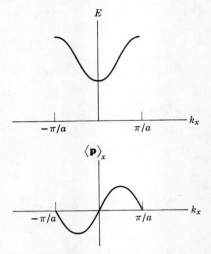

figure 8.7 Energy and $\langle \mathbf{p} \rangle_x$ as functions of k_x.

is about $10^{14}/cm^3$. In addition, thermally generated lattice vibrations occur, which disrupt the equal spacing between atoms. These factors cause electrons to be scattered or trapped at the site of an impurity.

When an electron encounters an impurity, two outcomes are possible:

1. The electron energy may remain unaltered but its momentum is changed. This is the condition for scattering, in which the electron energy stays within an energy band and the electron is free to move throughout the crystal.
2. The electron energy may fall outside an energy band, and the electron, no longer able to move through the lattice sites is trapped near the impurity atom. In this section we consider scattering and in Section 8.2.5 we shall study trapping.

We shall consider a one-dimensional array with an impurity located at the $m = 0$ position, as shown in Figure 8.8. The presence of the impurity is accounted for by assuming that at the $m = 0$ site the eigenvalue for the isolated atom will not be the same as the eigenvalue for all the other atoms. That is,

$$\mathcal{H}_0 |m\rangle = E_0 |m\rangle \qquad \text{for} \quad m \neq 0$$

and

$$\mathcal{H}_0 |m\rangle = (E_0 + \Delta) |m\rangle \qquad \text{for} \quad m = 0$$

where $\mathcal{H}_0$ is the Hamiltonian operator for an isolated atom and Δ is the difference between the eigenvalue of the isolated impurity atom and the eigenvalue for all other atoms. For $\mathcal{H}'$ we assume that all the atoms are identical and take $\mathcal{H}'_{mm} = \mathcal{H}'_{00}$ and $\mathcal{H}'_{m,m+1} = \mathcal{H}'_{01}$. An impurity will also alter the matrix elements of $\mathcal{H}'$

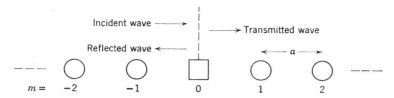

figure 8.8 An impurity atom is located at the $m = 0$ position, with incident, reflected, and transmitted waves propagating as shown.

in the vicinity of the impurity, but the primary features of the scattering process may be determined by assuming only a change in E_0.

If (8.11) is written out explicitly for $m = 0$, we have

$$c(0)[E_0 + \Delta + \mathcal{H}'_{00}] + [c(-1) + c(1)]\mathcal{H}'_{01} = c(0)E \tag{8.42}$$

If an electron is scattered by the impurity, it is necessary to consider reflected and transmitted waves as well as the incident wave. Let

$$\begin{aligned} c(m) &= e^{ikma} + Be^{-ikma} \quad \text{for} \quad m \leq 0 \\ c(m) &= Ce^{ikma} \quad \text{for} \quad m \geq 0 \end{aligned} \tag{8.43}$$

where the incident wave is e^{ikma}, the reflected wave is Be^{-ikma}, and the transmitted wave is Ce^{ikma}. The relationships among the c's, (8.11), are satisfied with the energy relationship given by (8.13) for all except the $m = 0$ equation of (8.42). If we substitute (8.13) and (8.43) into (8.42), then the $m = 0$ equation is also satisfied with the following expression for B:

$$B = \frac{-\Delta}{\Delta + 2i\mathcal{H}'_{01}\sin ka} \tag{8.44}$$

In the derivation of (8.44) it has been taken that $c(m)$ is continuous at $m = 0$, giving the relation

$$C = 1 + B$$

The probability of having a reflected electron is $|B|^2$, and the probability of a transmitted electron is $|C|^2$. Since the electron is either reflected or transmitted, it is necessary that

$$|B|^2 + |C|^2 = 1 \tag{8.45}$$

a condition that is satisfied by (8.44) in conjunction with $C = 1 + B$.

As one would expect, the reflected wave approaches zero as $\Delta \to 0$. For a given Δ, as ka approaches $n\pi$ the magnitude of the reflected wave approaches unity. This value for k corresponds to a band edge and the fact that there is complete

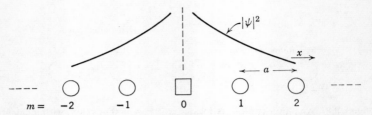

figure 8.9 A plot of $|\psi|^2$ for a trapped electron as a function of the distance from the impurity atom.

reflection, even for a vanishingly small discontinuity, is consistent with our previous discussion of Bragg reflection.

8.2.5 Trapping

An impurity may trap an electron so that it remains close to the location of the impurity. For this case we would expect $|\psi|^2$ to be of the form illustrated in Figure 8.9. The probability of finding the electron in an increment dx decreases as the distance from the impurity increases, in a manner that is symmetrical about the impurity site.

To analyze this problem we allow

$$k \to i\gamma$$

which results in decaying rather than propagating waves for real values of γ. Thus, we let

$$c(m) = e^{\gamma ma} \quad \text{for} \quad m \leqq 0$$
$$c(m) = e^{-\gamma ma} \quad \text{for} \quad m \geqq 0 \quad (8.46)$$

All the $c(m)$ equations, (8.11), are satisfied by the energy relationship (8.13), with the exception of the $m = 0$ equation of (8.42). For decaying waves, (8.13) becomes

$$E = E_0 + \mathscr{H}'_{00} + 2\mathscr{H}'_{01} \cosh \gamma a \quad (8.47)$$

where the cosine function has been replaced by the hyperbolic cosine. If we substitute (8.46) and (8.47) into (8.42), then with

$$\gamma = \frac{1}{a} \sinh^{-1}\left(\frac{\Delta}{2\mathscr{H}'_{01}}\right) \quad (8.48)$$

the $m = 0$ equation is also satisfied.

For $\mathscr{H}'_{01} < 0$, it is necessary that Δ be negative so that γ will be positive, whereupon $c(m)$ decreases away from the $m = 0$ site. Thus, to obtain trapping from an energy band that has a minimum at $k = 0$ the energy eigenvalue of the isolated impurity atom must be less than the eigenvalue for the isolated normal or intrinsic atom. Conversely, to obtain trapping from an energy band with a maximum at $k = 0$, the energy of the impurity atom must be greater than the energy of the intrinsic atom.

The substitution of (8.48) into (8.47) gives an expression for the energy of the trapped electron:

$$E = E_0 + \mathcal{H}'_{00} \pm [4(\mathcal{H}'_{01})^2 + \Delta^2]^{1/2} \qquad (8.49)$$

where the minus sign applies when $k = 0$ is a minimum in energy, and the plus sign applies when $k = 0$ is a maximum in energy. Figure 8.3 illustrates the case where $\mathcal{H}'_{01} < 0$, and therefore the minus sign in (8.49) applies. If $\Delta = 0$, then E given by (8.49) equals the value for E at $k = 0$ in Figure 8.3. Therefore, for a finite Δ the energy of an electron trapped by an impurity lies below the minimum energy of the energy band for the ideal crystal when $k = 0$ corresponds to the minimum energy.

8.3 INTRABAND EFFECTS

With the background material that has been presented in Section 8.2, we are now prepared to investigate the interaction between radiation and electrons in crystals. We shall first consider the response of an electron in a crystal to a signal whose photon energy is less than the energy required to raise an electron from one energy band to another by single-photon absorption. In semiconductors this includes signals from dc to the far infrared. If we choose our fundamental domain to include several hundred unit cells, the properties of this fundamental domain are almost indistinguishable from a crystal of infinite extent. A single wavelength for the signals we are considering includes at least several thousand unit cells, so that over the fundamental domain the field is approximately constant and the dipole approximation may be used.

Since the photon energy is not enough to produce band-to-band transitions, the primary effect is to change the energy state of an electron within a single energy band. If all the energy states in an energy band are filled, no change of state is possible within the band because of the "exclusion principle," which forbids more than one electron from occupying a given state. Therefore, intraband absorption, which refers to the absorption of radiation and the excitation of electrons within a band, can occur only with a partially filled energy band. In semiconductors this interaction generally involves electrons at the bottom of the conduction band, which are known as free carriers. Hence, intraband absorption is also referred to as free carrier absorption.

In Section 8.3 we shall first consider the motion of an electron in an ideal crystal, in both **k** space and coordinate space. An effective mass will be derived. Electron motion including scattering will be analyzed, and an expression will be obtained for electron current. The properties of the hole will be defined, and the current resulting from the motion of holes will be evaluated.

8.3.1 Intraband Motion in an Ideal Crystal

We shall find that under the influence of an applied field an electron moves through **k** space so that

$$\dot{\mathbf{k}}(t) = \frac{1}{\hbar}\mathbf{f} \tag{8.50}$$

where $\mathbf{k}(t)$ is the **k** value for the electron at any time t and $\mathbf{f}$ is the force applied on the particle by the field. Equation 8.50 may be derived from the equation of motion for the diagonal element of the density operator.

We are dealing with a continuum of basis functions since **k** is continuous, so that any state of the system may be represented as

$$\Psi(\mathbf{r}, t) = \int d^3k \sum_B c_B(\mathbf{k}, t)\psi_B(\mathbf{k}, \mathbf{r})$$

where $\psi_B(\mathbf{k}, \mathbf{r})$ is given by (8.22). The symbol d^3k represents the volume element in **k** space, $dk_x\, dk_y\, dk_z$, and $\sum_B$ is a sum over the different energy bands. In direct analogy with the discrete level system, the probability that an electron is found in energy band B and in a **k**-space volume increment d^3k is equal to $|c_B(\mathbf{k}, t)|^2\, d^3k$. (We suggest that Section 1.4.4, dealing with the density operator for a continuum of eigenstates, be reread at this time.) The matrix elements of the density operator are

$$\rho_{B'B}(\mathbf{k}', \mathbf{k}, t) = \langle \mathbf{k}', B| \rho(t) |\mathbf{k}, B\rangle$$

and in accordance with (1.37) the diagonal elements are

$$\rho_{BB}(\mathbf{k}, t) = |c_B(\mathbf{k}, t)|^2$$

where $\rho_{BB}(\mathbf{k}, \mathbf{k}, t) \equiv \rho_{BB}(\mathbf{k}, t)$.

The equation of motion for the diagonal element is given by (1.36), and for the ideal crystal (that is, no collisions), we take $T_1 \to \infty$. Thus, using the integral form of the commutator given by (1.39), we have

$$i\hbar \frac{\partial \rho_{BB}(\mathbf{k},t)}{\partial t} = [\mathscr{H}', \rho]_{\mathbf{k}B,\mathbf{k}B} = \sum_{B'} \int d^3k' [\mathscr{H}'_{\mathbf{k}B,\mathbf{k}'B'}\, \rho_{B'B}(\mathbf{k}', \mathbf{k}, t) - \text{c.c.}] \tag{8.51}$$

In going from (1.39) to (8.51), we have introduced a summation over different energy bands, since there may be many energy bands for each value of **k**. For intraband motion we consider transitions only within a band, and so for this case $B' = B$.

The interaction Hamiltonian $\mathscr{H}'$ to within an additive constant is $(e\mathbf{r} \cdot \mathscr{E})$. It is assumed that the wavelength of the radiation is appreciably greater than the maximum excursion of the electron, and **r** is measured from any fixed coordinate system in the crystal. The symbol $\mathscr{E}$ is used for the electric field in order to avoid confusion with the energy eigenvalue E. With this value for $\mathscr{H}'$, (8.51) becomes

$$i\hbar \frac{\partial \rho_{BB}(\mathbf{k}, t)}{\partial t} = e\mathscr{E} \cdot \int d^3k' \int dV [\rho_{BB}(\mathbf{k}', \mathbf{k}, t)\psi_B^*(\mathbf{k}, \mathbf{r})\mathbf{r}\psi_B(\mathbf{k}', \mathbf{r}) - \text{c.c.}] \quad (8.52)$$

From (8.22), we may write

$$\mathbf{r}\psi_B(\mathbf{k}', \mathbf{r}) = \mathbf{r}e^{i\mathbf{k}'\cdot\mathbf{r}}u_B(\mathbf{k}', \mathbf{r}) = -i\nabla_{\mathbf{k}'}\psi_B(\mathbf{k}', \mathbf{r}) + ie^{i\mathbf{k}'\cdot\mathbf{r}}\nabla_{\mathbf{k}'}u_B(\mathbf{k}', \mathbf{r}) \quad (8.53)$$

The substitution of (8.53) into (8.52) yields

$$i\hbar \frac{\partial \rho_{BB}(\mathbf{k})}{\partial t}$$

$$= \left[-ie\mathscr{E} \cdot \iint d^3k' \, dV \rho_{BB}(\mathbf{k}', \mathbf{k}, t)\psi_B^*(\mathbf{k}, \mathbf{r})\nabla_{\mathbf{k}'}\psi_B(\mathbf{k}', \mathbf{r}) - \text{c.c.} \right]$$

$$+ \left[ie\mathscr{E} \cdot \iint d^3k' dV \rho_{BB}(\mathbf{k}', \mathbf{k}, t)\psi_B^*(\mathbf{k}, \mathbf{r})e^{i\mathbf{k}'\cdot\mathbf{r}}\nabla_{\mathbf{k}'}u_B(\mathbf{k}', \mathbf{r}) - \text{c.c.} \right] \quad (8.54)$$

In Appendix 11 it is shown that the second bracket on the right-hand side of (8.54) is zero, and the first bracket reduces to $[ie\mathscr{E} \cdot \nabla_{\mathbf{k}}\rho_{BB}(\mathbf{k}, t)]$. Thus we have

$$\frac{\partial \rho_{BB}(\mathbf{k}, t)}{\partial t} = \frac{e}{\hbar}\mathscr{E} \cdot \nabla_{\mathbf{k}} \rho_{BB}(\mathbf{k}, t) \quad (8.55)$$

A general solution to (8.55) is

$$\rho_{BB}(\mathbf{k}, t) = \rho_{BB}\left(\mathbf{k} + \frac{e}{\hbar}\int_0^t \mathscr{E} \, dt\right) \quad (8.56)$$

as may be seen by the substitution of (8.56) into (8.55). If the electron distribution in **k** space is given by $\rho(\mathbf{k})$ in the absence of a field, then in the presence of the field the distribution is

$$\rho\left(\mathbf{k} - \frac{1}{\hbar}\int_0^t \mathbf{f} \, dt\right)$$

where $\mathbf{f} = -e\mathscr{E}$ is the force on an electron. In a time increment Δt it is necessary to have an increment in **k** given by

$$\Delta \mathbf{k} = \frac{1}{\hbar}\mathbf{f}\Delta t$$

to maintain the same probability for locating an electron in a given volume element in **k** space. Thus we may say that under the influence of a force **f** the electrons move in **k** space with the velocity

$$\dot{\mathbf{k}}(t) = \frac{1}{\hbar}\mathbf{f}$$

which is the same as (8.50).

8.3.2 Effective Mass

From (8.41) the expectation value of the canonical momentum for a particle in a given energy eigenstate is

$$\langle \mathbf{p} \rangle = \frac{m}{\hbar} \nabla_k E$$

For the dipole approximation the canonical momentum equals the particle momentum (Section 2.2.2), so that

$$\langle \mathbf{v} \rangle = \frac{1}{\hbar} \nabla_k E \tag{8.57}$$

where $\langle \mathbf{v} \rangle$ is the particle velocity. If we differentiate both sides of (8.57) with respect to time, we obtain

$$\langle \dot{\mathbf{v}} \rangle = \frac{1}{\hbar} (\dot{\mathbf{k}} \cdot \nabla_k) \nabla_k E \tag{8.58}$$

and from (8.50)

$$\langle \dot{\mathbf{v}} \rangle = \frac{1}{\hbar^2} (\mathbf{f} \cdot \nabla_k) \nabla_k E \tag{8.59}$$

where

$$\mathbf{f} \cdot \nabla_k = f_x \frac{\partial}{\partial k_x} + f_y \frac{\partial}{\partial k_y} + f_z \frac{\partial}{\partial k_z} \tag{8.60}$$

If we write out the α component of $\langle \dot{\mathbf{v}} \rangle$ as given by (8.59), we find that

$$\langle \dot{\mathbf{v}} \rangle_\alpha = \frac{1}{\hbar^2} \frac{\partial^2 E}{\partial k_\alpha \partial k_\beta} f_\beta \tag{8.61}$$

where α and β each refer to the x, y, or z coordinates, and the right-hand side of (8.61) is summed over the double subscript. For example, $\langle \dot{\mathbf{v}} \rangle_x$ is

$$\langle \dot{\mathbf{v}} \rangle_x = \frac{1}{\hbar^2} \left[\frac{\partial^2 E}{\partial k_x^2} f_x + \frac{\partial^2 E}{\partial k_x \partial k_y} f_y + \frac{\partial^2 E}{\partial k_x \partial k_z} f_z \right]$$

Equation 8.61 may be written as the matrix product

$$\begin{bmatrix} \langle \dot{\mathbf{v}} \rangle_x \\ \langle \dot{\mathbf{v}} \rangle_y \\ \langle \dot{\mathbf{v}} \rangle_z \end{bmatrix} = \frac{1}{\hbar^2} \begin{bmatrix} \dfrac{\partial^2 E}{\partial k_x^2} & \dfrac{\partial^2 E}{\partial k_x \partial k_y} & \dfrac{\partial^2 E}{\partial k_x \partial k_z} \\ \dfrac{\partial^2 E}{\partial k_y \partial k_x} & \dfrac{\partial^2 E}{\partial k_y^2} & \dfrac{\partial^2 E}{\partial k_y \partial k_z} \\ \dfrac{\partial^2 E}{\partial k_z \partial k_x} & \dfrac{\partial^2 E}{\partial k_z \partial k_y} & \dfrac{\partial^2 E}{\partial k_z^2} \end{bmatrix} \begin{bmatrix} f_x \\ f_y \\ f_z \end{bmatrix} \tag{8.62}$$

and is abbreviated as

$$\langle \dot{\mathbf{v}} \rangle = \frac{1}{m^*} \mathbf{f} \tag{8.63}$$

where $\langle \mathbf{v} \rangle$ and $\mathbf{f}$ are vectors and $1/m^*$ is a second-rank tensor.

The factor m^* is known as the effective mass for the electron in the crystal because of the similarity between (8.63) and Newton's second law of motion. The tensor elements of the effective mass are

$$\frac{1}{m^*_{\alpha\beta}} = \frac{1}{\hbar^2} \frac{\partial^2 E}{\partial k_\alpha \partial k_\beta} \tag{8.64}$$

Note that $1/m^*$ is a symmetric tensor because of the interchangeability of the order of differentiation in (8.64). The effective mass tensor has the same symmetry properties as the susceptibility tensor discussed in Section 3.3. It is always possible to diagonalize a second-rank, symmetric tensor by a rotation of coordinate axes, and directions for which the matrix is diagonal are termed the principal axes. We see from Table 3.2 that for all but two crystal classes the principal axes are along the direction of the crystal symmetry axes. Along the direction of a principal axis (8.61) becomes

$$\langle \dot{v} \rangle = \frac{1}{\hbar^2} \frac{\partial^2 E}{\partial k^2} f \tag{8.65}$$

and the effective mass is

$$m^* = \frac{\hbar^2}{\partial^2 E/\partial k^2} \tag{8.66}$$

where $\langle v \rangle$, f, and k are the components of the corresponding vectors parallel to a principal axis. In general, m^* has a different value along each of the three principal axes.

Figure 8.10 illustrates the behavior of m^* as a function of k. Near the top of an energy band the effective mass is negative, which means that if a free electron moves in a given direction in response to to an applied field, the electron in the crystal moves in the opposite direction. In this case the momentum transferred to the electron by the lattice is larger than the momentum contributed by the applied field and has an opposite direction to it. As we saw in Section 8.2.3, Bragg reflection occurs as the upper band edge is approached. Therefore a slight increase in the electron energy from an applied field could reduce the probability of penetration through the lattice because of the in-phase reflection from crystal layers. In other words, a small increase in the electron energy may produce a resonant response from the crystal that transfers a large amount of momentum to the electron in a direction that is opposite to that of the momentum produced by the field. This is why the effective mass is negative near the top of a band edge.

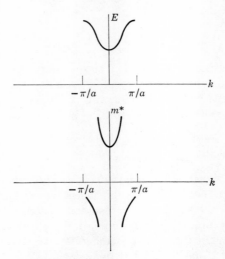

figure 8.10 The behavior of the effective mass as a function of k along a principal axis. Near the bottom of the energy band the effective mass is positive, and near the top of the band the effective mass is negative. At the inflection point in the energy curve the effective mass is infinite.

Let us now determine the energy, k, and velocity of an electron in a crystal as a function of time, in response to a constant force f_0 applied along a principal axis. We shall assume that, at $t = 0$, k is also zero. From (8.50) we see that k increases linearly with time as illustrated in Figure 8.11a. Since electron energy is a periodic function of k, energy is also a periodic function of time as shown in Figure 8.11b. From (8.57) we see that the electron velocity is proportional to the k derivative of energy, and so velocity is proportional to the time derivative of energy as shown in Figure 8.11c. Since the electron velocity is periodic in time, the electron position is also periodic in time, that is, the electron oscillates back and forth.

The results illustrated in Figure 8.11 neglect the effects of collisions, and in the following section we shall consider electron motion in both k and coordinate space allowing collisions to occur.

8.3.3 Electron Motion Including Collisions

As we saw in Section 8.2.4, an impurity may cause an incident electron to be reflected, and Figure 8.12a shows the motion of an electron, with collisions, in a

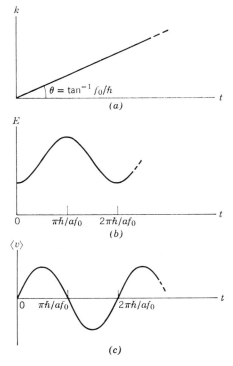

figure 8.11 Electron behavior in response to a dc field, neglecting collisions.

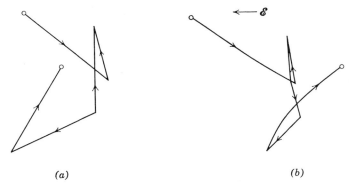

figure 8.12 Electron motion in a crystal with collisions, (a) in the absence, and (b) in the presence of an applied field.

crystal in the absence of an applied field. With the field present, Figure 8.12b, there is a drift along the direction of the field. For a given electron the time between collisions varies, and so the drift velocity must be determined from a time average of the velocity attained between collisions.

Consider an electron that has undergone its last collision at $t = t_0$. The probability increment dp that the electron is scattered in a succeeding time interval dt is proportional to dt and to the probability p that the electron has not collided in the time period $t - t_0$. Thus

$$dp = -\frac{1}{\tau} p \, dt \tag{8.67}$$

where $(-1/\tau)$ is the proportionality constant. From (8.67) it may be shown that τ is the average time between collisions. Upon integrating (8.67), we find

$$p = e^{-(t-t_0)/\tau} \tag{8.68}$$

where we have assumed $p = 1$ at $t = t_0$. If we are interested in the change in the probability that an electron has not collided at time t which is associated with considering a change in the time of its last collision, then we differentiate (8.68) with respect to t_0 and obtain

$$dp = \frac{1}{\tau} e^{-(t-t_0)/\tau} \, dt_0 \tag{8.69}$$

Between collisions, we see from (8.65) and (8.66) that the velocity along the direction of a principal axis is

$$\langle v \rangle = \langle v \rangle_0 + \int_{t_0}^{t} \frac{f}{m^*} \, dt \tag{8.70}$$

where $\langle v \rangle_0$ is the velocity at $t = t_0$. The increment in electron energy between collisions is generally sufficiently small that $\partial^2 E/\partial k^2$ does not change appreciably, and therefore m^* may be considered independent of time.

With an applied sinusoidal signal at frequency ω, f may be written as

$$f = -\frac{e\tilde{\mathscr{E}}}{2} e^{i\omega t} + \text{c.c.} \tag{8.71}$$

where e is the magnitude of the electron charge, and $\tilde{\mathscr{E}}$ is the time-independent field amplitude component parallel to a principal axis. The velocity of an electron that had its last collision at $t = t_0$ is given by the substitution of (8.71) into (8.70):

$$\langle v \rangle = \langle v \rangle_0 - \left[\frac{e\tilde{\mathscr{E}} e^{i\omega t}}{2i\omega m^*} [1 - e^{-i\omega(t-t_0)}] + \text{c.c.} \right] \tag{8.72}$$

To obtain the average electron velocity, it is necessary to average (8.72) over all possible collision times t_0. The average value of $\langle v \rangle$ is

$$\overline{\langle v \rangle} = \frac{\int_{-\infty}^{t} \langle v \rangle \, dp}{\int_{-\infty}^{t} dp} \tag{8.73}$$

which, from (8.69) and (8.72), has the value

$$\overline{\langle v \rangle} = \overline{\langle v \rangle}_0 - \left[\frac{e\tau}{m^*} \left[\frac{1 - i\omega\tau}{1 + (\omega\tau)^2} \right] \frac{\tilde{\mathscr{E}}}{2} e^{i\omega t} + \text{c.c.} \right] \tag{8.74}$$

Scattering by thermal vibrations is isotropic, so that for this case $\overline{\langle v \rangle}_0 = 0$ along any given direction. If a given scattering angle is preferred, such as scattering by ionized impurity centers, τ becomes modified accordingly.[1] Our discussions will be limited to isotropic scattering.

A typical range of values for τ is $10^{-12} - 10^{-13}$ sec, so that for frequencies well below 10^{12} rad/sec $\omega\tau \ll 1$, and from (8.74), with $\overline{\langle v \rangle} = \tfrac{1}{2}\overline{\langle \tilde{v} \rangle} e^{i\omega t} + \text{c.c.}$, we have that

$$\overline{\langle \tilde{v} \rangle} = -\frac{e\tau}{m^*} \tilde{\mathscr{E}}$$

$$\equiv -\mu \tilde{\mathscr{E}} \tag{8.75}$$

where

$$\mu = \frac{e\tau}{m^*} \tag{8.76}$$

The constant μ is known as the mobility, and for pure germanium at room temperature its value is 0.39 m²/V-sec.

Let us now consider the behavior of an electron in **k** space with collisions. In the absence of collisions we found that **k** increased linearly with time with an applied dc field:

$$\mathbf{k} = \mathbf{k}_0 + \frac{\mathbf{f}}{\hbar}(t - t_0) \tag{8.77}$$

where **f** is the amplitude of the dc force and $\mathbf{k}_0$ is the value for **k** at $t = t_0$. In a manner analogous to the method used for the evaluation of the average velocity of an electron, we find that the average value of **k** is

$$\overline{\mathbf{k}} = \frac{\mathbf{f}\tau}{\hbar} \tag{8.78}$$

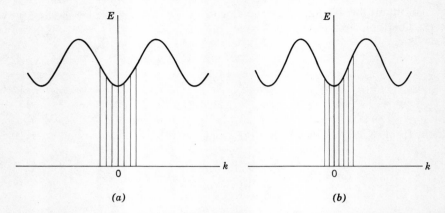

figure 8.13 The occupation of electron states: (a) without, and (b) with an applied constant force. A vertical line indicates that an eigenstate is occupied.

Figure 8.13 shows the occupation of a partially filled energy band without and with a constant applied force. There is a shift in the average value for **k** of the electrons by an amount $f\tau/\hbar$.

8.3.4 Current Density

Two approaches may be used to determine the current density that results from the application of a field. We may calculate the drift velocity of an electron in the presence of a field and thereby obtain the current density for a single electron as

$$\mathbf{J} = -\frac{e}{V}\overline{\langle \mathbf{v} \rangle} \tag{8.79}$$

where V is an arbitrary volume element over which we have normalized the wave function for an electron. The magnitude of electron charge is e, and the drift velocity is $\overline{\langle \mathbf{v} \rangle}$. Current resulting from having many electrons in the volume V is

$$\mathbf{J} = -\frac{e}{V}\sum_i \overline{\langle \mathbf{v} \rangle}_i \tag{8.80}$$

where the index i refers to the ith electron. We are generally interested in the electron current associated with the motion of electrons in the bottom of the conduction band, where each electron has approximately the same velocity. For this case,

$$\mathbf{J} = -N_e \overline{\langle \mathbf{v} \rangle} \tag{8.81}$$

where N_e is the number of conduction electrons per unit volume.

This approach has the advantage of simplicity and is particularly useful when the time between collisions, τ, is independent of the electron energy. However, when the variation of τ with energy is to be included, it is necessary to express an appropriate average value for τ. This average is obtained by using a second approach for the determination of **J**, namely, to consider a given increment in energy or velocity and to evaluate the change in charge density within that increment in the presence of an applied field. In Sections 8.3.4 and 8.3.5 we shall use the first method, and in Section 8.3.6 the second method will be employed.

From (8.74) and (8.81) we see that the current density $J = \frac{1}{2}\tilde{J}e^{i\omega t}$ + c.c. is

$$\tilde{J} = \frac{N_e e^2 \tau}{m^*}\left[\frac{1 - i\omega\tau}{1 + (\omega\tau)^2}\right]\tilde{\mathscr{E}}$$

$$\equiv \sigma^* \tilde{\mathscr{E}} \qquad (8.82)$$

$$\equiv (\sigma + i\omega\Delta\epsilon)\tilde{\mathscr{E}}$$

where σ^* is known as the complex conductivity, σ is the contribution to the real conductivity resulting from the electron motion, and $\Delta\epsilon$ is the contribution to the permittivity resulting from electron motion. We are using coordinate axes that correspond to the principal axes so that $\tilde{J}$ and $\tilde{\mathscr{E}}$ are the vector components parallel to a principal axis. In general, σ^* has three different values along the different principal axes. From (8.82), the expression for σ is

$$\sigma = \frac{N_e e^2}{m^*}\frac{\tau}{1 + (\omega\tau)^2} \qquad (8.83)$$

and Figure 8.14 is a plot of this function. When $\hbar\omega$ is greater than the energy required to make band-to-band transitions, the conductivity increases dramatically and the intraband or free-carrier absorption is generally negligible by comparison. This subject will be considered in Section 8.4.

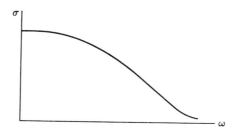

figure 8.14 The dependence of conductivity on frequency resulting from intraband or free-carrier absorption.

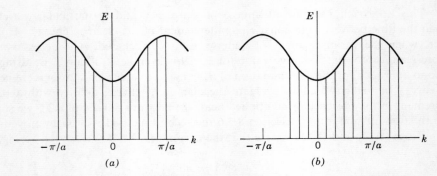

figure 8.15 The occupation of electron energy states in a filled band: (a) without a field; (b) with a field.

Let us now determine the current that results when a dc field is applied to a filled energy band. Figure 8.15a illustrates the filled band in the absence of a field. The occupation of states is shown only in the range of k from $-\pi/a$ to $+\pi/a$, since an electron with a k value given by $k_0 + 2\pi/a$ is indistinguishable from an electron with a k value equal to k_0. When a field is applied, the occupied states are shifted to the right as shown in Figure 8.15b. In the absence of a field there is no current. With the field present, the electrons in the newly occupied states between π/a and $2\pi/a$ are indistinguishable from the electrons in the formerly occupied states between $-\pi/a$ and 0. Therefore we conclude that there is no current when the field is turned on.

8.3.5 Holes

When a band is filled, we see from (8.80) that

$$\mathbf{J} = -\frac{e}{V}\sum_i \overline{\langle\mathbf{v}\rangle_i} = 0 \tag{8.84}$$

Suppose that a single electron is missing from the top of an energy band, as would be the case for the valence band when an electron makes a transition from the valence band to the conduction band. The current then is

$$\mathbf{J} = -\frac{e}{V}\sum_{i\neq p} \overline{\langle\mathbf{v}\rangle_i} \tag{8.85}$$

where the pth electron has been removed. Let us now add and subtract the pth electron to (8.85), giving

$$\begin{aligned}\mathbf{J} &= -\frac{e}{V}\sum_{i\neq p}\overline{\langle\mathbf{v}\rangle_i} - \frac{e}{V}\overline{\langle\mathbf{v}\rangle_p} + \frac{e}{V}\overline{\langle\mathbf{v}\rangle_p} \\ &= -\frac{e}{V}\sum_i\overline{\langle\mathbf{v}\rangle_i} + \frac{e}{V}\overline{\langle\mathbf{v}\rangle_p}\end{aligned} \tag{8.86}$$

Since the current for a filled band is zero, we see that the current for a band missing only the pth electron is

$$\mathbf{J} = \frac{e}{V} \overline{\langle \mathbf{v} \rangle}_p \tag{8.87}$$

The current given by (8.87) may be thought of as resulting from a particle with charge $+e$ and having the velocity and therefore the effective mass that the electron had before it was removed from the band. This particle is termed a hole. Since the effective mass is negative near the top of an energy band, the effective mass of holes, m_h^*, is defined as

$$\frac{1}{(m_h^*)_{\alpha\beta}} \equiv -\frac{1}{\hbar^2} \frac{\partial^2 E}{\partial k_\alpha \partial k_\beta} \tag{8.88}$$

which differs from the effective mass of an electron, (8.64), by a minus sign. From (8.74), therefore, we see that the drift velocity $\overline{\langle v \rangle}_h = \frac{1}{2}\langle v \rangle_h e^{i\omega t} + \text{c.c.}$ for a hole is

$$\widetilde{\langle v \rangle}_h = \left[\frac{1 - i\omega\tau_h}{1 + (\omega\tau_h)^2} \right] \frac{e\tau_h}{m_h^*} \widetilde{\mathscr{E}} \tag{8.89}$$

where the subscript h refers to a hole. In the derivation of (8.74) it has been assumed that the coordinate axes were the principal axes, so that $\widetilde{\langle v \rangle}_h$ and $\widetilde{\mathscr{E}}$ are the vector components along a principal axis. Similarly, m_h^* is the component of the effective mass tensor along the same axis.

The total current produced by an applied low frequency field, for which $\omega\tau_e, \omega\tau_h \ll 1$, when there are both electrons at the bottom of the conduction band and holes at the top of the valence band is

$$\widetilde{J} = \left(\frac{N_e \tau_e}{m_e^*} + \frac{N_h \tau_h}{m_h^*} \right) e^2 \widetilde{\mathscr{E}}$$

$$= (N_e \mu_e + N_h \mu_h) e \widetilde{\mathscr{E}} \tag{8.90}$$

where the subscript e refers to an electron, N_e is the number of conduction electrons per unit volume, N_h is the number of holes per unit volume, and e is the magnitude of the electronic charge.

8.3.6 Current Density from the Boltzmann Equation

In Sections 8.3.4 and 8.3.5 we determined current density by calculating the current for a single electron and summing over all the electrons. It had been assumed that the time between collisions, τ, was a constant with time. As the electron moves, however, its energy changes, and in general τ is a function of energy. To avoid this difficulty, we shall consider a fixed energy increment and determine the current that results from the change in charge density within this increment. The total current is obtained by integrating over energy and allowing τ to be a function of energy.

The charge density increment $d\xi$ resulting from conduction electrons in a small volume in **k** space is

$$d\xi = -N_V e \rho_{22}(\mathbf{k})\, d^3k \tag{8.91}$$

where

d^3k = volume element in **k** space = $dk_x\, dk_y\, dk_z$

$\rho_{22}(\mathbf{k})$ = diagonal element of the density operator in the conduction band (the subscript 2 refers to the conduction band)

= probability that an electron is to be found in a unit volume in **k** space in the conduction band

N_V = number of electrons per unit volume in coordinate space.

Current density increment $d\mathbf{J}$ that results from this charge element is

$$d\mathbf{J} = \langle \mathbf{v}(\mathbf{k}) \rangle\, d\xi \tag{8.92}$$

The function $\langle \mathbf{v}(\mathbf{k}) \rangle$ is the velocity for a given value of **k** as specified by (8.57).

Under the influence of a field, ρ_{22} changes because of motion in **k** space, and according to (8.55) we have

$$\frac{\partial \rho_{22}}{\partial t} = \frac{e}{\hbar} \mathscr{E} \cdot \nabla_\mathbf{k} \rho_{22} \tag{8.93}$$

where $\mathscr{E}$ is the applied electric field. Equation 8.55 was derived for electron motion in the absence of collisions, and resulted from letting $T_1 \to \infty$ in (1.36). With collisions, T_1 is finite so that (8.93) becomes

$$\frac{\partial \rho_{22}}{\partial t} + \frac{\rho_{22} - \rho_{22}^e}{T_1} = \frac{e}{\hbar} \mathscr{E} \cdot \nabla_\mathbf{k} \rho_{22} \tag{8.94}$$

where T_1 is the relaxation time and ρ_{22}^e is the equilibrium value. Later in this section we shall relate T_1 to the time constant τ, the time between collisions. With a low frequency field applied, (that is, $\omega T_1 \ll 1$) in the steady-state we have

$$\frac{\rho_{22} - \rho_{22}^e}{T_1} = \frac{e}{\hbar} \mathscr{E} \cdot \nabla_\mathbf{k} \rho_{22} \tag{8.95}$$

which is known as Boltzmann's equation. To first order in the applied field, ρ_{22} on the right-hand side of (8.95) is taken to be the equilibrium value since $\mathscr{E}$ is a factor of the right-hand-side term.

We may now obtain an expression for the current density in terms of the applied field. From (8.91) and (8.92) we have

$$\mathbf{J} = -eN_V \int d^3k \langle \mathbf{v}(\mathbf{k}) \rangle \rho_{22}(\mathbf{k}) \tag{8.96}$$

where, from (8.95),

$$\rho_{22}(\mathbf{k}) = \frac{eT_1}{\hbar} \mathscr{E} \cdot \nabla_\mathbf{k} \rho_{22}^e + \rho_{22}^e \tag{8.97}$$

In the absence of the applied field, **J** is zero, so that the ρ_{22}^e term on the right-hand side of (8.97) does not contribute to the current density and therefore may be dropped.

From (8.57) we have

$$\langle \mathbf{v}(\mathbf{k}) \rangle = \frac{1}{\hbar} \nabla_\mathbf{k} E \tag{8.57}$$

where E is the electron energy. A combination of (8.57), (8.96), and (8.97) yields

$$\mathbf{J} = -\frac{N_V e^2}{\hbar} \int d^3k (\nabla_\mathbf{k} E)(\mathscr{E} \cdot \nabla_\mathbf{k} \rho_{22}^e) T_1 \tag{8.98}$$

and T_1 is included inside the integral because, in general, T_1 is a function of E.

In Appendix 13 we show that (8.98) may be rewritten as

$$J_\alpha = \frac{N_e e^2}{m_\alpha^*} \langle T_1 \rangle \mathscr{E}_\alpha \tag{8.99}$$

where

$\mathscr{E}_\alpha$ = electric field along the α-direction, which is a principal axis,
m_α^* = effective mass at the bottom of the conduction band along the same direction,
N_e = number of conduction electrons per unit volume,

and

$$\langle T_1 \rangle = \frac{\int_0^\infty T_1 E^{3/2} \rho_{22}^e \, dE}{\int_0^\infty E^{3/2} \rho_{22}^e \, dE} \tag{8.100}$$

If we compare (8.82) and (8.99), we see that these two expressions are the same for a low frequency signal if we take the average time between collisions, τ, to be $\langle T_1 \rangle$. When T_1 is not a function of E, we have $\tau = T_1$. Equation 8.100 indicates the appropriate way to average the constant T_1 when T_1 is a function of the electron energy.

8.4 INTERBAND EFFECTS

We shall now consider photon energies that are large enough to produce transitions from the valence to the conduction band. In semiconductors this corresponds to wavelengths that are in the infrared or visible part of the spectrum. The dipole approximation is either shaky or invalid for these wavelengths, and so we shall use the Hamiltonian

$$\mathcal{H} = \frac{1}{2m}(\mathbf{p} - q\mathbf{A})^2 + \mathcal{V} \tag{8.101}$$

where we are taking $\nabla \cdot \mathbf{A} = 0$, and $\mathcal{V}$ is the potential field of the electron in the crystal. Current density for N_V charged particles per unit volume is

$$\mathbf{J} = qN_V \langle \mathbf{v} \rangle$$

and for the $\mathcal{H}$ given by (8.101) the velocity operator is related to the canonical momentum $\mathbf{p}$ by the relationship given in Section 2.2

$$\mathbf{v} = \frac{1}{m}(\mathbf{p} - q\mathbf{A}) \tag{8.102}$$

Therefore we see that the current density is

$$\mathbf{J} = q\frac{N_V}{m} \langle \mathbf{p} - q\mathbf{A} \rangle \tag{8.103}$$

The problem of direct transitions from the valence to conduction band resulting from absorption of radiation will be treated by means of the density operator, and equations will be derived similar to those in Chapter 2, which relate current and field. Indirect transitions, which involve interactions among the field, electrons, and lattice vibrations, will also be considered.

8.4.1 Direct Transitions

In this section we shall obtain an expression for the conductivity that results from direct transitions between energy bands. The direct transition is one in which the $\mathbf{k}$ value for the final state of the electron is approximately the same as the $\mathbf{k}$ value in the initial state, as illustrated in Figure 8.16. To obtain this transition, the photon energy must be equal to or greater than the band gap energy $\hbar\omega_0$. Therefore we would expect the conductivity to be zero for photon energies below $\hbar\omega_0$ and nonzero for energies above $\hbar\omega_0$. This is in contrast with the conductivity that results from free carrier absorption, which has a maximum at dc and decreases with increasing frequency. Figure 8.25 shows the infrared absorption for gallium arsenide, where below 1.45 eV we have free carrier absorption, and above 1.45 eV interband absorption predominates.

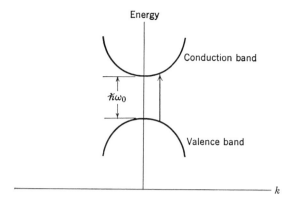

figure 8.16 A direct interband transition.

Our method of analysis will follow closely the procedure that was used in Section 2.4.2 to obtain the equation of motion for the electric dipole transition between two discrete levels. That is, we shall first express the current density in terms of the matrix elements of the density operator. This equation, in conjunction with the expression for the Hamiltonian, will yield a second-order differential equation for the current density as a function of time. The only significant difference between the present problem and the electric dipole transition considered in Chapter 2 is that now we are dealing with a continuum of eigenstates so that a **k** space integral will be necessary. From (1.41) and (8.103) we have that the current density involves a double integral in **k** space:

$$\mathbf{J} = \frac{qN_V}{m} \langle \mathbf{p} - q\mathbf{A} \rangle \propto \iint d^3k \, d^3k' \sum_{B,B'} \rho_{BB'}(\mathbf{kk'}) \, (\mathbf{p} - q\mathbf{A})_{\mathbf{k'}B',\mathbf{k}B} \qquad (8.104)$$

where

$$\rho_{BB'}(\mathbf{kk'}) = \langle \mathbf{k}B | \rho | \mathbf{k'}B' \rangle$$
$$(\mathbf{p} - q\mathbf{A})_{\mathbf{k'}B',\mathbf{k}B} = \langle \mathbf{k'}B' | \mathbf{p} - q\mathbf{A} | \mathbf{k}B \rangle \qquad (8.105)$$

In going from (1.41) to (8.104) we have added a summation over energy bands B and B', since there may be many different energy bands for a given value of **k**.

If we write out $\langle \mathbf{k'}, B' | \mathbf{p} | \mathbf{k}, B \rangle$ as an integral in spatial coordinates, from the form of the Bloch function we have

$$\langle \mathbf{k'}, B' | \mathbf{p} | \mathbf{k}, B \rangle = \int_F dV \, \phi(\mathbf{r}) e^{i(\mathbf{k}-\mathbf{k'})\cdot\mathbf{r}} \qquad (8.106)$$

where $\phi(\mathbf{r})$ is a function that is periodic with the periodicity of the lattice spacing. Since the exponential functions are orthogonal over the fundamental domain, this integral is zero unless $\mathbf{k} = \mathbf{k'}$.[2]

Similarly, if we consider the field to be a propagating wave of the form

$$\mathbf{A} = \frac{\tilde{\mathbf{A}}}{2} e^{i\omega t} + \text{c.c.} \tag{8.107}$$

where $\tilde{\mathbf{A}}$ varies as $e^{-\boldsymbol{\beta}\cdot\mathbf{r}}$ then $\langle \mathbf{k}', B'| \mathbf{A} |\mathbf{k}, B\rangle$ may be written as

$$\langle \mathbf{k}', B'| \tilde{\mathbf{A}} |\mathbf{k}, B\rangle = \int dV \phi'(\mathbf{r}) e^{i(\mathbf{k}-\mathbf{k}'+\boldsymbol{\beta})\cdot\mathbf{r}} \tag{8.108}$$

where $\phi'(\mathbf{r})$ has the periodicity of the lattice. We know that $|k|$ and $|k'|$ are of the order of some fractional part of $2\pi/a$, whereas $|\beta| \sim 2\pi/\lambda$ where λ is the wavelength of the signal. For optical frequency radiation, $\lambda \gg a$ so that the integral of (8.108) is also zero unless $\mathbf{k} \simeq \mathbf{k}'$.

We now see why the absorption of photons at optical frequencies leads to direct transitions. The constant $\boldsymbol{\beta}$, which is proportional to the momentum of the radiation, is small compared to the reciprocal of the lattice spacing, so that momentum conservation (that is, $\mathbf{k} - \mathbf{k}' + \boldsymbol{\beta} = 0$) leads to the condition that

$$\mathbf{k} \simeq \mathbf{k}' \tag{8.109}$$

Only if the photon wavelength is comparable to the lattice spacing, such as would be the case for soft X-rays, will photon absorption produce an appreciable change in electron momentum.

As a consequence of (8.109), the double integral in $\mathbf{k}$ space in (8.104) reduces to a single integral

$$\mathbf{J} = \frac{qN_V}{m} \int d^3k \sum_{BB'} (\rho_{BB'})(\mathbf{p} - q\mathbf{A})_{\mathbf{k}B', \mathbf{k}B} \tag{8.110}$$

where

$$\rho_{BB'} \equiv \rho_{BB'}(\mathbf{kk})$$

We shall only consider transitions between the valence band, corresponding to $B, B' = 1$, and the conduction band, corresponding to $B, B' = 2$. Thus (8.110) becomes

$$\mathbf{J} = \frac{qN_V}{m} \int d^3k \{\rho_{11}\langle \mathbf{k}1| \mathbf{p} - q\mathbf{A} |\mathbf{k}1\rangle + \rho_{22}\langle \mathbf{k}2| \mathbf{p} - q\mathbf{A} |\mathbf{k}2\rangle$$
$$+ \rho_{21}\langle \mathbf{k}1| \mathbf{p} - q\mathbf{A} |\mathbf{k}2\rangle + \rho_{12}\langle \mathbf{k}2| \mathbf{p} - q\mathbf{A} |\mathbf{k}1\rangle\} \tag{8.111}$$

The first two terms in the integrand of (8.111) give the currents resulting from the electron velocity associated with a given energy eigenstate, and the last two terms apply to interband transitions. If the photon energy is greater than the band gap energy, then ρ_{21} and ρ_{12} have resonant denominators and the last two terms will predominate.

Equation 8.111 may be further simplified by noting that for $\lambda \gg a$,

$$\langle \mathbf{k}1| \mathbf{A} |\mathbf{k}2\rangle \simeq 0$$

because of the orthogonality of the eigenfunctions. Therefore, (8.111) may be rewritten as

$$\mathbf{J} = \frac{qN_V}{m}\int d^3k(\rho_{21}\mathbf{M} + \rho_{12}\mathbf{M}^*) \tag{8.112}$$

where

$$\mathbf{M} \equiv \langle \mathbf{k}1 | \mathbf{p} | \mathbf{k}2 \rangle \tag{8.113}$$

It is convenient to introduce a current density per unit volume in **k** space, **j**, so that

$$\mathbf{J} = \int d^3k \mathbf{j} \tag{8.114}$$

From (8.112) and (8.114) we have

$$\mathbf{j} = \frac{qN_V}{m}(\rho_{21}\mathbf{M} + \rho_{12}\mathbf{M}^*) = \frac{qN_V}{m}(\rho_{21}\mathbf{M} + \text{c.c.}) \tag{8.115}$$

Equation 8.115 represents the first half of the analysis, in which we have obtained an expression for current density in terms of the matrix elements of the density operator. The next step is to find the form of the interaction Hamiltonian. From (8.101) we have

$$\mathscr{H}' = -\frac{q}{m}\mathbf{p}\cdot\mathbf{A} + \frac{q^2}{2m}\mathbf{A}^2 \tag{8.116}$$

The conservation of momentum condition (8.109) applies to

$$\mathscr{H}'_{\mathbf{k}'B',\mathbf{k}B} = \langle \mathbf{k}'B' | \mathscr{H}' | \mathbf{k}B \rangle \tag{8.117}$$

so that the only nonzero matrix elements for $\mathscr{H}'$ occur when $\mathbf{k} = \mathbf{k}'$. If we define $\mathscr{H}'_{BB'}$ as

$$\mathscr{H}'_{BB'} \equiv \mathscr{H}'_{\mathbf{k}.B,\mathbf{k}B'} \tag{8.118}$$

the $\mathscr{H}'$ matrix may be written as

$$\mathscr{H}' = \begin{pmatrix} \mathscr{H}'_{11} & \mathscr{H}'_{12} \\ \mathscr{H}'_{21} & \mathscr{H}'_{22} \end{pmatrix} \tag{8.119}$$

where the subscript 1 refers to the valence band and the subscript 2 refers to the conduction band. The diagonal terms in the $\mathscr{H}'$ matrix cause a shift in the transition frequency that is generally small and will be neglected. The off-diagonal terms are responsible for interband transitions.

The matrix element $\mathscr{H}'_{21}$ is found from (8.116) to be

$$\mathscr{H}'_{21} = -\frac{q}{m}\mathbf{A}\cdot\mathbf{M}^* \tag{8.120}$$

where **M** is defined by (8.113). The **A**² term in $\mathcal{H}'$ does not contribute to this matrix element for $\lambda \gg a$ because of the orthogonality of the eigenfunctions. Therefore the matrix representation for the $\mathcal{H}'$ of interest is

$$\mathcal{H}' = \begin{pmatrix} 0 & -\dfrac{q}{m} A_\alpha M_\alpha \\ -\dfrac{q}{m} A_\alpha M_\alpha^* & 0 \end{pmatrix} \tag{8.121}$$

where there is a summation over the repeated subscript α.

Equations 8.115 and 8.121 are identical in form to (2.28) and (2.27), respectively. Therefore the solution to (2.28) and (2.27), which is (2.35), applies directly to the present problem:†

$$\ddot{j}_\alpha + \frac{2}{\tau_{21}} \dot{j}_\alpha + \omega_{21}^2 j_\alpha = -\frac{2\omega_{21} q^2 N_V}{m^2 \hbar} (\rho_{22} - \rho_{11}) M_\alpha M_\beta^* A_\beta \tag{8.122}$$

where α and β refer to the coordinate directions, and the right-hand side of (8.122) is summed over the repeated subscript β. The product $M_\alpha M_\beta^*$ is a second-rank tensor, and if we choose our axes to be the principal axes for this tensor, then in (8.122) we allow

$$M_\alpha M_\beta^* A_\beta \rightarrow |M_\alpha|^2 A_\alpha \tag{8.123}$$

The diagonal elements of the density operator give the occupation probability. For example, the probability that an electron is in the valence band in a k-space volume element d^3k is $\rho_{11} d^3k$, where ρ_{11} is given by (A13.8) in Appendix 13. At room temperature, in a semiconductor that is not too heavily doped with an impurity the probability of occupation of an eigenstate is high in the valence band and low in the conduction band. Thus, from (A13.8) and (A13.9) in Appendix 13 we find that

$$(\rho_{22} - \rho_{11}) = -\frac{2 N_V^{-1}}{(2\pi)^3} \tag{8.124}$$

From (8.122)–(8.124) we see that the steady-state expression $\tilde{j}_\alpha$ is

$$\tilde{j}_\alpha = \frac{ie^2 |M_\alpha|^2 \tilde{\mathcal{E}}_\alpha}{2\pi^3 m^2 \hbar \omega} \frac{\omega_{21}(\omega_{21}^2 - \omega^2 - 2i\omega/\tau_{21})}{(\omega_{21}^2 - \omega^2)^2 + (2\omega/\tau_{21})^2} \tag{8.125}$$

† We observe that the right hand side of (8.122) is proportional to $(\rho_{22} - \rho_{11})$. Because of the exclusion principle a $2 \rightarrow 1$ transition probability is proportional to $\rho_{22}(1 - \rho_{11})$. That is, if $\rho_{11} = 1$ then there cannot be any transitions to state 1. However, if we include transitions in both directions we have

$$\rho_{22}(1 - \rho_{11}) - \rho_{11}(1 - \rho_{22}) = \rho_{22} - \rho_{11}$$

so that (8.122) is correct as given.

where

$$j_\alpha = \frac{\tilde{j}_\alpha}{2} e^{i\omega t} + \text{c.c.}$$

$$\mathscr{E}_\alpha = \frac{\tilde{\mathscr{E}}_\alpha}{2} e^{i\omega t} + \text{c.c.}$$

$$= -\partial A_\alpha/\partial t$$

$$= -i\omega \frac{\tilde{A}_\alpha}{2} e^{i\omega t} + \text{c.c.}$$

The real part of (8.125) contributes to the conductivity and the imaginary part to the dielectric constant. To evaluate conductivity we need only consider the real part of (8.125).

Let us now determine the conductivity resulting from interband transitions for the specific example of a semiconductor, where the photon energy of the incident radiation is close to the band gap transition energy. Figure 8.16 illustrates the type of transition under consideration. The transition is shown as a vertical line because the initial and final values for k are equal.

Conductivity is the ratio of electric field to current density. Current density J_α is given by

$$J_\alpha = \int d^3k \, j_\alpha$$

and since j_α is given in terms of ω_{21} it is necessary to express d^3k in terms of ω_{21} so as to be able to perform the integration. In Appendix 10 we see that

$$d^3k = 4\pi \left(\frac{2\mu_x^* \mu_y^* \mu_z^*}{\hbar^3}\right)^{1/2} (\omega_{21} - \omega_0)^{1/2} \, d\omega_{21} \tag{8.126}$$

where $\hbar\omega_0$ is the minimum energy separation between bands,

$$\frac{1}{\mu_\alpha^*} \equiv \frac{1}{m_\alpha^*(2)} - \frac{1}{m_\alpha^*(1)} \tag{8.127}$$

and

$m_\alpha^*(2) = \alpha$ component of effective mass in the conduction band, where the α direction is a principal axis.
$m_\alpha^*(1) = \alpha$ component of effective mass in the valence band.

From (8.126) and the real part of (8.125) we obtain

$$J_\alpha = \frac{2}{\pi} G_\alpha \mathscr{E}_\alpha \int \frac{(2\omega_{21}/\tau_{21})(\omega_{21} - \omega_0)^{1/2}}{(\omega_{21}^2 - \omega^2)^2 + (2\omega/\tau_{21})^2} \, d\omega_{21} \tag{8.128}$$

where

$$G_\alpha \equiv \frac{e^2 |M_\alpha|^2}{\pi^2 m^2 \hbar^{5/2}} (2\mu_x^* \mu_y^* \mu_z^*)^{1/2} \tag{8.129}$$

In (8.128) the limits of ω_{21} are from ω_0 to the maximum frequency difference between conduction and valence bands. However, the integrand is sharply peaked near $\omega_{21} = \omega$, and thus the integral is not appreciably altered by extending the limits of integration from $-\infty$ to $+\infty$. We then obtain

$$J_\alpha = G_\alpha \frac{(\omega - \omega_0)^{1/2}}{\omega} \mathscr{E}_\alpha \quad \omega \geqq \omega_0$$
$$= 0 \quad \omega \leqq \omega_0 \tag{8.130}$$

From (8.130),

$$\sigma_\alpha = G_\alpha \frac{(\omega - \omega_0)^{1/2}}{\omega} \quad \omega \geqq \omega_0$$
$$= 0 \quad \omega \leqq \omega_0 \tag{8.131}$$

where σ_α is the conductivity along the α direction.

A consequence of a finite conductivity is that electromagnetic radiation decays exponentially with distance. An absorption constant Γ is defined so that the Poynting vector decays to $1/e$ of its initial value in a distance equal to Γ^{-1}. For a field polarized in the α direction, the relationship between Γ and σ_α is

$$\Gamma = \frac{\sigma_\alpha}{\eta \epsilon_0 c} \tag{8.132}$$

where η is the refractive index along the α direction, c is the free-space velocity of light, and ϵ_0 is the free-space permittivity. The general shape of the absorption constant resulting from direct interband transitions is shown in Figure 8.17. In the neighborhood of $\omega \simeq \omega_0$, the absorption coefficient varies approximately as $\sqrt{\omega - \omega_0}$.

An estimate may be obtained for the matrix element $|M_\alpha|^2$ and hence for G_α. Just as there is a sum rule for the dipole moment operator, as given by (3.21), there is also a sum rule for the momentum operator. An oscillator strength f_B

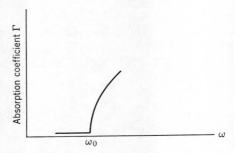

figure 8.17 The absorption coefficient Γ for direct interband transitions.

for an interband transition from the valence band to the Bth energy band is defined as

$$f_B \equiv \frac{2\,|\mathbf{M}|_B^2}{m\hbar\omega} \qquad (8.133)$$

where $\hbar\omega$ is the photon energy of the radiation producing the transition and

$$|\mathbf{M}|_B^2 \equiv |\langle \mathbf{k}, B|\,\mathbf{p}\,|\mathbf{k}, 1\rangle|^2 \qquad (8.134)$$

For an electromagnetic field polarized in the α direction the sum rule is[3]

$$\sum_{B \neq 1} (f_B)_\alpha = 1 - \frac{m}{m_\alpha^*(1)} \qquad (8.135)$$

where $m_\alpha^*(1)$ is the α component of electron effective mass in the valence band. In general, transitions to the conduction band predominate so that only one term need be included in the right-hand side of (8.135). If we take the electron effective mass near the top of the valence band to be approximately the negative of the free-space mass, then from (8.133) and (8.135) we have

$$|M_\alpha|^2 \simeq m\hbar\omega \qquad (8.136)$$

From (8.129) and (8.131), the conductivity then has the value

$$\sigma_\alpha \simeq \frac{e^2 m^{1/2}}{2\pi^2 \hbar^{3/2}} \sqrt{\omega - \omega_0} \qquad \omega \geqq \omega_0 \qquad (8.137)$$

$$= 1.1 \times 10^{-3} \sqrt{\omega - \omega_0} \quad \text{mhos/m}$$

where we have taken the effective mass to be the free-space mass, so that $\mu_x^* = \mu_y^* = \mu_z^* = m/2$. For a semiconductor with $\eta = 4$, from (8.132) we obtain $\Gamma \simeq 10^6$ m^{-1} for a photon energy that is 0.05 eV greater than the minimum band gap energy. This means that incident radiation decays to $1/e$ of its initial value in a distance equal to 10^{-6} m.

In the derivation of the expression for conductivity it had been assumed that $|M_\alpha|^2$ was independent of frequency or slowly varying for $\omega \simeq \omega_0$. In general, we may write

$$|M_\alpha|^2 \simeq |M_\alpha|_0^2 + \frac{\partial |M_\alpha|^2}{\partial \omega}(\omega - \omega_0) \qquad (8.138)$$

where we have kept the first two terms in the power series expansion near $\omega = \omega_0$. If $|M_\alpha|_0^2 = 0$, then we define the transition as a forbidden transition, and for this case we find from (8.125) and (8.126) that

$$J_\alpha = G'_\alpha \frac{(\omega - \omega_0)^{3/2}}{\omega} \mathcal{E}_\alpha \qquad \omega \geqq \omega_0 \qquad (8.139)$$
$$= 0 \qquad \qquad \omega \leqq \omega_0$$

where

$$G'_\alpha \equiv \frac{1}{|M_\alpha|^2} \frac{\partial |M_\alpha|^2}{\partial \omega} G_\alpha \qquad (8.140)$$

Thus, for a forbidden transition the absorption constant varies as the 3/2 power of the difference between the photon energy and the minimum band gap energy.

8.4.2 Indirect Transitions

In Section 8.4.1 we discussed the case where the **k** value for the final state in the conduction band equaled the **k** value for the initial state in the valence band. Direct transitions predominate when the top of the valence band has the same **k** value as the bottom of the conduction band, as illustrated in Figure 8.16. This is the case for such semiconductors as gallium arsenide (GaAs) and indium antimonide (InSb), but for Si and Ge the appropriate form of the energy diagram is as shown in Figure 8.18. If the frequency of the radiation is in the range $\omega_0 \leqq \omega < \omega'_0$, then transitions occur only with a change in the **k** between initial and final states. If $\omega \simeq \omega_0$, then the transition that may occur is shown by the arrow in Figure 8.18.

An operator of the form $Qe^{i\boldsymbol{\beta}\cdot\mathbf{r}}$, where Q is spatially invariant, may have a nonzero matrix element $\langle \mathbf{k}'B'| Qe^{i\boldsymbol{\beta}\cdot\mathbf{r}} |\mathbf{k}B\rangle$ only if

$$\mathbf{k} - \mathbf{k}' + \boldsymbol{\beta} = 0 \qquad (8.141)$$

As discussed with regard to (8.108), $|\boldsymbol{\beta}| \ll |\mathbf{k}|$ for electromagnetic fields so that $\mathbf{k}' \simeq \mathbf{k}$ and only direct transitions occur. However, if $\boldsymbol{\beta} = \mathbf{k}_v$, corresponding to

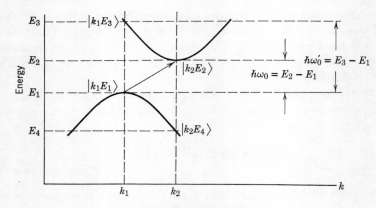

figure 8.18 Indirect transitions.

a phonon, then $\mathbf{k}'$ may differ appreciably from $\mathbf{k}$ since the wavelength for the phonon is comparable to the lattice spacing a. In Section 7.4 it was indicated that (8.141) is known as the conservation of wave vector condition.

Let us consider the ways a transition may occur from $|k_1 E_1\rangle$ to $|k_2 E_2\rangle$. In Chapter 5 we noted with regard to multiple-photon effects that energy must be conserved between the initial and final states for a resonance denominator to exist. However, energy is not necessarily conserved for each matrix element involved in the process, and the same condition holds for indirect transitions in semiconductors. Momentum, on the other hand, must be conserved for each matrix element, or else the element is zero.

Consider the Hamiltonian to be of the form

$$\mathcal{H} = \mathcal{H}_0 + \mathcal{H}_1 + \mathcal{H}_2 \tag{8.142}$$

where $\mathcal{H}_1$ and $\mathcal{H}_2$ are the interaction terms:

$\mathcal{H}_1 = $ interaction energy between an electron and a photon
$\mathcal{H}_2 = $ interaction energy between an electron and a phonon.

The term $\mathcal{H}_1$ is associated only with vertical transitions, whereas $\mathcal{H}_2$ produces a change in $\mathbf{k}$. If one photon and one phonon are involved, then $|k_1 E_1\rangle \rightarrow |k_2 E_2\rangle$ may result from the product

$$\langle k_2 E_2 | \mathcal{H}_2 | k_1 E_3 \rangle \langle k_1 E_3 | \mathcal{H}_1 | k_1 E_1 \rangle \tag{8.143}$$

or the product

$$\langle k_2 E_2 | \mathcal{H}_1 | k_2 E_4 \rangle \langle k_2 E_4 | \mathcal{H}_2 | k_1 E_1 \rangle \tag{8.144}$$

Expressions 8.143 and 8.144 represent the only possible ways for achieving the desired transitions with a single photon and a single phonon, as long as transitions to other energy bands are excluded. The transition probability from $|k_1 E_1\rangle$ to $|k_2 E_2\rangle$ is proportional to the sum of the squares of the magnitudes of (8.143) and (8.144).

Let us now determine the frequency dependence of the conductivity resulting from indirect transitions. If we carry out the details of the perturbation procedure as was done in Chapter 5, we find that a resonance denominator of the form

$$\frac{1}{\omega_{21} \pm \omega_v - \omega} \tag{8.145}$$

will appear in the expression for the off-diagonal elements of the density operator and therefore j_α will have this term as a factor. In (8.145) the following definitions hold:

$$\hbar \omega_{21} = E_2 - E_1$$
$$\hbar \omega_v = \text{phonon energy}$$
$$\hbar \omega = \text{photon energy}$$

The transition $|k_1 E_1\rangle \to |k_2 E_2\rangle$ takes place with the absorption of a photon and either the absorption or emission of a phonon. In (8.145), the plus sign corresponds to the emission of a phonon and the minus sign corresponds to phonon absorption. If a phonon is absorbed, resonance occurs when the transition energy equals the sum of the photon and phonon energies. If a phonon is emitted, resonance occurs when the transition energy equals the difference in energy between the photon and phonon. The physical significance of (8.145) is that whenever energy is conserved in the overall process, the current component j_α is large.

The form of j_α that is valid near resonance is

$$j_\alpha = \frac{C}{\omega_{21} \pm \omega_v - \omega} \mathcal{E}_\alpha \tag{8.146}$$

where C is a function of the occupation probability and the matrix elements given by (8.143) and (8.144),† and the field $\mathcal{E}_\alpha$ is present because it is a factor in $\mathcal{H}_1$. Current density J_α is given by the **k**-space integral of j_α:

$$J_\alpha = \iint d^3k_1 \, d^3k_2 \, j_\alpha \tag{8.147}$$

where d^3k_1 is a **k**-space volume in the conduction band and d^3k_2 is a **k**-space volume in the valence band. For direct transitions, (8.147) reduces to a single integral because the initial and final **k** values are the same.

Since j_α is expressed in terms of frequency, it is necessary to write d^3k_1 and d^3k_2 also in terms of frequency. This derivation is very similar to the derivation given in Appendix 10 for the case of direct transitions, and indeed by using (A10.3) rather than (A10.4) to relate d^3k to frequency we find that

$$\begin{aligned} d^3k_1 &= a(\omega_a - \omega_1)^{1/2} \, d\omega_1 \\ d^3k_2 &= b(\omega_2 - \omega_b)^{1/2} \, d\omega_2 \end{aligned} \tag{8.148}$$

$$a \equiv 4\pi \left(\frac{2m_x^*(1) \, m_y^*(1) \, m_z^*(1)}{\hbar^3} \right)^{1/2}$$

$$b \equiv 4\pi \left(\frac{2m_x^*(2) \, m_y^*(2) \, m_z^*(2)}{\hbar^3} \right)^{1/2}$$

where ω_1, ω_2, ω_a, and ω_b are as shown in Figure 8.19, and $m_x^*(1)$ is the effective mass in the valence band along the x direction. From Figure 8.19 we have

$\hbar(\omega_a - \omega_1)$ = energy increment below the bottom of the valence band

$\hbar(\omega_2 - \omega_b)$ = energy increment above the top of the conduction band.

† For an evaluation of these matrix elements see, for example, Section 13.5.2 in Ref. 3 listed at the end of the chapter.

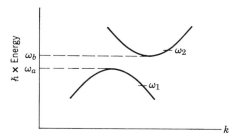

figure 8.19 The frequencies involved in indirect transitions.

The combination of (8.146)–(8.148) yields

$$J_\alpha = abC\mathscr{E}_\alpha \int^{\omega_a} (\omega_a - \omega_1)^{1/2} d\omega_1 \int_{\omega_b} \frac{(\omega_2 - \omega_b)^{1/2}}{(\omega_2 - \omega_1) - \omega \pm \omega_v} d\omega_2 \quad (8.149)$$

where, for ω_{21} in (8.146) we have substituted the equivalent expression, $\omega_{21} = \omega_2 - \omega_1$. From Figure 8.19 we see that the lower limit on ω_2 is ω_b and the upper limit on ω_1 is ω_a. The integrand in the ω_2 integral is sharply peaked near $\omega_2 = \omega_1 + \omega \mp \omega_v$, and we need only consider the contribution to the integral from the pole at $\omega_2 = \omega_1 + \omega \mp \omega_v$. In performing this integral it is necessary for us to note that as $\omega_2 \to \infty$, $C \to 0$ since C is proportional to the Fermi-Dirac distribution, and thus this integral is bounded as $\omega_2 \to \infty$. Therefore, integration with respect to ω_2 gives

$$J_\alpha = abC\mathscr{E}_\alpha \int^{\omega_a} (\omega_a - \omega_1)^{1/2} (\omega_1 + \omega \mp \omega_v - \omega_b)^{1/2} d\omega_1 \quad \omega_1 + \omega \mp \omega_v \geqq \omega_b$$

$$= 0 \quad \omega_1 + \omega \mp \omega_v \leqq \omega_b \quad (8.150)$$

Since J_α is zero for $\omega_1 \leqq \omega_b - \omega \pm \omega_v$, the lower limit on the ω_1 integral may be specified to be $(\omega_b - \omega \pm \omega_v)$, so that J_α can be written as

$$J_\alpha = abC\mathscr{E}_\alpha \int_{\omega_b - \omega \pm \omega_v}^{\omega_a} (\omega_a - \omega_1)^{1/2} (\omega_1 + \omega \mp \omega_v - \omega_b)^{1/2} d\omega_1 \quad (8.151)$$

The integration of (8.151) yields the desired relationship between J_α and $\mathscr{E}_\alpha$:

$$J_\alpha = \frac{\pi}{8} abC\mathscr{E}_\alpha (\omega - \omega_0 \mp \omega_v)^2 \quad \omega \geqq \omega_0 \pm \omega_v$$

$$= 0 \quad \omega \leqq \omega_0 \pm \omega_v \quad (8.152)$$

where $\omega_0 \equiv \omega_b - \omega_a$, and $\hbar\omega_0$ is the minimum band gap energy.

The constant C is not the same for both phonon emission and phonon absorption, and so it is necessary to consider these separate intervals in frequency:

$$\begin{aligned} J_\alpha &= 0 & \omega &\leq \omega_0 - \omega_v \\ &= C_a \mathscr{E}_a (\omega - \omega_0 + \omega_v)^2 & \omega_0 - \omega_v &\leq \omega \leq \omega_0 + \omega_v \\ &= C_a \mathscr{E}_a (\omega - \omega_0 + \omega_v)^2 \\ &\quad + C_e \mathscr{E}_a (\omega - \omega_0 - \omega_v)^2 & \omega_0 + \omega_v &\leq \omega \end{aligned} \quad (8.153)$$

In thermal equilibrium the ratio of the phonon emission constant C_e to the phonon absorption constant C_a is given by the Boltzmann factor

$$\frac{C_e}{C_a} = e^{-\hbar \omega_v / \kappa T} \tag{8.154}$$

Emission is the deexcitation of a phonon, and absorption is the excitation; thus the ratio of the probability of these events is determined by the ratio of the occupation of the phonon state to the ground state, which is the Boltzmann factor.

Figure 8.20 illustrates the absorption coefficient associated with indirect, interband transitions. Absorption begins when the photon energy equals the difference between the band gap energy $\hbar \omega_0$ and the phonon energy $\hbar \omega_v$. In the range $\omega_0 - \omega_v \leq \omega \leq \omega_0 + \omega_v$, the transition results from the absorption of a photon and a phonon. In the range $\omega_0 + \omega_v \leq \omega$, the transition may occur either from the absorption of a photon or phonon, or from the absorption of a photon and the emission of a phonon.

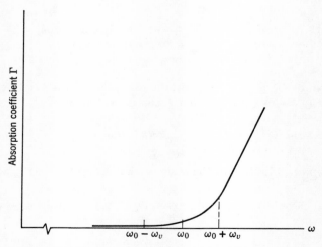

figure 8.20 The absorption coefficient Γ for indirect, interband transitions.

8 ELECTRONS IN CRYSTALS

In general, phonon frequency ω_v is a function of k_v, as for the case of acoustic phonons. The value for ω_v to be used in (8.153) is that which corresponds to $k_v = k_2 - k_1$, where $k_2 - k_1$ is the change in k between final and initial states, as illustrated in Figure 8.18.

8.5 PHOTOCONDUCTIVITY

In Section 8.3 we considered some consequences of intraband motion, and in Section 8.4 we studied interband transitions. Photoconductivity is a phenomenon that combines both these effects, for it is the change in the dc or low-frequency conductivity that results from band-to-band transitions.

Consider the current density that is produced by an applied dc field as given by (8.90):

$$J = (N_e \mu_e + N_h \mu_h) e \mathscr{E} \tag{8.90}$$

where e is the magnitude of electronic charge, N_e is the number of conduction electrons per unit volume, N_h is the number of valence holes per unit volume, μ_e and μ_h are electron and hole mobilities, respectively, measured along a direction that is a principal axis for the effective mass tensor, $\mathscr{E}$ is a dc electric field, and J is the current density. The parameters N_e and N_h depend upon the frequency and intensity of an infrared or visible signal that impinges on the crystal because of interband transitions. We shall proceed to calculate the dependence of dc conductivity on the power density of an incident optical-frequency radiation, and the dc photocurrent will also be determined. Photocurrent is that component of the current that flows in response to an applied dc or low-frequency field when light is incident on the crystal.

We shall evaluate N_e by equating the rate of photon absorption to the rate of change of conduction electrons. That is, a quantum efficiency, (q.e.), of unity is assumed, in which each absorbed photon results in the transition of an electron from the valence to the conduction band. Quantum efficiency is generally close to unity, but if we wish to include (q.e.) $\neq 1$, it is merely necessary to multiply the rate of photon absorption by (q.e.).

The net rate of change of N_e that results from applied radiation is

$$\frac{\partial N_e}{\partial t} + \frac{N_e - N_e^e}{T_e} \tag{8.155}$$

where N_e^e is the steady-state density of conduction electrons in the absence of incident radiation and T_e is the free carrier lifetime, or lifetime of electrons in the conduction band. The value for T_e is determined by the recombination rate for electrons and holes and also by the rate at which electrons are trapped by impurities as discussed in Section 8.2.5. This parameter is a function of temperature and may vary over many orders of magnitude. In cadmium sulfide, for example, T_e

varies from 10^{-2}–10^{-10} sec. At room temperature, a typical value for T_e is 10^{-6} sec. If electron-hole recombination is the sole mechanism for losing conduction electrons, then $T_h = T_e$, where T_h is the lifetime of holes in the valence band. In the presence of traps, however, T_h does not necessarily equal T_e, because the trapping cross-section for holes may differ from the capture cross-section for electrons. An extensive discussion of lifetime is presented by Rose.[4]

With an incident power density I resulting from a homogeneous irradiation of the crystal, and an absorption constant Γ, the power absorbed per unit volume is ΓI. Therefore the number of photons absorbed per unit time per unit volume is

$$\frac{\Gamma I}{\hbar \omega} \tag{8.156}$$

With unity quantum efficiency we may equate (8.155) and (8.156), and in the steady-state we have

$$\frac{N_e - N_e^e}{T_e} = \frac{\Gamma I}{\hbar \omega} \tag{8.157}$$

Similarly, the absorption of a photon creates a hole in the valence band so that

$$\frac{N_h - N_h^e}{T_h} = \frac{\Gamma I}{\hbar \omega} \tag{8.158}$$

The substitution of (8.157) and (8.158) into (8.90) yields

$$\tilde{J} = (N_e^e \mu_e + N_h^e \mu_h) e \tilde{\mathscr{E}} + (T_e \mu_e + T_h \mu_h) \left(\frac{\Gamma I}{\hbar \omega} \right) e \tilde{\mathscr{E}} \tag{8.159}$$

where the first term on the right-hand side results from the equilibrium concentrations and the second term, the photocurrent, is produced by the incident light. From (8.159) we see that the increment in conductivity $\Delta \sigma$ resulting from the photoconductive effect is

$$\Delta \sigma = (T_e \mu_e + T_h \mu_h) \frac{\Gamma I}{\hbar \omega} e \tag{8.160}$$

The frequency dependence of $\Delta \sigma$ depends on the frequency dependence of Γ, which, in turn, is determined by whether or not the transition is direct or indirect. If (q.e.) $\neq 1$, the right-hand side of (8.160) is to be multiplied by (q.e.).

Photocurrent I_p is often written in the form

$$I_p = eGF \tag{8.161}$$

where

$G \equiv$ photoconductive gain
$F =$ electron-hole pairs created per unit time by absorption of light
$=$ photons absorbed per unit time (for (q.e.) $= 1$).

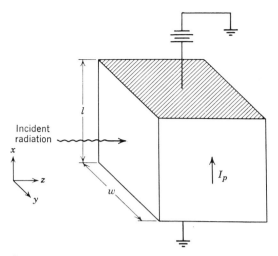

figure 8.21 Homogeneous radiation incident on a semiconductor crystal with an applied dc bias.

The ratio I_p/e is the number of charge carriers, electrons and holes, that pass between the electrodes per unit time, and thus G is the ratio of the number of charge carriers to the number of created electron-hole pairs. To evaluate G, let us consider the configuration illustrated in Figure 8.21. Current I_p is related to current density J by the relationship

$$I_p = w \int_0^\infty J \, dz \qquad (8.162)$$

so that from (8.159) we find that the photocurrent is

$$I_p = (T_e \mu_e + T_h \mu_h) \frac{e\mathscr{E} w \Gamma}{\hbar \omega} \int_0^\infty I \, dz \qquad (8.163)$$

Power density varies as

$$I = I_0 e^{-\Gamma z} \qquad (8.164)$$

where I_0 is the power density on the front surface. For a z dimension of the crystal that is much greater than Γ^{-1} so that all the radiation is absorbed, we have

$$I_p = (T_e \mu_e + T_h \mu_h) \frac{e\mathscr{E} w}{\hbar \omega} I_0 \qquad (8.165)$$

The transit time T_{re} for electrons between electrodes is

$$T_{re} = \frac{l}{\mu_e \mathscr{E}} \qquad (8.166)$$

and the transit time for holes is

$$T_{rh} = \frac{l}{\mu_h \mathscr{E}} \tag{8.167}$$

A combination of (8.165)–(8.167) yields

$$I_p = \left(\frac{T_e}{T_{re}} + \frac{T_h}{T_{rh}}\right) e \frac{(wlI_0)}{\hbar\omega} \tag{8.168}$$

The factor (wlI_0) is the incident power, and thus the rate of photon absorption is $wlI_0/\hbar\omega$. Therefore, on the basis of unity quantum efficiency

$$F = \frac{wlI_0}{\hbar\omega} \tag{8.169}$$

A comparison of (8.168), (8.169), and (8.161) gives

$$G = \frac{T_e}{T_{re}} + \frac{T_h}{T_{rh}} \tag{8.170}$$

The photoconductive gain depends on the ratio of lifetime to transit time and may be greater or less than unity.

8.6 SEMICONDUCTOR INJECTION LASERS

In Chapter 4 we considered various types of lasers for which population inversion was achieved between discrete energy levels. We shall now study population inversion and laser action as it may occur between the conduction and valence bands in a semiconductor. A general description and some of the operating characteristics of the semiconductor laser will be discussed in Section 8.6.1, and the threshold requirements will be evaluated in Section 8.6.2.

8.6.1 A General Description

The semiconductor injection laser involves the flow of electrons across a junction from n-type to p-type material. Figure 8.22 illustrates the configuration. Electron flow is in the positive x direction, and the surfaces of the semiconductor that are normal to the z direction are polished flat and parallel so as to form an interferometer with its axis along the z direction.

An n-type material is a semiconductor that has been doped with a donor impurity, which is an impurity that contributes electrons to the conduction band. The presence of the donor atom results in energy eigenstates just below the level of the conduction band. This means that almost all the impurity atoms are ionized, and they thereby release electrons to the conduction band. A p-type material is one that has been doped with an acceptor impurity, which is an impurity that

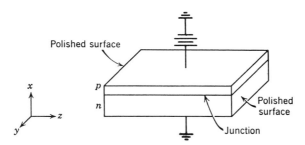

figure 8.22 The configuration for the semiconductor injection laser. Electron flow is in the positive x direction, and an interferometer is formed by the polished surfaces.

produces holes in the valence band. The presence of the acceptor atom results in energy eigenstates just above the top of the valence band in which electrons are trapped from the valence band.

At thermal equilibrium in a semiconductor the probability that an eigenstate is occupied by an electron is given by the Fermi-Dirac distribution

$$f = \frac{1}{1 + e^{(E-E_F)/\kappa T}} \tag{8.171}$$

where E is the energy of the eigenstate, E_F is the Fermi energy, κ is Boltzmann's constant, and T is temperature in degrees Kelvin. If the eigenstate energy is more than one κT energy increment below E_F, then $f \simeq 1$. For degenerate doping in n-type material, E_F is in the conduction band, and for degenerate doping in p-type material E_F is in the valence band. When a junction is formed between n-type and p-type, E_F is continuous throughout the medium, and thus the occupation of states in the neighborhood of a junction is as shown in Figure 8.23.

If a voltage Φ is applied across the junction so that the p-type material is made positive relative to the n-type, an electron that traverses the junction from n-type to p-type loses an amount of potential energy equal to $e\Phi$. Thus, in the presence of the bias it is necessary to modify Figure 8.23 so that the energy on the p-side is reduced in relation to the n-side energy, by an amount equal to $e\Phi$. Figure 8.24 illustrates the energy diagram for a forward bias, where the p type is made positive in relation to the n type.

Conduction electrons move from the n-type to the p-type material in the presence of the bias, so that there is a region of 1–10 microns in the p-type material where the conduction band is occupied and the valence band has holes. This constitutes a population inversion and allows for laser action. Population inversion means that $(\rho_{22} - \rho_{11})$ is positive, so that the absorption coefficient given by (8.132) becomes negative. A negative absorption coefficient implies that there

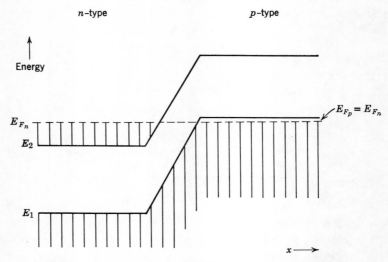

figure 8.23 The vertical lines indicate the regions in which the energy eigenstates are occupied. This energy diagram applies in thermal equilibrium when there is no bias voltage applied across the junction. The energy at the top of the valence band is E_1 and the energy at the bottom of the conduction band is E_2.

is gain, and when this gain exceeds the total losses of the system, self-sustained oscillations occur.

Gallium arsenide, GaAs, was the first semiconductor to exhibit laser action.[5,6] The interband transition for this material is direct, with a fairly large matrix element, and the loss resulting from free carrier absorption is relatively small at

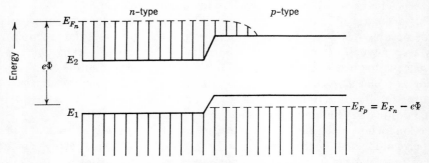

figure 8.24 The occupation of eigenstates for a forward-bias voltage Φ. Note that in the p-type material there is population inversion in the immediate vicinity of the junction.

the band gap transition frequency. Laser oscillation in GaAs occurs primarily in the *p*-type material within 1–10 μ of the junction, and at a wavelength of 0.84 μ. The overall efficiency of the device, defined as the ratio of the optical output power to the power delivered by the bias supply, can be on the order of 30%. More than 10 W cw has been obtained at liquid helium temperature.

In Sections 8.6.1 and 8.6.2 we are considering laser action resulting from population inversion between the conduction and valence bands. There are other ways to achieve laser oscillation in semiconductors, and these possibilities are enumerated and discussed in the literature.[7]

8.6.2 Threshold Current

Oscillations occur when the gain resulting from the population inversion exceeds losses. Losses include the following.

1. Free carrier absorption at the laser frequency resulting in the excitation of electrons from the bottom of the conduction band to higher energy states in the conduction band, and the excitation of electrons from within the valence band to the top of the valence band where holes exist.
2. Transmission losses through the mirrors or surfaces forming the interferometer.
3. Diffraction losses within the interferometer.

At 77°K, liquid nitrogen temperature, with impurity concentrations of about 10^{18} cm^{-3}, the losses give a typical equivalent attenuation constant of approximately 30 cm^{-1}. This means that the gain associated with the population inversion must exceed 30 cm^{-1} for oscillations to occur.

Figure 8.25 shows the absorption constant in GaAs near the band edge at 77°K with an acceptor concentration of 1.6×10^{18} cm^{-3}. The contribution to the

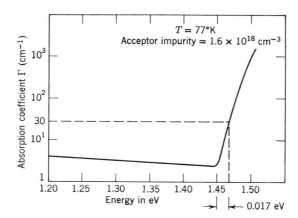

figure 8.25 Absorption coefficient in *p*-type GaAs at liquid nitrogen temperature.

absorption below 1.45 eV results primarily from free carrier absorption, and above 1.45 eV there is the additional contribution from interband transitions. If Figure 8.25 corresponded to the case of a filled valence band and empty conduction band, then with population inversion the contribution to Γ from interband transitions would be the negative of the value given in Figure 8.25. To obtain a gain of 30 cm^{-1} it is necessary to have filled states at approximately 0.017 eV above the band gap energy. This means that if enough electrons are injected into the p-type material to fill energy eigenstates 0.017 eV above the bottom of the conduction band, there will be a gain of 30 cm^{-1}. It is assumed that the acceptor doping is sufficiently high that the valence band is empty at the valence band energy corresponding to direct transitions.

We shall now estimate the current density that must be injected across the junction to reach threshold. From (1.40) we have

$$\rho_{22} = n(\mathbf{k})PN_e^{-1} \tag{1.40}$$

where

$n(\mathbf{k})$ = number of eigenstates per unit volume in $\mathbf{k}$ space per unit volume in coordinate space
P = probability that an eigenstate is occupied by an electron
N_e = number of conduction electrons per unit volume in coordinate space.

Since $\int d^3k_2 \rho_{22} = 1$, where d^3k_2 is the $\mathbf{k}$-space volume in the conduction band (the probability that a conduction electron is somewhere in the conduction band is unity), from (1.40) we obtain

$$N_e = \int d^3k_2 n(\mathbf{k})P \tag{8.172}$$

With the aid of (8.148), we are able to write (8.172) as

$$N_e = 4\pi \left[\frac{2m_x^*(2)\,m_y^*(2)\,m_z^*(2)}{\hbar^3} \right]^{1/2} \int n(\mathbf{k})P\sqrt{(\omega_2 - \omega_b)}\,d\omega_2 \tag{8.173}$$

where $\hbar\omega_b$ is the electron energy at the bottom of the conduction band and $\hbar\omega_2$ is the energy within the conduction band. Let us approximate the Fermi-Dirac distribution in the presence of the injected current so that

$$\begin{aligned} P &= 1 \quad \text{for} \quad \hbar(\omega_2 - \omega_b) \leq \Delta E \\ &= 0 \quad \text{for} \quad \hbar(\omega_2 - \omega_b) > \Delta E \end{aligned} \tag{8.174}$$

where ΔE is the energy increment above the bottom of the conduction band at which threshold occurs.

From (A13.9) in Appendix 13, we have $n(\mathbf{k}) = 2/(2\pi)^3$, and we find from (8.173) and (8.174) that

$$N_e = 3.5 \times 10^{27} [2m_{rx}^*(2)\,m_{ry}^*(2)\,m_{rz}^*(2)]^{1/2} (\Delta E)^{3/2} \tag{8.175}$$

where (ΔE) is measured in eV, N_e is in m^{-3}, and

$$m_{rx}^*(2) \equiv \frac{m_x^*(2)}{m}; \quad m_{ry}^*(2) \equiv \frac{m_y^*(2)}{m}; \quad m_{rz}^*(2) \equiv \frac{m_x^*(2)}{m}$$

We have assumed an occupation probability that is unity below a given energy and zero above that energy. One should use the Fermi-Dirac distribution in (8.173) to determine N_e, particularly for higher temperatures. For the present calculation this refinement is not necessary, and we may approximate the existence of a Fermi-Dirac distribution by requiring (ΔE) to be $(0.017 + \kappa T)$ eV, in order to have filled eigenstates 0.017 eV above the bottom of the conduction band. At 77°K, $\kappa T \simeq 0.006$ eV, so that in (8.175) we let $\Delta E = 0.023$ eV. For GaAs we have

$$m_{rx}^*(2) \simeq m_{ry}^*(2) \simeq m_{rz}^*(2) \simeq 0.07$$

whereupon (8.175) yields

$$N_e = 3.1 \times 10^{23} \text{ m}^{-3} \tag{8.176}$$

If a current density J flows from the n-type to the p-type region, then

$$\begin{array}{l}\text{The number of electrons} \\ \text{injected per unit time} \\ \text{per unit volume into the} \\ p \text{ region}\end{array} = \frac{J}{ed} \tag{8.177}$$

where d is the depth of penetration of electrons in the p region. In the steady-state the density of conduction electrons produced per unit time by the injected current is obtained by equating (8.155) to (8.177):

$$\frac{N_e}{T_e} = \frac{J}{ed} \text{(q.e.)} \tag{8.178}$$

where (q.e.) = quantum efficiency, which is the ratio of the number of electrons injected to the number of injected electrons that arrive in the conduction band in the p type material. In (8.155), we have assumed that $N_e^e = 0$, since in the absence of the injected current there are very few conduction electrons.

We wish to compare the theoretical threshold current with measurements made on a material with an impurity concentration of 5×10^{17} cm^{-3}. Our absorption data, Figure 8.25, is for a different concentration, but it can be assumed that the slope of the absorption curve is relatively insensitive to doping. For GaAs with a doping concentration of 5×10^{17} cm^{-3}, $T_e \simeq 2.2 \times 10^{-9}$ sec.[8] With the parameter values

$$\text{(q.e.)} = 1$$
$$d = 1\,\mu$$

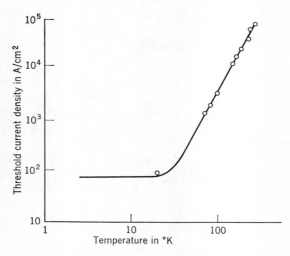

figure 8.26 Threshold current density for a GaAs injection laser as a function of temperature.

and for the value of N_e given by (8.176), from (8.178) we obtain a threshold current density of

$$J = 2.3 \times 10^3 \text{ amps/cm}^2$$

With dimensions $200\,\mu \times 200\,\mu$ transverse to the direction of current flow, this means that an injection current of 0.92 amps is required. Figure 8.26 illustrates the experimental data obtained by Burns, Dill, and Nathan[9] for threshold current as a function of temperature. At 77°K the measured current density is approximately 2×10^3 amps/cm², which is very close to the calculated value. At temperatures below 20°K the threshold current reduces to 80 amps/cm².

REFERENCES

1. See R. A. Smith, *Semiconductors*, Cambridge University Press, London, 1961, pp. 95–96.
2. A proof of this statement is presented in E. Spenke, *Electronic Semiconductors*, McGraw-Hill, New York, 1958, Appendix I, 381–383.
3. This sum rule is derived in R. A. Smith, *Wave Mechanics of Crystalline Solids*, Chapman and Hall, London, 1963, Appendix 2, pp. 464–466.
4. A. Rose, *Concepts in Photoconductivity and Allied Problems*, Interscience Tracts on Physics and Astronomy, Number 19, Interscience Publishers, New York, 1963.
5. M. I. Nathan, W. P. Dumke, G. Burns, F. H. Dill, Jr., and G. J. Lasher, "Stimulated Emission of Radiation from GaAs *p-n* Junctions," *Appl. Phys. Letters*, **1**, 62–64, 1962.

6. R. N. Hall, G. E. Fenner, J. D. Kingsley, T. J. Soltys, and R. O. Carlson, "Coherent Light Emission From GaAs Junctions," *Phys. Rev. Letters*, **9**, 366–368, 1962.
7. See, for example, G. Birnbaum, *Optical Masers*, Academic Press, New York, 1964, p. 155.
8. This value for T_e was obtained from Eqs. (7-21) and (7-24) on p. 384 of: W. V. Smith and P. P. Sorokin, *The Laser*, McGraw-Hill, New York, 1966.
9. G. Burns, F. H. Dill, Jr., and M. I. Nathan, "The Effect of Temperature on the Properties of GaAs Lasers," *Proc. IEEE*, **51**, 947–948, 1963.

Some additional references, listed under the section to which the referenced material applies:

Sections 8.2–8.3

(a) E. Spenke, *Electronic Semiconductors*, McGraw-Hill, New York, 1958, Chapter 7.
(b) C. Kittel, *Introduction to Solid State Physics*, 3rd ed., Wiley, New York, 1966, Chapter 9.
(c) R. P. Feynman, R. B. Leighton, and M. Sands, *The Feynman Lectures on Physics*, Addison-Wesley, Reading, Mass., 1965, Vol. III, Chapter 13.

Section 8.4

(d) R. A. Smith, *Semiconductors*, Cambridge University Press, London, 1961, Chapter 7.
(e) R. A. Smith, *Wave Mechanics of Crystalline Solids*, Chapman and Hall, London, 1963, Chapter 13.

Section 8.5

(f) R. H. Bube, *Photoconductivity of Solids*, Wiley, New York, 1960.
(g) A. Rose, *Concepts in Photoconductivity and Allied Problems*, Interscience Tracts on Physics and Astronomy, Number 19, Interscience, New York, 1963.

Section 8.6

(h) G. Birnbaum, *Optical Masers*, Academic Press, New York, 1964, Chapter XI.
(i) W. V. Smith and P. P. Sorokin, *The Laser*, McGraw-Hill, New York, 1966, Chapter 7.
(j) B. A. Lengyel, *Introduction to Laser Physics*, Wiley, New York, 1966, pp. 136–157.
(k) A. Yariv, *Quantum Electronics*, Wiley, New York, 1967, Chapter 17.

PROBLEMS

8.1 Derive the three-dimensional form for the Bragg reflection condition given by (8.15).

8.2 Prove that the orthogonality condition (8.25) holds for periodic boundary conditions.

8.3 From (8.27) and (8.28), show that for a free electron

$$E = \frac{\hbar^2 k^2}{2m}$$

8.4 We wish to determine the probability that an electron will be reflected or transmitted at the site of an impurity. Using (8.44), with $\Delta = 0.2 \mathcal{H}'_{01}$, plot the probabilities for an electron to be reflected and for an electron to be transmitted as a function of ka.

8.5 We wish to locate the energy levels for trapped electrons. If, in (8.49), $\Delta = 0.2\mathcal{H}'_{01}$, and

(a) For an energy band with $E = 1$ eV for $k = 0$ and $E = 2$ eV for $k = \pi/a$, locate the energy level for the trapped electron. (Answer: Energy is 1/400 eV below the band.)

(b) With $E = 1$ eV for $k = \pi/a$ and $E = 2$ eV for $k = 0$, locate the energy level for the trapped electron.

8.6 Show that (8.55) satisfies the condition for charge conservation. That is, prove that

$$\frac{d\rho}{dt} = 0 = \frac{\partial \rho}{\partial t} + \dot{\mathbf{k}} \cdot \nabla_{\mathbf{k}}\rho$$

8.7 Fill in the missing steps in going from (8.57) to (8.58).

8.8 In Figure 8.11 we have plotted k, E, and $\langle v \rangle$ as functions of time when a dc field is applied and there are no collisions. From the equations derived in Section 8.3.3, plot these same variables as functions of time when a dc field is applied and collisions are included.

8.9 Derive (A13.9) in Appendix 13 for the density of states in **k** space.

8.10 Derive (A13.2) in Appendix 13.

8.11 Show that the real and imaginary parts of the complex conductivity defined by (8.82) conform to the Kramers-Kronig relations.

8.12 (a) Obtain an expression for the change in conductivity, (8.160), when the quantum efficiency is not equal to unity.

(b) What is the percentage change in conductivity when a neodymium laser with a power intensity of 100 W/cm² is incident on an intrinsic GaAs crystal at room temperature?

8.13 Let us consider certain aspects of the behavior of the photoconductive gain constant G defined by (8.161) and (8.170).

(a) Show that G is independent of the quantum efficiency.

(b) What keeps G, as given by (8.170), from becoming arbitrarily large?

(c) Draw a curve showing the temperature dependence of G. (Reference: R. H. Bube, *Photoconductivity of Solids*, Wiley, New York, 1960.)

8.14 A semiconductor injection laser is operated at 77°K, and with current flowing across the junction the equivalent Fermi level is 0.023 eV above the bottom of the conduction band. If the junction is used as an amplifier rather than as an oscillator, sketch the shape of the gain curve as a function of frequency for a photon energy range 1.2 to 1.6 eV. Assume that the absorption constant given in Figure 8.25 is the gain constant for an inverted population.

THE EFFECT OF $\mathcal{H}^r$ ON ρ_{11}

appendix
1

We wish to solve (1.25) and (1.26) for $\rho_{11}(t)$ by means of the Laplace transform. The Laplace transform is defined as

$$\rho_{ij}(s) = L\{\rho_{ij}(t)\} = \int_0^\infty e^{-st}\rho_{ij}(t)\,dt \tag{A1.1}$$

If we take the transform of (1.25) and (1.26) and solve for $\rho_{11}(s)$, we find that

$$\rho_{11}(s) = \frac{1}{s + (1/\hbar^2)\sum_k (2s\,|\mathcal{H}^r_{1k}|^2)/(s^2 + \omega_{1k}^2)} \tag{A1.2}$$

where $\omega_{1k} \equiv (E_1 - E_k)/\hbar$. The inverse transform of (A1.2) is

$$\rho_{11}(t) = \frac{1}{2\pi i}\int_{\epsilon-j\infty}^{\epsilon+j\infty} \frac{e^{st}\,ds}{s + (1/\hbar^2)\sum_k (2s\,|\mathcal{H}^r_{1k}|^2)/(s^2 + \omega_{1k}^2)} \tag{A1.3}$$

where ϵ is a small, positive number.

The term $\mathcal{H}^r_{1k}$ may represent, for example, the matrix element between two electronic states of the molecule that are coupled by a vibrational mode of the lattice. Vibrational modes are closely spaced so that the summation in (A1.3) may be replaced by an integral

$$\sum_k \frac{2s\,|\mathcal{H}^r_{1k}|^2}{s^2 + \omega_{1k}^2} \to \int_{-\infty}^\infty g(\omega_{1k})\,d\omega_{1k}\,\frac{2s\,|\mathcal{H}^r_{1k}|^2}{s^2 + \omega_{1k}^2} \tag{A1.4}$$

where $g(\omega_{1k})$ is the number of vibrational states per unit frequency interval.

If the coupling is small (that is, $\mathcal{H}^r_{1k} \simeq 0$) then (A1.3) is evaluated from the residue of the pole at $s = 0$, and $\rho_{11}(t)$ is a constant. Therefore for small coupling we evaluate the integral in (A1.4) for $s \simeq 0$. Whereupon (A1.4) becomes

$$\int_{-\infty}^{\infty} g(\omega_{1k}) \, d\omega_{1k} \frac{2s |\mathscr{H}^r_{1k}|^2}{s^2 + \omega_{1k}^2} = 2\pi g(\omega_{1k}) |\mathscr{H}^r_{1k}|^2 \bigg|_{\omega_{1k}=0} \tag{A1.5}$$

Thus we evaluate the density of states $g(\omega_{1k})$ and the matrix element $|\mathscr{H}^r_{1k}|^2$ for vibrational modes that exist at a vanishingly small energy increment from the electronic state.

The substitution of (A1.5) into (A1.3) yields

$$\rho_{11}(t) = \frac{1}{2\pi i} \int_{\epsilon-j\infty}^{\epsilon+j\infty} \frac{e^{st} \, ds}{s + 1/\tau} = e^{-t/\tau} \tag{A1.6}$$

where

$$\frac{1}{\tau} \equiv \frac{2\pi g(\omega_{1k}) |\mathscr{H}^r_{1k}|^2}{\hbar^2} \bigg|_{\omega_{1k}=0} \tag{A1.7}$$

TRANSITION RATES IN EQUILIBRIUM

appendix

2

In equilibrium we require that ρ_{jj} be independent of time and that there be no net emission or absorption of energy at any of the transition frequencies. For a two-level system this means that transitions from $|2\rangle \to |1\rangle$ per unit time must equal the number of transitions per unit time from $|1\rangle \to |2\rangle$:

$$\rho_{22}^e W_{21} = \rho_{11}^e W_{12} \tag{A2.1}$$

For a three-level system it is possible to satisfy the condition that ρ_{jj} is constant by having

$$\rho_1^e W_{12} = \rho_{22}^e W_{23} = \rho_{33}^e W_{31} \tag{A2.2}$$

This condition is illustrated in Figure A2.1. However, the system illustrated in Figure A2.1 requires the absorption of energy at frequencies ω_{21} and ω_{32} and the emission of energy at ω_{31}. This is not an equilibrium condition. Equilibrium is maintained for a three-level system only if the transition rate from the j state to

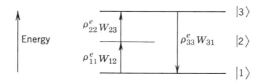

figure A2.1 A possible transition rate system for maintaining constant ρ_{jj}.

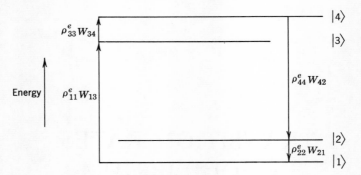

figure A2.2 A possible transition rate system for maintaining equilibrium in a four-level system.

the k state is the same as the transition rate from the k to the j state:

$$\rho^e_{kk} W_{kj} = \rho^e_{jj} W_{jk} \tag{A2.3}$$

For a four-level system it is possible to maintain equilibrium either by (A2.3) or, for example, by satisfying the following conditions:

$$\omega_{21} = \omega_{43}$$
$$\rho^e_{11} W_{13} = \rho^e_{33} W_{34} = \rho^e_{44} W_{42} = \rho^e_{22} W_{21} \tag{A2.4}$$
$$W_{12} = W_{14} = W_{23} = W_{24} = W_{31} = W_{32} = W_{41} = W_{43} = 0$$

The situation described by (A2.4) is illustrated in Figure A2.2. However, there are many more conditions to be satisfied by (A2.4) than by (A2.3), and so it would be far less likely to have equilibrium satisfied by (A2.4). Therefore, even for multilevel systems the equilibrium condition is taken to be (A2.3). Equation A2.3 is known as the principle of detailed balance. (See Ref. 5 at the end of Chapter 1.)

TO PROVE THAT $\mu_\alpha \mu_\beta^*$ IS REAL

appendix 3

From the definition of the dipole matrix element we have

$$\mu_\alpha \mu_\beta^* = \langle u_1|\mu_\alpha|u_2\rangle\langle u_1|\mu_\beta|u_2\rangle^* = \langle u_1|\mu_\alpha|u_2\rangle\langle u_2|\mu_\beta|u_1\rangle$$

where the second equality follows from the Hermitian property of the operator μ_α. Then we add and subtract the term $\langle u_1|\mu_\alpha|u_1\rangle\langle u_1|\mu_\beta|u_1\rangle$, yielding

$$\mu_\alpha \mu_\beta^* = \sum_{i=1}^{2} \langle u_1|\mu_\alpha|u_i\rangle\langle u_i|\mu_\beta|u_1\rangle - \langle u_1|\mu_\alpha|u_1\rangle\langle u_1|\mu_\beta|u_1\rangle$$

Upon removal of the identity relation $I = \sum_{i=1}^{2} |u_i\rangle\langle u_i|$, the above further reduces to

$$\mu_\alpha \mu_\beta^* = \langle u_1|\mu_\alpha\mu_\beta|u_1\rangle - \langle u_1|\mu_\alpha|u_1\rangle\langle u_1|\mu_\beta|u_1\rangle \tag{A3.1}$$

Furthermore, by the rule for Hermitian conjugation of operators and the fact that the operators μ_α and μ_β are individually Hermitian, we have for the operators the relationship

$$(\mu_\alpha\mu_\beta)^\dagger = \mu_\beta^\dagger\mu_\alpha^\dagger = \mu_\beta\mu_\alpha$$

However, μ_α and μ_β are both functions of **r** alone and, hence, commuting observables. Therefore the order of the operators may be reversed, yielding

$$(\mu_\alpha\mu_\beta)^\dagger = \mu_\alpha\mu_\beta$$

that is, the product is also Hermitian.

Since the dipole moment and its product are Hermitian, they are represented by Hermitian matrices, the diagonal elements of which are real. Therefore the product $\mu_\alpha \mu_\beta^*$ given by (A3.1), which is expressed entirely in terms of diagonal elements, must be real, and this proves the assertion.

The quantity $\mu_\alpha \mu_\beta^*$ appears as a coupling coefficient that relates the α component of induced dipole moment to the β component of applied electric field. The three-by-three array of coefficients that relates the two vectors constitutes a second-rank tensor. Since $\mu_\alpha \mu_\beta^* = \mu_\alpha^* \mu_\beta = \mu_\beta \mu_\alpha^*$, the tensor is symmetric in the indices α and β.

LORENTZ LOCAL FIELD CORRECTION FACTOR

appendix

4

The field equations for an isotropic polarizable medium in the absence of free charge are

$$\nabla \cdot \mathbf{B} = 0 \qquad \nabla \cdot \mathbf{D} = 0$$
$$\nabla \times \mathbf{H} = \mathbf{J} + \frac{\partial \mathbf{D}}{\partial t} \qquad \nabla \times \mathbf{E} = -\frac{\partial \mathbf{B}}{\partial t} \qquad (A4.1)$$
$$\mathbf{B} = \mu_0 \mathbf{H} \qquad \mathbf{D} = \epsilon_0 \mathbf{E} + \mathbf{P}^{tot}$$
$$\mathbf{J} = \sigma \mathbf{E}$$

Of special interest is the polarization term $\mathbf{P}^{tot}$, which appears in the expression for the electric flux density $\mathbf{D}$. In the discussion of a particular process that takes place on an atomic or molecular level, such as the electric dipole transition, it is useful to separate the total polarization into two parts,

$$\mathbf{P}^{tot} = \mathbf{P} + \mathbf{P}' \qquad (A4.2)$$

$\mathbf{P}$ is taken to be the polarization associated with the molecular process of interest that is explicitly accounted for, and $\mathbf{P}'$ is the polarization resulting from all other transitions.

We now assume that the polarization $\mathbf{P}'$ caused by other transitions is related to the local electric field $\mathbf{E}^{loc}$ by

$$\mathbf{P}' = \sum_i N_i \bar{\alpha}_i \mathbf{E}^{loc} \qquad (A4.3)$$

where N_i is the number of molecules per unit volume of type i, and $\bar{\alpha}_i$ is the spatially averaged linear polarizability of molecules of type i. For transitions whose resonant frequencies are far from the frequencies of the applied fields, the polarizability $\bar{\alpha}_i$ is simply a real constant. The total polarization $\mathbf{P}^{tot}$ can therefore be expressed from (A4.2) and (A4.3) as

$$\mathbf{P}^{tot} = \mathbf{P} + \sum_i N_i \bar{\alpha}_i \mathbf{E}^{loc} \tag{A4.4}$$

Standard elementary physics texts show that in an isotropic medium the macroscopic field $\mathbf{E}$ differs from the local electric field $\mathbf{E}^{loc}$ seen by a given atom or molecule, because of the influence of nearby polarizable matter. The two fields are related by†

$$\mathbf{E}^{loc} = \mathbf{E} + \frac{1}{3\epsilon_0} \mathbf{P}^{tot} \tag{A4.5}$$

Combination of (A4.4) and (A4.5) yields

$$\mathbf{P}^{tot} = \frac{1}{1 - \sum_i N_i \bar{\alpha}_i / 3\epsilon_0} \mathbf{P} + \frac{\sum_i N_i \bar{\alpha}_i}{1 - \sum_i N_i \bar{\alpha}_i / 3\epsilon_0} \mathbf{E} \tag{A4.6}$$

Substitution of (A4.6) into the expression for $\mathbf{D}$ in (A4.1) yields

$$\mathbf{D} = \epsilon_0 \mathbf{E} + \mathbf{P}^{tot} = \left(\epsilon_0 + \frac{\sum_i N_i \bar{\alpha}_i}{1 - \sum_i N_i \bar{\alpha}_i / 3\epsilon_0} \right) \mathbf{E} + \frac{1}{1 - \sum_i N_i \bar{\alpha}_i / 3\epsilon_0} \mathbf{P} \tag{A4.7}$$

The above expression for $\mathbf{D}$ contains two terms. The first is independent of the process of interest accounted for by $\mathbf{P}$; the second is the contribution resulting from $\mathbf{P}$. If we characterize the dielectric properties of the medium *excluding the effects of the process of interest* by a permittivity ϵ so that in the absence of $\mathbf{P}$ we have $\mathbf{D} = \epsilon \mathbf{E}$, we find that (A4.7) can be written in the form

$$\mathbf{D} = \epsilon \mathbf{E} + \mathbf{P}^s \tag{A4.8}$$

Comparison of (A4.7) and (A4.8) indicates that (*a*) the dielectric properties of the medium excluding the effects of $\mathbf{P}$ are described by what is known as the Lorentz-Lorenz relation,

$$\frac{\eta^2 - 1}{\eta^2 - 2} = \frac{1}{3\epsilon_0} \sum_i N_i \bar{\alpha}_i \tag{A4.9}$$

where $\eta = \sqrt{\epsilon/\epsilon_0}$ is the refractive index of the medium excluding the effects of $\mathbf{P}$; and (*b*) the polarization source term $\mathbf{P}^s$ to be used in the macroscopic field equations to represent the effects of the process of interest is related to the actual polarization $\mathbf{P}$ by

$$\mathbf{P}^s = \left(\frac{\eta^2 + 2}{3} \right) \mathbf{P} \tag{A4.10}$$

† See, for example, C. Kittel, *Introduction to Solid-State Physics*, 3rd ed., Wiley, New York, 1966, p. 375 ff.

Furthermore, if Equations A4.6 through 4.10 are substituted into (A4.5), we find that the local electric field $\mathbf{E}^{loc}$ is expressed in terms of the macroscopic field $\mathbf{E}$ by

$$\mathbf{E}^{loc} = \left(\frac{\eta^2 + 2}{3}\right)\mathbf{E} + \frac{1}{3\epsilon_0}\mathbf{P}^s \approx \left(\frac{\eta^2 + 2}{3}\right)\mathbf{E} \qquad (A4.11)$$

where the approximation is generally good for all but the strongest polarization source terms $\mathbf{P}^s$. The factors $(\eta^2 + 2)/3$ in (A4.10) and (A4.11) are known as Lorentz local field correction factors.

A wave equation in the variable $\mathbf{E}$ is derived by taking the curl of the $\nabla \times \mathbf{E}$ equation in (A4.1) and substituting from the other equations as appropriate with $\mathbf{D}$ given by (A4.8). The result is

$$\nabla \times (\nabla \times \mathbf{E}) + \frac{\eta \mathscr{A}}{c}\frac{\partial \mathbf{E}}{\partial t} + \frac{\eta^2}{c^2}\frac{\partial^2 \mathbf{E}}{\partial t^2} = -\mu_0 \frac{\partial^2 \mathbf{P}^s}{\partial t^2} \qquad (A4.12)$$

where $c^2 = 1/\mu_0\epsilon_0$, $\mathscr{A}$ is a damping coefficient given by $\mathscr{A} = \mu_0\sigma c/\eta$, and $\eta = \sqrt{\epsilon/\epsilon_0}$ is the refractive index of the medium excluding the effects of the process of interest that is accounted for in the driving term $\mathbf{P}^s$.

In summary, then, we have shown that for an isotropic medium the polarization $\mathbf{P}$ resulting from a macroscopic process of interest must be modified by a Lorentz local field correction factor as shown in (A4.10) before its effects on macroscopic fields can be determined as in (A4.12). In addition, the macroscopic field $\mathbf{E}$ must be similarly modified as shown in (A4.11) to obtain the local field $\mathbf{E}^{loc}$, which may appear in equations describing the microscopic process. The above arguments hold regardless of whether $\mathbf{P}$ results from linear or nonlinear processes. The derivation presented here for an isotropic medium can be extended to cover anisotropic media, in which case the Lorentz correction factors are tensors.†

† N. Bloembergen, *Nonlinear Optics*, W. A. Benjamin, New York, 1965, pp. 68, 178.

KRAMERS-KRONIG RELATIONSHIPS
appendix
5

According to (3.5) the polarization $\mathbf{P} = \tfrac{1}{2}\tilde{\mathbf{P}}e^{i\omega t}$ + c.c. and electric field $\mathbf{E} = \tfrac{1}{2}\tilde{\mathbf{E}}e^{i\omega t}$ + c.c. are related in a linear isotropic medium by

$$\tilde{\mathbf{P}} = \epsilon_0 \chi(\omega) \tilde{\mathbf{E}} \qquad (A5.1)$$

The quantity $\chi(\omega)$ is a linear susceptibility that has real and imaginary parts,

$$\chi(\omega) = \chi'(\omega) + i\chi''(\omega) \qquad (A5.2)$$

We shall now show that $\chi'(\omega)$ and $\chi''(\omega)$ are related in such a way that if one is known, the other can be determined by a set of integral relationships known as the Kramers-Kronig relations,

$$\chi'(\omega) = -\frac{1}{\pi} \text{PP} \int_{-\infty}^{\infty} \frac{\chi''(\omega')}{\omega' - \omega} d\omega' \qquad (A5.3)$$

$$\chi''(\omega) = \frac{1}{\pi} \text{PP} \int_{-\infty}^{\infty} \frac{\chi'(\omega')}{\omega' - \omega} d\omega' \qquad (A5.4)$$

where PP stands for the principal part, defined following (3.29).

Let us consider the integration of the function $\chi(\omega')/(\omega' - \omega)$ around the contour shown in Figure A5.1. For a passive medium the function $\chi(\omega')$ is analytic within the region enclosed by the contour, and so $[\chi(\omega')]/(\omega' - \omega)$ is also analytic within this region. Therefore, by Cauchy's integral theorem the integral around the closed path is zero,[†]

$$\int_{c'} \frac{\chi(\omega')}{\omega' - \omega} d\omega' + \int_{-R}^{\omega-\epsilon} \frac{\chi(\omega')}{\omega' - \omega} d\omega' + \int_{\omega+\epsilon}^{R} \frac{\chi(\omega')}{\omega' - \omega} d\omega' + \int_{c} \frac{\chi(\omega')}{\omega' - \omega} d\omega' = 0 \qquad (A5.5)$$

[†] See, for example, T. M. Macrobert, *Functions of a Complex Variable*, 4th ed., Macmillan London, 1954, p. 51.

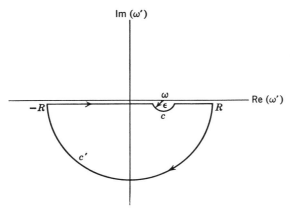

figure A5.1 Contour in the complex plane over which (A5.5) is evaluated.

We now consider the integrals in the limits $R \to \infty$, $\epsilon \to 0$. The first integral vanishes because with $\chi(\omega')$ given by (3.6) the numerator of the integrand is of degree 2 lower in the variable ω' than the denominator, and thus integration around an infinite semicircular arc yields zero. The sum of the second and third integrals in the limit $\epsilon \to 0$ is by definition the principal part of the integral. The fourth integral around the simple pole at $\omega' = \omega$ yields $i\pi\chi(\omega)$. We therefore have in the limits $R \to \infty$, $\epsilon \to 0$,

$$\text{PP} \int_{-\infty}^{\infty} \frac{\chi(\omega')}{\omega' - \omega} \, d\omega' + i\pi\chi(\omega) = 0 \tag{A5.6}$$

or

$$\chi(\omega) = \frac{i}{\pi} \text{PP} \int_{-\infty}^{\infty} \frac{\chi(\omega')}{\omega' - \omega} \, d\omega' \tag{A5.7}$$

By substituting (A5.2) into the above and equating real and imaginary parts, we then obtain the Kramers-Kronig relationships, Equations A5.3 and A5.4.

SPATIAL AVERAGING

appendix
6

We wish to consider the error introduced by the assumption that (2.39) holds when there is saturation, for the case when all molecular orientations are equally likely. This assumption is that

$$\overline{N_V(\mu_\alpha \mu_\beta^*)(\rho_{11} - \rho_{22})} = \overline{(\mu_\alpha \mu_\beta^*)}(N_1 - N_2) \qquad (2.39)$$

where the overbar indicates a spatial average. If all molecular orientations are equally likely, then $|\boldsymbol{\mu}_{12}|^2$ along any given direction has a $\cos^2 \theta$ dependence, where θ is the angle between the dipole vector and the coordinate direction, as shown in Figure A6.1.

The gain constant given in (3.67) was obtained by assuming (2.39) to hold, and this leads to a factor in the gain constant of the form

$$\frac{1}{1 + I/I_{\text{sat}}} \qquad (A6.1)$$

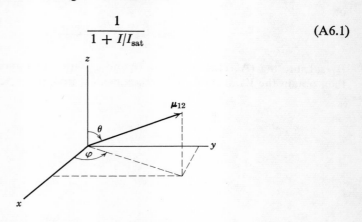

figure A6.1 Coordinate directions for calculating spatial averages.

However, if an unaveraged equation of the type of (3.66) is substituted into an unaveraged (3.65), the factor that appears in the gain constant is

$$\frac{1}{4\pi} \int \frac{3\cos^2\theta}{1 + (I/I_{sat})3\cos^2\theta} \, d\Omega \tag{A6.2}$$

where $d\Omega$ is an element of solid angle. The factor 3 that appears in (A6.2) results from the fact that in the derivation of (A6.1) the average of $\cos^2\theta$ over all directions gives a $\frac{1}{3}$ factor.

The ratio of (A6.2) to (A6.1) is always greater than unity. For very small and very large values of I/I_{sat} this ratio approaches unity, and the maximum value of this ratio is approximately 1.3.

DERIVATION OF PHOTON RATE EQUATION

appendix

7

In order to evaluate the right-hand side of (6.49), we begin with (6.44) for the diagonal element $\rho_{1,n;1,n}$. When we evaluate the commutator on the right-hand side of (6.44), we have from (6.40),

$$[\mathcal{H}', \rho]_{1,n;1,n} = \mathcal{H}'_{1,n;2,n-1}\rho_{2,n-1;1,n} + \mathcal{H}'_{1,n;2,n+1}\rho_{2,n+1;1,n} - \text{c.c.} \quad (A7.1)$$

where c.c. is the complex conjugate. The discussion presented in the next paragraph will indicate that the second term on the right-hand side of (A7.1) may be dropped.

In solving equations of motion (6.43) and (6.44) for the density operator, we note that certain simplifying assumptions can be made. The homogeneous solution for $\rho_{i,n;j,m}$ is an exponentially damped sinusoid at the resonant frequency $\omega_{i,n;j,m}$. Therefore, $\rho_{i,n;j,m}$ responds most strongly to driving terms that vary at or near the natural resonant frequency. Therefore we shall retain only those terms on the right-hand side of (6.43) that have a sinusoidal time dependence given by $\omega_{i,n;j,m}$. From (6.40) we see that the matrix elements of $\mathcal{H}'$ are not functions of time, which means that the frequencies of the driving terms in (6.43) are those of the density matrix elements $\rho_{k,p;l,q}$ that appear. The frequency associated with $\rho_{k,p;l,q}$ is $\omega_{k,p;l,q}$ whereas the resonant frequency in (6.43) is $\omega_{i,n;j,m}$. For these two frequencies to be nearly the same, we find from (6.45) that only those expressions in $\rho_{k,p;l,q}$ that satisfy

$$E_k - E_l + (p - q)\hbar\omega \approx E_i - E_j + (n - m)\hbar\omega \quad (A7.2)$$

are kept. Any term for which (A7.2) is not satisfied does not drive the equation near its resonant frequency and therefore is not retained. This condition applies equally well to the diagonal elements given by (6.44), which constitute a special case in which the resonant frequency is near zero; that is,

$$E_k - E_l + (p - q)\hbar\omega \approx 0 \qquad (A7.3)$$

With the aid of (A7.3) we find that only the first term in (A7.1) which contains the $\rho_{2,n-1;1,n}$ term is to be retained, whereupon combination of (6.44) and (A7.1) yields

$$\dot{\rho}_{1,n;1,n} + \frac{1}{\tau_{1,n;1,n}}(\rho_{1,n;1,n} - \rho_{1,n;1,n}^e) = \frac{1}{i\hbar}(\mathcal{H}'_{1,n;2,n-1}\rho_{2,n-1;1,n} - \text{c.c.}) \qquad (A7.4)$$

By a similar procedure we find that

$$\dot{\rho}_{2,n;2,n} + \frac{1}{\tau_{2,n;2,n}}(\rho_{2,n;2,n} - \rho_{2,n;2,n}^e) = \frac{1}{i\hbar}(\mathcal{H}'_{2,n;1,n+1}\rho_{1,n+1;2,n} - \text{c.c.}) \qquad (A7.5)$$

The off-diagonal terms on the right-hand side of (A7.4) and (A7.5) may be evaluated from (6.43). For the element $\rho_{2,n-1;1,n}$ we have $\omega_{2,n-1;1,n} = \Omega - \omega \approx 0$, and for $\rho_{1,n+1;2,n}$ the corresponding natural frequency is $\omega - \Omega \approx 0$. Therefore these matrix elements will be slowly varying functions of time, and for typical optical transitions we have $\partial/\partial t \ll 1/\tau_{i,n;j,m}$. It will be apparent from the final form of the equations that we shall derive that $\tau_{i,n;j,m}$ becomes the transverse relaxation time constant T_2. This is the same condition (that is, $\partial/\partial t \ll 1/T_2$) that was used to derive the laser rate equations, (3.78) and (3.79). Under this condition, which we assume at the outset, the first term may be neglected with respect to the third in (6.43). The equation for the off-diagonal elements of interest is therefore given by

$$\rho_{2,n-1;1,n} = \frac{\mathcal{H}'_{2,n-1;1,n}/\hbar}{(\omega - \Omega) + i/\tau_{2,n-1;1,n}}(\rho_{1,n;1,n} - \rho_{2,n-1;2,n-1}) \qquad (A7.6)$$

and

$$\rho_{1,n+1;2,n} = \frac{\mathcal{H}'_{1,n+1;2,n}/\hbar}{(\Omega - \omega) + i/\tau_{1,n+1;2,n}}(\rho_{2,n;2,n} - \rho_{1,n+1;1,n+1}) \qquad (A7.7)$$

where once again we have retained in the driving terms only those terms that satisfy (A7.2).

Combining (A7.4)–(A7.7) and substituting the result for the diagonal elements of the density matrix into (6.49), after some simplification we obtain

$$\frac{\partial}{\partial t}\langle n \rangle + \frac{\langle n \rangle - \langle n \rangle^e}{\tau_c} = \frac{2\pi}{\hbar^2} g_L(\omega, \Omega)$$

$$\times \sum_n n[|\mathcal{H}'_{1,n;2,n-1}|^2(\rho_{2,n-1;2,n-1} - \rho_{1,n;1,n})$$

$$+ |\mathcal{H}'_{2,n;1,n+1}|^2(\rho_{1,n+1;1,n+1} - \rho_{2,n;2,n})], \qquad (A7.8)$$

where $g_L(\omega, \Omega)$ is the Lorentzian lineshape function given by (3.14),

$$g_L(\omega, \Omega) = \frac{1}{\pi} \frac{1/T_2}{(\Omega - \omega)^2 + (1/T_2)^2} \tag{3.14}$$

In the derivation of (A7.8) identification of the time constant as τ_c on the left-hand side of (A7.8) is in correspondence with the condition that τ_c is the cavity lifetime that accounts for the loss of photons from the cavity from causes other than interaction with the molecule, for example, mirror losses. Similarly, the identification of the time constant which appears in $g_L(\omega, \Omega)$ as the transverse relaxation time constant for the medium, T_2, insures that the final result is consistent with that obtained in the semiclassical analysis presented in Chapter 3, where a more detailed consideration of the time constants was given.

The matrix elements of the interaction Hamiltonian required in (A7.8) are given by (6.40), and upon substitution (A7.8) reduces to

$$\frac{\partial}{\partial t} \langle n \rangle + \frac{\langle n \rangle - \langle n \rangle^e}{\tau_c} = -G \sum_n n(\rho_{1,n;1,n} - \rho_{2,n;2,n}) + G\rho_{22} \tag{A7.9}$$

where

$$G \equiv \frac{\pi \omega}{\hbar \epsilon} (\mu_\alpha \mu_\beta^*)(E_a)_\alpha (E_a)_\beta g_L(\omega, \Omega) \tag{A7.10}$$

SUBSTANCES FOUND TO EXHIBIT STIMULATED RAMAN EFFECT
appendix
8

SUBSTANCES FOUND TO EXHIBIT STIMULATED RAMAN EFFECT

Liquids	frequency shift (cm^{-1})	References
Bromoform	222	C
Tetrachloroethylene	447	J
Carbon Tetrachloride	460	L
Hexafluorobenzene	515	L
Bromoform	539	C
Trichlorethylene	640	C
Carbon Disulfide	656	C
Chloroform	667	C
Orthoxylene	730	B
α Dimethyl Phenethylamine	836	I
Dioxan	836	C
Morpholine	841	L
Thiophenol	916	L
Nitro Methane	927	L
Deuterated Benzene	944	A
Cumene	990	L
1:3 Di Bromobenzene	990	J
Benzene	992	A, C
Pyridine	992	A
Aniline	997	H
Styrene	998	D, H
m Toluidine	999	L
Bromobenzene	1000	H

SUBSTANCES FOUND TO EXHIBIT STIMULATED RAMAN EFFECT (contd.)

Liquids	frequency shift (cm^{-1})	References
Chlorobenzene	1001	L
Benzonitrile	1002	H
T: Butyl Benzene	1002	J
Ethyl Benzene	1002	B
Toluene	1004	A, C
Fluorobenzene	1012	E
γ Picoline	1016	L
m Cresol	1029	L
Meta-Dichlorobenzene	1030	L
1: Fluoro 2: Chlorobenzene	1030	J
Iodo Benzene	1070	L
Benzoyl Chloride	1086	L
Benzaldehyde	1086	L
Anisole	1097	L
Pyrrole	1178	L
Furan	1180	L
Styrene	1315	D, H
Nitrobenzene	1344	A, C
1: Bromonaphthalene	1368	A
1: Chloronaphthalene	1368	H
2: Ethyl Naphthalene	1381	J
m Nitro Toluene	1389	L
Quinolene	1427	L
Furan	1522	L
Methyl Salicylate	1612	L
Cinnamaldehyde	1624	H
Styrene	1629	D, H
3: Methyl Butadiene	1638	G
Pentadiene	1655	G
Isoprene	1792	I
1: Hexyne	2116	L
Ortho-Dichlorobenzene	2202	L
Benzonitrile	2229	H
1:2 Dimethyl Aniline	2292	L
Methyl Cyclohexane	2817	L
Methanol	2831	C
cis trans 1:3 Dimethyl Cyclohexane	2844	J
Tetrahydrofuran	2849	H
Cyclohexane	2852	A, C
cis 1:2 Dimethyl Cyclohexane	2854	J
α Dimethyl Phenethylamine	2856	I
Dioxan	2856	C
Cyclohexane	2863	A, C
Cyclohexanone	2863	B
cis trans 1:3 Dimethyl Cyclohexane	2870	J

SUBSTANCES FOUND TO EXHIBIT STIMULATED RAMAN EFFECT (contd.)

Liquids	frequency shift (cm^{-1})	References
cis 1:4 Dimethyl Cyclohexane	2873	J
Cyclohexane	2884	A, C
Dichloro Methane	2902	L
Morpholine	2902	L
2: Octene	2908	J
2:3 Dimethyl 1:5 Hexadiene	2910	J
Limonene	2910	I
Orthoxylene	2913	B
1: Hexyne	2915	L
cis 2: Heptene	2920	J
Mesitylene	2920	I
2: Bromopropane	2920	J
Acetone	2921	B, C
Ethanol	2921	C
Carvone	2922	I
cis 1:2 Dimethyl Cyclohexane	2927	J
Dimethyl Formamide	2930	C
2: Chloro 2 Methyl Butane	2931	J
2: Octene	2931	J
cis trans 1:3 Dimethyl Cyclohexane	2931	J
Metaxylene	2933	B
1:2 Diethyl Tartrate	2933	I
Orthoxylene	2933	B
Piperidine	2933	B
1:2 Diethyl Benzene	2934	J
2 Chloro 2 Methyl Butene	2935	J
1: Bromopropane	2935	J
Piperidine	2936	B
Tetrahydrofuran	2939	H
Piperidine	2940	B
Cyclohexanone	2945	B
2: Nitropropane	2948	J
1:2 Diethyl Carbonate	2955	L
1:2 Dichloro Ethane	2956	L
Trans Dichloroethylene	2956	C
1: Bromopropane	2962	J
2 Chloro 2 Methyl Butane	2962	J
α Dimethyl Phenethylamine	2967	I
Dioxan	2967	C
Cyclohexanol	2982	L
Cyclopentane	2982	L
Cyclopentanol	2982	L
Bromo-Cyclopentane	2982	L
Ortho-Dichlorobenzene	2982	L
Para Chlorotoluene	2982	L

SUBSTANCES FOUND TO EXHIBIT STIMULATED RAMAN EFFECT (contd.)

Liquids	frequency shift (cm^{-1})	References
α Picoline	2982	L
Paraxylene	2988	B
Orthoxylene	2992	B
Di Butyl n Phthalate	2992	L
1:1:1 Trichloroethane	3018	C
Ethylene Chlorhydrin	3022	L
Iso Phorone	3022	L
Nitroso Dimethylamine	3022	L
Propylene Glycol	3022	L
Cyclohexane	3038	L
Styrene	3056	D, H
Benzene	3064	A, C
T: Butyl Benzene	3064	J
1: Fluoro 2: Chlorobenzene	3084	J
Turpentine	3090	L
Pseudo Cumene	3093	L
Acetic Acid	3162	L
Acetonyl Acetone	3162	L
Methyl Methacrylate	3162	L
γ Picoline	3182	L
Aniline	3300	H
Water	3651	L

Solids		
Quartz	128	K
Lithium Niobate	152	O
α Sulphur	216	H
Lithium Niobate	248	N, O
Quartz	466	K
α Sulphur	470	H
Lithium Niobate	628	N, O
Calcium Tungstate	911	H
Stilbene	997	I
Polystyrene	1001	C
Calcite	1084	C
Diamond	1332	H
Naphthalene	1380	H
Stilbene	1591	I
Tryglycine Sulphate	2422	L
Tryglycine Sulphate	2702	L
Tryglycine Sulphate	3022	L
Polystyrene	3054	C

Gases		
Oxygen	1552	H
Potassium Vapor	2721	M
Methane	2916	F
Deuterium	2991	F
Hydrogen	4155	F

SUBSTANCES FOUND TO EXHIBIT STIMULATED RAMAN EFFECT (contd.)

References

A. G. Eckhardt, R. W. Hellwarth, F. J. McClung, S. E. Schwarz, and D. Weiner, "Stimulated Raman Scattering from Organic Liquids," *Phys. Rev. Letters*, **9**, 455-457, December 1962.
B. M. Geller, D. P. Bortfeld, and W. R. Sooy, "New Woodbury-Raman Laser Materials," *Appl. Phys. Letters*, **3**, 36–40, August 1963.
C. S. Kern and B. Feldman, "Stimulated Raman Emission," *M.I.T. Lincoln Lab. Solid-State Res. Rept.*, **3**, 18, 1964.
D. D. P. Bortfeld, M. Geller, and G. Eckhardt, "Combination Lines in the Stimulated Raman Spectrum of Styrene," *J. Chem. Phys.*, **40**, 1770–1771, March 15, 1964.
E. J. A. Calviello and Z. H. Heller, "Raman Laser Action in Mixed Liquids," *Appl. Phys. Letters*, **5**, 112–113, September 1964.
F. R. W. Minck, R. W. Terhune, and W. G. Rado, "Laser-Stimulated Raman Effect and Resonant Four-Photon Interactions in Gases H_2, D_2, and CH_4," *Appl. Phys. Letters*, **3**, 181–184, November 15, 1963.
G. V. A. Subov, M. M. Sushchinskii, and I. K. Shuvalton, "Investigation of the Excitation Threshold of Induced Raman Scattering," *J. Exptl. Theoret. Phys.* (USSR), **47**, 784–786, August 1964.
H. G. Eckhardt, "Selection of Raman Laser Materials," *IEEE J. Quant. Electr.*, **QE-2**, 1–8, January 1966.
I. D. L. Weinberg, "Stimulated Raman Emission in Crystals and Organic Liquids," *M.I.T. Lincoln Lab. Solid-State Res. Rept.*, **2**, 31, 1965.
J. J. J. Barrett and M. C. Tobin, "Stimulated Raman Emission Frequencies in 21 Organic Liquids," *J. Opt. Soc. Amer.*, **56**, 129–130, January 1966.
K. P. E. Tannenwald and J. B. Thaxter, "Stimulated Brillouin and Raman Scattering in Quartz at 2.1° to 293° Kelvin," *Science*, **134**, 1319–1320, December 9, 1966.
L. M. D. Martin and E. L. Thomas, "Infrared Difference Frequency Generation," *IEEE J. Quant. Electr.*, **QE-2**, August 1966.
M. M. Rokni and S. Yatsiv, "Resonance Raman Effect in Free Atoms of Potassium," *Phys. Letters*, **24a**, 277, 1967.
N. S. K. Kurtz and J. A. Giordmaine, "Stimulated Raman Scattering by Polaritons," *Phys. Rev. Letters*, **22**, 192, February 3, 1969.
O. J. Gelbwachs, R. H. Pantell, H. E. Puthoff, and J. M. Yarborough, A Tunable Stimulated Raman Oscillator" *Appl. Phys. Letters*, **14**, May 1, 1969.

RELATIONSHIP BETWEEN POLARIZABILITY AND DISPLACEMENT

appendix
9

From (7.96) we have that the electronic polarizability at the mth lattice site has a fluctuating component because of the passage of an acoustic wave q_m that takes the form

$$\alpha'_m = \sum_l \left(\frac{\partial \alpha_m}{\partial (q_{l+1} - q_l)}\right)_0 (q_{l+1} - q_l) \tag{A9.1}$$

where

$$q_l = \frac{\tilde{q}}{2} e^{i(\omega_k t - kla)} + \text{c.c.} \tag{A9.2}$$

and $l = 0, \pm 1, \pm 2, \ldots$ labels the lattice sites along the crystal. That is, the polarizability at site m is affected by the relative displacements of atoms at other lattice sites labeled by l. We now show that the summation over lattice sites l in (A9.1) reduces to a simple expression so that

$$\alpha'_m = C_b(q_{m+1} - q_m) \tag{A9.3}$$

Substituting (A9.2) into (A9.1), we have

$$\alpha'_m = \text{Re } F \tag{A9.4}$$

where

$$F = \sum_l \left(\frac{\partial \alpha_m}{\partial (q_{l+1} - q_l)}\right)_0 \tilde{q} e^{i(\omega_k t - kla)} (e^{-ika} - 1) \tag{A9.5}$$

and Re stands for "the real part of."

Since the crystal lattice is uniform in structure and considered to be infinitely large, it is readily apparent that a change in polarizability at lattice site m caused by a relative atomic displacement at lattice site l and a change in polarizability at lattice site $m + 1$ resulting from a similar atomic displacement at lattice site $l + 1$ must be identical. Therefore

$$\left(\frac{\partial \alpha_m}{\partial(q_{l+1} - q_l)}\right)_0 = \left(\frac{\partial \alpha_{m+1}}{\partial(q_{l+2} - q_{l+1})}\right)_0 \tag{A9.6}$$

Equation A9.5 can therefore be written

$$F = \sum_l \left(\frac{\partial \alpha_m}{\partial(q_{l+1} - q_l)}\right)_0 \tilde{q} e^{i(\omega_k t - kla)} [e^{-ika} - 1]$$

$$= \sum_{l'} \left(\frac{\partial \alpha_{m+1}}{\partial(q_{l'+2} - q_{l'+1})}\right)_0 \tilde{q} e^{i(\omega_k t - kl'a)} [e^{-ika} - 1]$$

$$= \sum_l \left(\frac{\partial \alpha_{m+1}}{\partial(q_{l+1} - q_l)}\right)_0 \tilde{q} e^{i(\omega_k t - kla)} e^{ika} [e^{-ika} - 1] \tag{A9.7}$$

where we let $l' = l - 1$ in rearranging the right-hand side of (A9.7).

Equating exponential expressions term by term in (A9.7), we obtain the recursion relationship

$$c_m = c_{m+1} e^{ika} \tag{A9.8}$$

where $c_m = [\partial \alpha_m / \partial(q_{l+1} - q_l)]_0$. Equation A9.8 is a difference equation that can be solved by an assumed solution of the form $c_m = A e^{pm}$. The result is

$$c_m = c_0 e^{-ikma} \tag{A9.9}$$

where $c_0 = [\partial \alpha_0 / \partial(q_{l+1} - q_l)]_0$ is the value of c_m at lattice site $m = 0$.

Combination of (A9.4), (A9.5) and (A9.9) yields

$$\alpha'_m = \operatorname{Re} \tilde{q} e^{i(\omega_k t - kma)} [e^{-ika} - 1] \sum_l c_0 e^{-ikla}$$

$$= \left[\frac{\tilde{q}}{2} e^{i[\omega_k t - k(m+1)a]} - \frac{\tilde{q}}{2} e^{i[\omega_k t - kma]}\right] \sum_l c_0 e^{-ikla} + \text{c.c.} \tag{A9.10}$$

Since l ranges over positive and negative values and c_0 is real, (A9.10) reduces to

$$\alpha'_m = C_b(q_{m+1} - q_m) \tag{A9.11}$$

where C_b is a constant,

$$C_b = \sum_l \left(\frac{\partial \alpha_0}{\partial(q_{l+1} - q_l)}\right)_0 e^{-ikla}$$

Thus the polarizability at lattice site m, α'_m, can be expressed simply as a constant C_b times the acoustic displacement at lattice site m.

AN EXPRESSION FOR d^3k IN TERMS OF ω_{21}

appendix

10

In the neighborhood of $\mathbf{k} = 0$, the conduction band energy is

$$E(\mathbf{k}, 2) \simeq E(0, 2) + \frac{1}{2} \sum_{\alpha,\beta} \frac{\partial^2 E(\mathbf{k}, 2)}{\partial k_\alpha \partial k_\beta} k_\alpha k_\beta \tag{A10.1}$$

where we have kept the first two nonzero terms in the power series expansion. From the definition of effective mass, (A10.1) may be rewritten as

$$E(\mathbf{k}, 2) = E(0, 2) + \frac{\hbar^2}{2} \sum_{\alpha,\beta} \frac{1}{m^*_{\alpha\beta}(2)} k_\alpha k_\beta \tag{A10.2}$$

where $m^*_{\alpha\beta}(2)$ is the effective mass at the bottom of the conduction band. If the coordinate axes are chosen along the directions of the principal axes, (A10.2) becomes

$$E(\mathbf{k}, 2) = E(0, 2) + \frac{\hbar^2}{2} \sum_\alpha \frac{1}{m^*_\alpha(2)} k_\alpha^2 \tag{A10.3}$$

where the α direction is parallel to a principal axis.

An expression similar to (A10.3) is obtained for the valence band, so that we have

$$\hbar\omega_{21} = E(\mathbf{k}, 2) - E(\mathbf{k}, 1) = \hbar\omega_0 + \frac{\hbar^2}{2} \sum_\alpha \frac{1}{\mu^*_\alpha} k_\alpha^2 \tag{A10.4}$$

where

$$\hbar\omega_0 = E(0, 2) - E(0, 1) = \text{minimum band gap energy}$$

and

$$\frac{1}{\mu^*_\alpha} \equiv \frac{1}{m^*_\alpha(2)} - \frac{1}{m^*_\alpha(1)}$$

Let us now define the variable q as

$$q_\alpha \equiv \frac{k_\alpha}{\sqrt{\mu_\alpha^*}}$$

so that

$$d^3k = dk_x\, dk_y\, dk_z = \sqrt{\mu_x^* \mu_y^* \mu_z^*}\, dq_x\, dq_y\, dq_z \tag{A10.5}$$

From (A10.4) we have

$$\omega_{21} - \omega_0 = \frac{\hbar}{2}(q_x^2 + q_y^2 + q_z^2) \equiv \frac{\hbar}{2}\rho^2 \tag{A10.6}$$

where

$$\rho^2 \equiv q_x^2 + q_y^2 + q_z^2$$

The variable ρ is the radius in q space, so that the volume integral in q space is

$$dq_x\, dq_y\, dq_z = 4\pi\rho^2\, d\rho \tag{A10.7}$$

A combination of (A10.5)–(A10.7) yields

$$d^3k = 4\pi\left(\frac{2\mu_x^* \mu_y^* \mu_z^*}{\hbar^3}\right)^{1/2}(\omega_{21} - \omega_0)^{1/2}\, d\omega_{21} \tag{A10.8}$$

which is the desired expression for d^3k in terms of ω_{21}.

EVALUATION OF THE INTEGRALS IN EQUATION 8.54

appendix
11

First, let us consider the integral

$$I_1 \equiv i \int dV \rho_{BB}(\mathbf{k}', \mathbf{k}, t) \, \psi_B^*(\mathbf{k}, \mathbf{r}) \, e^{i\mathbf{k}'\cdot\mathbf{r}} \, \nabla_{\mathbf{k}'} u_B(\mathbf{k}', \mathbf{r}) - \text{c.c.} \tag{A11.1}$$

Since

$$\psi_B^*(\mathbf{k}, r) = e^{-i\mathbf{k}\cdot\mathbf{r}} u_B^*(\mathbf{k}, \mathbf{r})$$

there is a factor $e^{i(\mathbf{k}'-\mathbf{k})\cdot\mathbf{r}}$ in the integrand of (A11.1), which causes the integral in coordinate space to be zero† unless $\mathbf{k} = \mathbf{k}'$. Thus we may write I_1 as

$$I_1 = i \int dV \rho_{BB} u_B^*(\mathbf{k}, \mathbf{r}) \, \nabla_{\mathbf{k}} u_B(\mathbf{k}, \mathbf{r}) - \text{c.c.}$$

$$= i\rho_{BB} \int dV [u_B^* \nabla_{\mathbf{k}} u_B + u_B \nabla_{\mathbf{k}} u_B^*]$$

$$= i\rho_{BB} \nabla_{\mathbf{k}} \int dV \, |u_B|^2 \tag{A11.2}$$

where $\rho_{BB} \equiv \rho_{BB}(\mathbf{k}, \mathbf{k}, t)$. Since $\int dV \, |u_B|^2 = 1$, $\nabla_{\mathbf{k}} \int dV \, |u_B|^2 = 0$, and so we obtain the result that $I_1 = 0$. This means that the second bracket on the right-hand side of (8.54) is zero.

Second, let us consider

$$I_2 \equiv i \iint d^3k' \, dV \rho_{BB}(\mathbf{k}', \mathbf{k}, t) \, \psi_B^*(\mathbf{k}, \mathbf{r}) \, \nabla_{\mathbf{k}'} \psi_B(\mathbf{k}', \mathbf{r}) - \text{c.c.} \tag{A11.3}$$

† Refer to Reference 2 at the end of Chapter 8.

If we integrate (A11.3) by parts with respect to $\mathbf{k}'$, we obtain

$$I_2 = -i \iint d^3k' \, dV \, \psi_B^*(\mathbf{k}, \mathbf{r}) \, \psi_B(\mathbf{k}', \mathbf{r}) \, \nabla_{\mathbf{k}'} \rho_{BB}(\mathbf{k}', \mathbf{k}, t) - \text{c.c.} \quad \text{(A11.4)}$$

From (8.26) we have $\int \psi_\beta^*(\mathbf{k}, \mathbf{r}) \psi_B(\mathbf{k}', \mathbf{r}) \, dV = \delta(\mathbf{k}' - \mathbf{k})$ so that I_2 may be rewritten as

$$I_2 = -i \lim_{\mathbf{k}' \to \mathbf{k}} [\nabla_{\mathbf{k}'} \rho_{BB}(\mathbf{k}', \mathbf{k}, t) + \nabla_{\mathbf{k}'} \rho_{BB}(\mathbf{k}, \mathbf{k}', t)] \quad \text{(A11.5)}$$

The matrix element of ρ is given by (1.37):

$$\rho_{BB}(\mathbf{k}', \mathbf{k}, t) = c_B(\mathbf{k}', t) c_B^*(\mathbf{k}, t) \quad \text{(A11.6)}$$

where $c_B(\mathbf{k}, t)$ is the coefficient of the term $|\mathbf{k}B\rangle$ in the expansion for the state vector. From (A11.6) we find that

$$\lim_{\mathbf{k}' \to \mathbf{k}} [\nabla_{\mathbf{k}'} \rho_{BB}(\mathbf{k}', \mathbf{k}, t) + \nabla_{\mathbf{k}'} \rho_{BB}(\mathbf{k}, \mathbf{k}', t)] = \nabla_{\mathbf{k}} \rho_{BB}(\mathbf{k}, \mathbf{k}, t) \equiv \nabla_{\mathbf{k}} \rho_{BB}(\mathbf{k}, t) \quad \text{(A11.7)}$$

Thus we have

$$I_2 = -i \nabla_{\mathbf{k}} \rho_{BB}(\mathbf{k}, t) \quad \text{(A11.8)}$$

which is the relationship that leads to (8.55).

THE DERIVATION OF

$$\left(\frac{\partial E}{\partial k_\alpha}\right)^2 = \frac{2\hbar^2}{3m_\alpha^*} E$$

appendix

12

In Appendix 10 Equation (A10.3), it is shown that

$$E = \frac{\hbar^2}{2} \sum_{\alpha=x,y,z} \frac{1}{m_\alpha^*} k_\alpha^2 \qquad (A12.1)$$

where α refers to the coordinate directions. The energy at the bottom of the conduction band has been taken to be zero, and m_α^* is the effective mass at the bottom of the conduction band for a principal axis coordinate system. From (A12.1) we may write

$$E = \frac{\hbar^2}{2} \sum_\alpha q_\alpha^2 \qquad (A12.2)$$

where $q_\alpha \equiv k_\alpha (m_\alpha^*)^{-1/2}$. Thus

$$\left(\frac{\partial E}{\partial k_\alpha}\right)^2 = \frac{1}{m_\alpha^*} \left(\frac{\partial E}{\partial q_\alpha}\right)^2 \qquad (A12.3)$$

Since

$$E = \frac{\hbar^2}{2}(q_x^2 + q_y^2 + q_z^2) \qquad (A12.4)$$

in $\mathbf{q}$ space the constant energy surfaces are spheres and therefore E has the same dependence on each of the q coordinates, so that

$$\left(\frac{\partial E}{\partial q_x}\right)^2 + \left(\frac{\partial E}{\partial q_y}\right)^2 + \left(\frac{\partial E}{\partial q_z}\right)^2 = 3\left(\frac{\partial E}{\partial q_\alpha}\right)^2 = 2\hbar^2 E \qquad (A12.5)$$

where the last equality was obtained from (A12.2). Hence, from (A12.5) and (A12.3) we obtain

$$\left(\frac{\partial E}{\partial k_\alpha}\right)^2 = \frac{2\hbar^2}{3m_\alpha^*} E \qquad (A12.6)$$

THE DERIVATION OF (8.99) FROM (8.98)

appendix 13

We start with the equation

$$\mathbf{J} = -\frac{N_V e^2}{\hbar} \int d^3k (\nabla_\mathbf{k} E)(\mathscr{E} \cdot \nabla_\mathbf{k} \rho_{22}^e) T_1 \tag{8.98}$$

If we define N_e as the number of conduction electrons per unit volume we see that

$$\int d^3k \rho_{22}^e = \frac{N_e}{N_V} \tag{A13.1}$$

where N_V is the number of electrons per unit volume. Also, we may write

$$\mathscr{E} \cdot \nabla_\mathbf{k} \rho_{22}^e = \left(\frac{\partial \rho_{22}^e}{\partial E}\right) \mathscr{E} \cdot \nabla_\mathbf{k} E \tag{A13.2}$$

A combination of (A13.1), (8.98), and (A13.2) enables us to rewrite the current density as

$$\mathbf{J} = -\frac{e^2 N_e}{\hbar^2} \frac{\int d^3k (\nabla_\mathbf{k} E)(\partial \rho_{22}^e/\partial E)(\mathscr{E} \cdot \nabla_\mathbf{k} E) T_1}{\int d^3k \rho_{22}^e} \tag{A13.3}$$

If the electric field is applied along the α direction we have

$$J_\alpha = -\frac{e^2 N_e}{\hbar^2} \frac{\int d^3k (\partial E/\partial k_\alpha)^2 (\partial \rho_{22}^e/\partial E) T_1}{\int d^3k \rho_{22}^e} \mathscr{E}_\alpha \tag{A13.4}$$

In Appendix 10 it is shown that

$$d^3k \text{ is proportional to } E^{1/2} \, dE \tag{A13.5}$$

and from Appendix 12 we have

$$\left(\frac{\partial E}{\partial k_\alpha}\right)^2 = \frac{2\hbar^2}{3m_\alpha^*} E \tag{A12.6}$$

where m_α^* is the effective mass at the bottom of the conduction band for a principal axis coordinate system. The substitution of (A12.6) and (A13.5) into (A13.4) yields

$$J_\alpha = -\frac{2e^2 N_e}{3m_\alpha^*} \frac{\int E^{3/2} (\partial \rho_{22}^e/\partial E) T_1 \, dE}{\int E^{1/2} \rho_{22}^e \, dE} \mathcal{E}_\alpha \tag{A13.6}$$

As given by (1.40), $\rho_{22}(\mathbf{k})$ may be written as

$$\rho_{22}(\mathbf{k}) = n(\mathbf{k}) P N_V^{-1} \tag{A13.7}$$

where

$n(\mathbf{k})$ = number of eigenstates per unit volume in $\mathbf{k}$ space, per unit volume in coordinate space

P = probability that an eigenstate in the conduction band is occupied by an electron

At thermal equilibrium, P is given by the Fermi-Dirac distribution, so that the diagonal elements of ρ^e are

$$\rho_{ii}^e = \frac{n(\mathbf{k}) N_V^{-1}}{1 + e^{(E-E_F)/\kappa T}} \tag{A13.8}$$

where E_F is known as the Fermi energy, κ is Boltzmann's constant, and T is temperature in degrees Kelvin. Equation 8.24 relating n and k leads to the following expression for $n(\mathbf{k})$:

$$n(\mathbf{k}) = \frac{2}{(2\pi)^3} \tag{A13.9}$$

where an extra factor of 2 has been included to account for the two spin states of an electron. From (A13.8) we find that

$$\frac{\partial \rho_{22}^e}{\partial E} = -\frac{1}{\kappa T} \rho_{22}^e (1 - \rho_{22}^e) \simeq -\frac{1}{\kappa T} \rho_{22}^e, \quad \text{for} \quad \rho_{22}^e \ll 1 \tag{A13.10}$$

In the conduction band $\rho_{22}^e \ll 1$, and so the approximation given in (A13.10) would apply. Since ρ_{22}^e decays exponentially with increasing energy, the limits on the integral in (A13.6) may be approximated as $\int_0^\infty$ because the integrand is small at the upper limit. The actual limits on the integral are from zero to the maximum energy of the conduction band.

In the conduction band, where ρ_{22}^e is approximately proportional to $e^{-E/\kappa T}$, we find that the following relationship holds:

$$\int_0^\infty E^{1/2} \rho_{22}^e \, dE = \tfrac{2}{3} \kappa T \int_0^\infty E^{3/2} \rho_{22}^e \, dE \tag{A13.11}$$

The substitution of (A13.10) and (A13.11) into (A13.6) yields

$$J_\alpha = \frac{N_e e^2}{m_\alpha^*} \langle T_1 \rangle \mathscr{E}_\alpha \tag{8.99}$$

where

$$\langle T_1 \rangle \equiv \frac{\int_0^\infty T_1 E^{3/2} \rho_{22}^e \, dE}{\int_0^\infty E^{3/2} \rho_{22}^e \, dE} \tag{8.100}$$

INDEX

Absorption, 58–63
   m-photon, 142
   three-photon, 133, 142
   two-photon, 132, 139–141, 156
Absorption constant, 38, 327
   direct transitions in a crystal, 300–302
   electric dipole transition, 60
   indirect transitions, 306
   in terms of spontaneous emission time, 190
   paramagnetic resonance, 96
   photoconductivity, 308
   saturated, 73
Absorption losses, 39
Acoustic modes, 191–207
see also Vibrational modes
Adjoint operator, 6
AgBr, 223
AgCl, 223
Alkali halides, 217
Ammonia maser, 102–103, 128
Ammonia molecule, 60–62, 74
Analytic function, 328
Anaxial crystal, 82–83
Angular momentum, classical, 24, 49
   quantum mechanical, 44
   spin, 41–44
Anisotropic media, local field correction factor, 78
   polarization equation, 42
   tensor properties of susceptibility, 77
Annihilation operator, 162
   acoustic mode, 204–205
   matrix elements, 166

Annihilation operator (*continued*)
   optical mode, 208
Anomalous dispersion, 59

Band-to-band transitions—*see*
   Interband effects
Basis vector, 4, 10, 14, 280
Benzene (C_6H_6), 236–237, 241–243, 335, 338
Biaxial crystal, 83
Blackbody radiation, 186
Bleaching, 74
Bloch equations, 48, 95, 156
Bloch function, 271–273, 295
Bohr magneton, 44
Bohr radius, 21
Boltzmann distribution, 12, 18, 77, 186–187, 306
Boltzmann equation, 291–292
Bragg reflection, 245, 269, 273–274, 277–278, 283, 317
bra vector, 2, 5, 10
Brillouin effect, 192, 243–255
   conservation conditions (energy and momentum), 212–214
   equation of motion, 249
   gain, 252
   Hamiltonian, 247
   susceptibility, 251–252, 254
Brillouin zones, 195, 271

Canonical momentum, 21, 282, 294
Carbon disulfide, (CS_2), 259, 335
Cauchy's integral theorem, 328

Cavity, decay of energy in
  quantized mode, 167
 lifetime, 39, 53, 167–168, 334
 mode density, 179, 183, 185, 189
 normal modes, 38–39, 159
 optical lifetime, 39, 53
 optical losses, 39, 111–112, 313
Center of symmetry, 27, 82, 156
C_6H_6, 236–237, 241–243
Chromium ion in sapphire (Al_2O_3), 103–105
Collisions, average time between, 286–287, 289, 291–293
 electron motion with, 284–288
Commutator operator, 10, 15, 135–136
Complex plane, 329
Compton scattering, 189, 263
Conductivity, 37
 complex, 289
 interband, 299–301
 intraband, 289
 photo-, 307–310, 318
Conservation of energy, Brillouin scattering, 212–214
 phonon processes, 209–217
 Raman scattering, 214–217
Coordinate transformation matrix, 79
Correspondence principle, 64
Coupled modes, phonon-photon, 223–224
Creation operator, 162
 acoustic mode, 204–205
 matrix elements, 166
 optical mode, 208
Cross-section, absorption, 62–63, 98
 classical absorption, 64
 multiple-photon processes, 143–145
 Raman, 239–243
Crystal, anaxial, 82
 biaxial, 83
 classes, 83
 energy bands in, 266–272
 symmetry, 82
 uniaxial, 83
Crystalline field, 28
Crystal vibrations—*see* Vibrational modes
CS_2, 259, 335
CsCl, 223
Cubic crystal, 83, 192
Current density, Boltzmann equation, 291–293
 hole, 290–291

Current density (*continued*)
 indirect transitions, 304–306
 interband transitions, 295–307
 intraband transitions, 288–290, 307
 photoconductivity, 307–308
 threshold for semiconductor laser, 313–316

De Broglie relations, 213
Debye, 62
Degeneracy, 27, 75, 110, 130
Density of modes—*see* Mode density
Density of states, semiconductor, 314, 318
Density operator, 6–15, 280
 diagonal matrix elements of, 9
 electrons in a crystal, 292–293
 multiple-photon analysis, 139–142
 nonlinear analysis, 135–136
 off-diagonal matrix, elements of, 12
 properties of, 9
 relaxation terms for, 11–14
 resonant Raman effect, 154
 time dependence of, 10
Detailed balance, 322
Diamagnetism, 24
Diamond, 338
Diamond lattice structure, 197, 217
Dielectric constant, complex (optical crystal modes), 219
 optical modes in MgF_2, 226
 α-quartz, 252
 table of, for crystals, 223
Diffraction-limited beam, 256
Diffraction losses, 39, 313
Dirac delta function, 14
Dirac notation, 2, 5, 10, 268, 273
Direct transitions, 294–302
Dispersion, 58
Dispersion characteristic, Brillouin scattering, 215
 determination by neutrons, 200
 determination by X-rays, 196
 GaP optical mode, 222
 $LiNbO_3$ optical mode, 263–264
 MgF_2 optical modes, 226
 number of branches, 200
 optical and acoustic modes, diatomic crystal, 199
 Raman scattering, 216

INDEX

Dispersion characteristic (*continued*)
   transverse and longitudinal
      acoustic modes, 194–195
Dispersion vector, 214
   Brillouin scattering, 215
   Compton scattering, 263
   Raman scattering, 216
Doppler broadening, 66–70, 122, 129
Doppler effect, 246

Effective mass, 282–284, 293, 299, 304, 342, 346, 348
Ehrenfest's theorem, 1
Eigenfunction, 3–4
   continuum, 14–15
   crystal, 267–273
   degeneracy, 27
Eigenvalue, 3, 18
   in crystal, 267–270
Einstein A and B coefficients, 185, 187
Elasto-optical coefficient, 251–252
Electric dipole, Hamiltonian for, 23
   operator, 28, 31
   permanent, 28
Electric dipole transition, absorption constant, 60
   bleaching, 74
   equations of motion, 40
   matrix elements, 26
   matrix elements (degeneracy), 75
   parity, 28
   rate equation for, 107
   saturation, 71
   single photon process, 44
   Stark shift, 150–153, 157
   susceptibility, 57
   tensor properties, 80
   transition probability, 94
Electric quadrupole—*see* Quadrupole
Electronic resonances, contribution to dielectric constant, 223
Electron motion, with collisions, 284–288
   without collisions, 280–284
Electrostrictive effect, 256
Energy bands in crystals, 266–272
   filled, 290
Energy conservation—*see* Conservation of energy
Entropy, 18
Equation of motion, Brillouin effect, 249

Equation of motion (*continued*)
   current density in a crystal, 298
   electric dipole transition, 40
   electron in a crystal, 280–281, 318
   energy decay in quantized cavity mode, 167
   expectation value of an operator (general case), 15–17
   magnetic dipole spin ½ system (paramagnetic resonance), 48
   optical mode, 218
   quantized cavity field, 168
   quantized plane wave, 188
   Raman effect, 232–233, 264
Even parity, 27
Exclusion principle, 279, 298
Expectation value, 6, 10, 15
   equations of motion for, 15–17
Extinction coefficient, 224–225
"Extra photon," 179, 232, 239, 247, 254
Extraordinary refractive index, 83

Fabry-Perot interferometer, 253
Fermi-Dirac distribution, 305, 311, 314–315, 348
Fermi energy, 311, 348
Field equations, isotropic polarizable medium, 37
   paramagnetic medium, 51
Field quantization—*see* Quantization of electromagnetic field
Filling factor, 39, 110, 131
Fluorescence, 182–183
Forbidden transitions, 27, 301–302
Free carriers, 279
   absorption, 289, 313
   lifetime, 307
Frequency pulling, 90, 130
Fundamental domain, 273, 279

Gain, Brillouin effect, 252
   photoconductive, 308–310, 318
   Raman effect, 235
   traveling-wave laser, 88
Gallium arsenide (GaAs) 312–313, 315–316
GaP, 220–225
   dielectric constant,
      frequency dependence of, 220
   dispersion characteristic (optical mode), 222
   extinction coefficient, optical mode, 225

GaP (*continued*)
  reflectivity, 221
  refractive index, optical mode, 225
Gas laser, 105–106, 121–127, 131
Gaussian lineshape function, 66–70, 138, 141
  helium-neon laser, 122–125, 129
Gyromagnetic ratio, 44

Hamiltonian, 1–4
  acoustic mode, 206
  atom in electromagnetic field, 20
  Brillouin effect, 247
  electric dipole, 23
  electron-radiation interaction in crystals, 297–298
  harmonic oscillator, 160
  indirect transitions, 303
  interaction, 4
  interaction of radiation with optical mode, 218
  interband effects, 294, 297–298
  intraband effects, 280
  magnetic dipole, 24
  optical mode, 208
  perturbation analysis, 135
  quadrupole, 24
  quantized cavity mode, 166
  quantized matter-radiation interaction, 172–173
  quantized plane waves (e.m. field), 171
  Raman effect, 231–232
  randomizing, 11–13
  spin ½ magnetic dipole system, 45
Hamilton's equations, 159, 204
Harmonic generation, 145–150
  parity considerations, 146
Harmonic oscillator, eigenvalues, 162, 164
  eigenvectors, 162
  Hamiltonian, 160
  quantization, 160–164
Helium-neon laser, 105–106, 121–127, 131
Hermitian operator, 6, 10, 12, 17–19, 274, 323–324
Hexagonal crystal, 83
Hole burning, 124
Holes, 290–291
Homogeneous broadening, 65, 108, 124, 129
Hooke's law, 194, 204
Hydrogen atom, Bohr radius, 21
  oscillator strength, 63

Identity operator, 6
Impurities, acceptor, 310
  donor, 310
  reflection by, 275–278, 317
  trapping by, 278–279, 318
Index of refraction, 38, 326
  contribution of electronic resonances, 223
  GaP (optical mode), 225
  intensity-dependent, 255
  ordinary, extraordinary, 83
Indices, crystal directions, 192
Indirect transitions, 302–307
Infrared-active optical modes, 217–225
  see also Vibrational modes
Inhomogeneous broadening, 66–70
  helium-neon laser, 122–125, 129
Interband effects, 294–307
  direct transitions, 294–302
  indirect transitions, 302–307
Intraband effects, 279–293
Inversion symmetry, 82
Ionic bonds, 217, 231

KCl, 223
Kerr cell, 118, 228
Kerr effect, 256
ket vector, 2, 5, 10
KI, 223
Kramers-Kronig relations, 70, 318, 328–329
Kronecker delta function, 3
Kuhn-Thomas sum rule, 64

Lagrangian, 21
Laplace transform, 319
Laser, 101–131
  cavity, 91
  circuit losses, 111–112
  four-level, 120–121, 131
  helium-neon, 105–106, 121–127, 131
  hole burning, 124
  neodymium, 120–121, 131
  pump power, 110–111, 126
  Q-switching, 117–120
  rate equations, 92, 107
  ruby, 103–120
  semiconductor, 310–316, 318
  steady-state emission from, 111–112
  three-level, 103–120
  threshold requirements, 107–111, 126
  transient behavior, 112–117

Laser (*continued*)
  traveling-wave, 86
Lattice modes—*see* Vibrational modes
LiF, 223
Lifetime—*see* Relaxation time constants
LiNbO$_3$, 230, 263–264, 338
Linewidth, Gaussian line, 68
  Lorentzian line, 60
  saturated, 73
Lithium niobate (LiNbO$_3$), 230, 262–264, 338
Local field, as seen by molecule, 33, 325–327
Local field correction factor, 40–41, 219, 234, 325–327
  anisotropic media, 78
Longitudinal relaxation, T_1, 13–14, 36, 50
  measurement of, 74
Lorentz correction factor—*see* Local field correction factor
Lorentzian lineshape function, 57–60, 65, 122, 129, 137–141, 334
Lorentz-Lorenz relation, 326
Losses, optical, 39
Lyddane - Sachs - Teller relation, 224

Magnetic dipole, Hamiltonian for, 24
Magnetic dipole spin ½ transition— *see* Paramagnetic resonance
Magnetic moment, 24
Magnetic susceptibility, 96
Magnetization, 48
Magnon, 158
Maser, ammonia, 102–103, 128
  rate equations, 93
Matrix, diagonal elements of, 5, 9, 16
  element, degeneracy, 75
  notation, 4–5
  off-diagonal elements of, 12, 16
  trace of, 8–9, 15, 18–19
Maxwellian velocity distribution, 67
Maxwell's equations, 37
Metastable state, 104
MgF$_2$, 226
MgO, 223
Mirror plane, 82
Mixed state, 7–9
Mobility, 287, 307
Mode density, cavity modes, 179, 183, 185, 189
  traveling-wave modes, 181, 241

Momentum, angular—*see* Angular momentum
  canonical, 21, 282, 294
  electron in crystal, 273–275
  free electron, 273
  particle, 21, 282, 294
Momentum matching—*see* Wave vector conservation
Monoclinic crystal, 83
Multiparticle systems, 25
Multiple photon processes, 132–145
  cross section, 143–145
  parity for, 141–143
Multipole expansion, 21

NaBr, 223
NaCl, 197, 217, 223
NaF, 223
Neodymium laser, 120–121, 131
Neutrons, 200
NH$_3$ molecule, 60–62, 74
Normal modes, acoustic, 200–204
  cavity, 38–39, 159
Normalization constants for rate equations, energy density, 92–93
  population difference, 92–93, 109
Nuclear magnetic resonance, 54
Number operator, 165–166

Occupation probability, 9
  in a crystal, 298
Odd parity, 27
Operator, 5–6
  adjoint, 6
  commutator, 10, 15, 135–136
  density, 6–15
  expectation value of, 6, 10, 15
  Hermitian, 6, 10, 12, 17–19, 274, 323–324
  identity, 6
  time derivatives of, 15–17
  trace of, 15, 18
Optical losses—*see* losses, optical
Optical modes, 191, 196–200, 207–209, 216–226
  see also Vibrational modes
Optical resonator—*see* Cavity
Ordinary refractive index, 83
Orientational averaging, 34–35
Orthogonality, cavity modes, 38, 170
  traveling wave modes (acoustic), 202
  traveling wave modes (e.m.), 171

Orthorhombic crystal, 83
Oscillator strength, 63, 300–301

Paramagnetic resonance, 41–44, 95
Parity, 26–30, 52, 133
   electric dipole transition, 28
   electric quadrupole transition, 29
   harmonic generation, 146
   magnetic dipole transition, 30
   multiple-photon processes, 141–143
Pauli exclusion principle, 279, 298
Pauli spin operators, 44
Periodic boundary conditions, 169, 181, 200–204, 272–273, 317
Permanent dipole moment, 28
Perturbation analysis, 134–136
Phase matching—see Wave vector conservation
Phonon (see also Vibrational modes), 191, 200–209
   indirect transitions, 302–304, 306
Photoconductive gain, 308–310, 318
Photoconductivity, 307–310, 318
Photoelastic effect, 251–252
Photon, 158, 164
   "density," 92
   location, 177
   rate equation (quantized fields), 176–178, 183, 332–334
π pulse, 53
$\pi/2$ pulse, 54
Piezoelectric tensor, 98
Planck constant, 2, 159
Planck's law, 186–187
Plasmon, 158
Point group, 82
Polarizability, 212, 227, 229–230, 246–247, 325–327, 340–341
Polarization, 33
   anisotropic medium, 42
   equation of motion, 34
   gas laser, 125
   harmonic generation, 146–147
   source term, 37, 41, 326
   Stark shift, 151
Population difference, 33
   equation of motion, 36
   harmonic generation, 146–147
   Stark shift, 151

Population inversion, 101–107, 118–121, 314
   semiconductor, 311
Precession, 49–50
Principal axes, 42, 283, 298, 342, 346, 348
Principal part, PP, 70, 328
Probability, occupation, 32–34
Propagation constant, absorbing medium, 59
   amplifying medium, 102
Pure state, 7–9, 14

Q-switching of a laser, 117–120, 228
Quadrupole focuser, 102, 128
Quadrupole transition, Hamiltonian, 24
   parity, 29
   sum rule, 83, 98
Quantization of electromagnetic field, 158–190
   annihilation operator, 162
   cavity fields, 159–166
   creation operator, 162
   decay of energy in quantized cavity mode, 167
   energy eigenvalues, 164
   equation of motion for quantized cavity field, 168
   equation of motion for quantized plane wave, 188
   matrix elements for important operators, 166
   number operator, 165
   plane wave quantization, 169–171
Quantization of vibrational modes—see Vibrational modes
Quantized matter-radiation interaction, photon rate equation, 176–178, 183, 332–334
   population difference equation, 177–178, 183–185
Quantum efficiency, 307–308, 315, 318
α-Quartz, 252

Radiation, blackbody, 186
   broadband, interaction with matter, 184
Raman effect, 132–133, 192, 225–243
   anti-Stokes, 227, 231
   benzene, 236–237, 241–243
   cross section, 239–243
   equation of motion, 232–233, 264
   gain, 235
   Hamiltonian, 231–232, 264

Raman effect (*continued*)
  resonant, 153–155
  spontaneous, 228, 239–243
  stimulated, 228, 335–339
  Stokes emission, 227, 231
  susceptibility, 235, 265
  threshold condition, 236–237
Rate equations, laser (semiclassical), 92, 107
  maser, 93
  normalization constants, energy
    density, 92–93
    population difference, 92–93, 109
  photons, (quantized fields), 176–178, 183, 332–334
  population difference (quantized fields), 177–178, 183–185
  steady-state solution, 108–110
  traveling-wave laser, 87
RbI, 223
Reciprocal lattice vector, 214
Reflection by impurities, 275–278, 317
Reflectivity, 221
Refractive index—*see* Index of refraction
Relaxation time constants, 11–14, cavity
    lifetime, 39, 53
  energy in a resonator, 111–112, 126–127
  free carrier, 307–308
  longitudinal, T_1, 13–14, 36, 50
  saturation measurement, 74
  transverse, T_2, 13–14, 35, 50, 333
Residual ray, 221
Resonance, 55
Resonator—*see* Cavity
Reststrahl reflection, 221
Rotating prism, 118–119
Rotation axis, 82
Ruby laser, 103–120

Saturable dye, 119
Saturation, 71–74
  laser, 129
Scattering cross-section—*see* Cross-section
Scattering losses, 39
Schrodinger wave equation, 1–5
Second quantization, 160
Selection rule, 28–29
Self-focusing, 255–260
Semiclassical analysis, 20

Semiconductor, n-type, 310
  p-type, 310
Semiconductor injection laser, 310–316, 318
  threshold, 313–316
Single-photon transition probability, 135–138
Snell's law, 256–257
Sodium chloride, 197, 217, 223
Solid angle, 181
Space group, 82
Spatial averaging, 330–331
Spectroscopic splitting factor, 44
Spin ½, 42–51, 95–97, 156, 348
Spin angular momentum, 41-44
Spin eigenstates, 43
Spin-lattice relaxation time constant, 50
  see also Transverse relaxation
Spin operators, 44
Spin-orbit coupling, 44–45
Spin packets, 66
Spin paramagnetism—*see* Paramagnetic
    resonance
Spinning top, 49
Spontaneous emission, 14, 108, 172, 176, 179
  "extra photon," 179, 232, 239, 247, 254
  fluorescence, 182–183
  lifetime, 183
Stark shift, 150–153, 157
Stern-Gerlach experiment, 43
Stimulated Brillouin effect—
    see Brillouin effect
Stimulated emission, 176, 178
Stimulated Raman effect—*see* Raman effect
Sum rule, Kuhn-Thomas, 64
  momentum operator in a crystal, 300–301
  quadrupole operator, 98
Susceptibility, 57–58, 328–329
  anisotropic media, 77
  Brillouin, 251–252, 254
  inhomogeneously broadened line, 69
  magnetic, 96
  nonlinear, 85
  Raman, 235, 265
  resonant Raman, 154
  tensor properties, 77–85
Symmetry elements (crystal), 82

Tensor properties of the electric dipole
    transition, 80
Tetragonal crystal, 83

Thermal equilibrium, 12, 18, 306
  matter-radiation interaction, 186
  transition rates, 321
Three-photon absorption, 133, 142
Threshold, 150
  laser, 107–111, 126
  Raman effect, 236, 260
  resonant Raman effect, 153–155
  semiconductor injection laser, 313–316
TlBr, 223
TlCl, 223
Top, classical, 49
Torque, 49, 53
Total internal reflection, 256
Trace, 8–9, 15, 18–19
Transition probability, 12, 94, 185
  m-photon absorption, 142–143
  single-photon absorption, 135–138
  three-photon absorption, 142
  two-photon absorption, 139–141
Transverse relaxation time, T_2, 13–14, 35, 50, 333
Trapping by impurities, 278–279, 318
Traveling-wave laser, 86
Triclinic crystal, 83
Trigonal crystal, 83
Two-photon, absorption, 132, 139–141, 156, 188–189
  emission, 188–189

Umklapp processes, 214
Uncertainty principle, 92, 177
Uniaxial crystal, 83

Vector, basis, 4, 10, 14
Vibrational modes, acoustic, 191–207
  coupling to electronic state, 319–320
  diatomic crystal lattice, 196–200
  monatomic crystal lattice, 192–196
  normal modes, 200–204
  optical, 191, 196–200, 207–209, 216–226
    crystal parameters for, 221
    GaP, 220–225
    infrared properties of optical
      phonons, 217–225
    $LiNbO_3$, 263–264
    MgF_2, 226
  quantization, 191, 200–209
    acoustic, 204–207
    optical, 207–209

Wave equation, 1–5
  paramagnetic medium, 51
  polarizable medium, 38
Wave vector conservation, 209–217
  Brillouin scattering, 212–214, 244, 250
  direct transitions, 296
  general case, 211
  harmonic generation, 149
  indirect transitions, 302–303
  Raman scattering, 214–217, 234

X-rays, 196

Zero point, energy, 164, 167, 188
  fluctuations, 188

Physical Constants

Electronic charge, e	1.60210×10^{-19} coulombs
Electron rest mass, m	9.1091×10^{-31} kg
Planck constant, h	6.6256×10^{-34} joule-sec
(h—"bar"), $\hbar$	1.054×10^{-34} joule-sec
Velocity of light in vacuum, c	2.997925×10^{8} m/sec
Charge to mass ratio for electron, e/m	1.758796×10^{11} coulombs/kg
Boltzmann constant, κ	1.38054×10^{-23} joules/°K
Bohr magneton, β	9.2732×10^{-24} joules/Tesla
Wavelength associated with 1 eV, λ	12396.44×10^{-10} m
Energy associated with 1 eV, E	1.60210×10^{-19} joules
Permittivity of vacuum, ϵ_0	8.854×10^{-12} farads/m
Permeability of vacuum, μ_0	$4\pi \times 10^{-3}$ henries/m

Use of the Nomograph

A. To convert from a wavelength to an energy-like unit:
  1. Find the wavelength on the vertical wavelength scale.
  2. Go horizontally over to the appropriate curve for the quantity you want and vertically down to read the desired value on the horizontal scale. Be sure to note the power of 10 for that particular quantity from the appropriate scale at the bottom of the nomograph.
  Examples:
  5000 Å → 20,000 cm^{-1}
  3 microns → 0.41 eV.

B. To convert from an energy-like quantity to a wavelength:
  1. Find the value of the quantity on the appropriate horizontal scale (check for correct power of 10).
  2. Go vertically to the appropriate curve and horizontally over to read the wavelength on the vertical scale.
  Examples:
  5×10^{-12} erg → 4000 Å
  10,000 °K → 1.44 micron.

C. To convert from one energy-like quantity to another energy-like quantity:
  1. Convert from the first energy-like quantity to wavelength.
  2. Convert the wavelength to the second energy-like quantity. (Note that it is not necessary actually to read the value of the wavelength—just go up to the first curve, then over to the second curve and down. The curves must be used, however, because the scales at the bottom of the nomograph are otherwise independent and give no information. Thus *do not* interpret the nomograph as indicating that 1 eV = 10^{15} Hz, which is, of course, incorrect.)

  Examples:
  1 eV → (12,400 Å) → 8000 cm^{-1}
  2×10^{-12} erg → (1 μ) → 3×10^{14} Hz.

Extensions

The nomograph may be extended to other regions of the spectrum by extrapolating the linear curves or simply by applying scale factors of some power of 10 to the log scales. If the wavelength scale values are multiplied by some power of 10, the energy scale values should be *divided* by the same power of 10.

Examples:
10^9 Hz = $10^{-6} \times 10^{15}$ Hz → $10^{+6} \times 3000$ Å = 3×10^9 Å = 0.3 meter
300 °K = $10^{-2} \times 30,000$ °K → $10^{-2} \times 2.6$ eV = 0.026 eV.

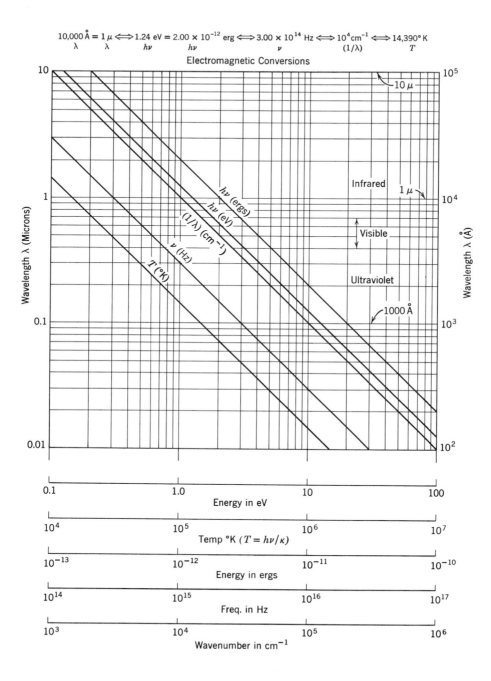